The Dynamic Nature of Mitochondria

... a novel and amazing view into mitochondrial research revealing the dynamic nature of this organelle and how this is linked to numerous fundamental biological processes ...

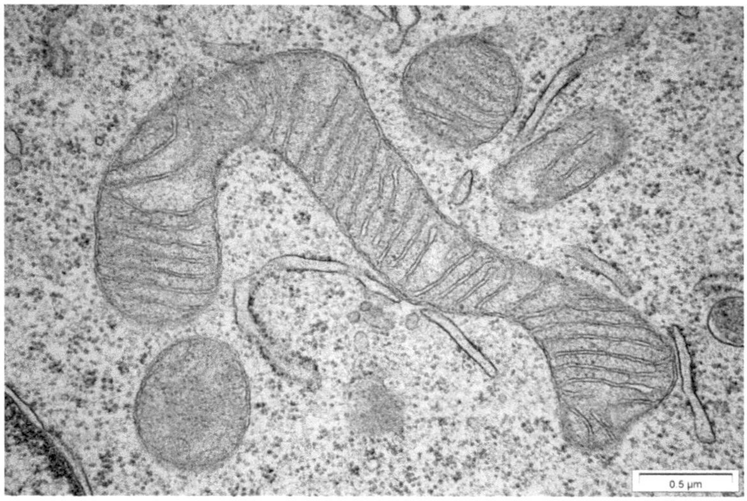

FIGURE 0.1 Electron micrograph of wild-type mitochondria in HEK cells. Tubular mitochondrial section with regular lamellar cristae is shown. Close association of mitochondria to the endoplasmatic reticulum is nicely visible at multiple regions. (Image with courtesy of Andrea Borchardt and Ruchika Anand.)

Mitochondrial research has exploded over the last ~150 years. This book gives an amazing view on a conceptual change in our understanding of mitochondrial biology. It becomes clear that mitochondria are extremely dynamic in nature, controlling life at multiple levels. Mitochondria rule energy conversion, adapt cells well to changing stress and nutrient conditions, and regulate many cellular processes including immunity. The dynamic nature of mitochondria occurs at an intramitochondrial level but also includes its ability to interact with other organelles and to modulate multiple signalling pathways. It is thus not surprising that alterations or inabilities to ensure this dynamic behaviour is linked to ageing and human diseases.

The following sections give an updated view on mitochondria:

- Mitochondrial ultrastructure: molecular mechanisms shaping the inner membrane
- Mitochondrial cristae and lipid dynamics: from super-resolution microscopy to lipid-OXPHOS interplay
- Mitochondrial control of cellular homeostasis: From redox signalling to interorganellar contact sites
- Mitochondria in health and disease: from mtDNA release to Complex I assembly
- Advanced methods in mitochondrial biology and metabolism research
- Integrative view on mitochondrial research and outlook

The field of mitochondrial research has always been full of surprises and has helped science to advance tremendously. It developed hand in hand with landmark developments in technology, such as super-resolution microscopy (nanoscopy), and is currently influencing an increasing number of scientific disciplines. There is still much 'new' to find out about this 'old' organelle, and I think that you can find interesting and also unexpected aspects of mitochondrial biology in this book. I hope the book will enhance your scientific curiosity and inspire your own research.

Oxidative Stress and Disease

Series Editors
Helmut Sies, MD
Enrique Cadenas, MD, PhD
University of Southern California School of Pharmacy
Los Angeles, California

The Dynamic Nature of Mitochondria: From Ultrastructure to Health and Disease, *edited by Andreas Reichert*

Environmental Stressors and OxInflammatory Tissue Responses, *edited by Giuseppe Valacchi and Andreas Daiber*

Peroxiporins: Redox Signal Mediators In and Between Cells, *edited by Iria Medraño-Fernandez, Gerd Patrick Bienert, and Roberto Sitia*

Physiological Consequences of Brain Insulin Action, *edited by André Kleinridders*

Oxidative Eustress in Exercise Physiology, *edited by James N. Cobley and Gareth W. Davison*

Redox Regulation of Differentiation and De-differentiation, *edited by Carsten Berndt and Horst Lillig*

Oxidative Stress in Cancer, AIDS, and Neurodegenerative Diseases, *edited by Luc Montagnier, René Olivier, and Catherine Pasquier*

Understanding the Process of Aging: The Roles of Mitochondria, Free Radicals, and Antioxidants, *edited by Enrique Cadenas and Lester Packer*

Redox Regulation of Cell Signaling and Its Clinical Application, *edited by Lester Packer and Junji Yodoi*

Antioxidants in Diabetes Management, *edited by Lester Packer, Peter Rösen, Hans J. Tritschler, George L. King, and Angelo Azzi*

Free Radicals in Brain Pathophysiology, *edited by Giuseppe Poli, Enrique Cadenas, and Lester Packer*

Nutraceuticals in Health and Disease Prevention, *edited by Klaus Krämer, Peter-Paul Hoppe, and Lester Packer*

Environmental Stressors in Health and Disease, *edited by Jürgen Fuchs and Lester Packer*

Handbook of Antioxidants: Second Edition, Revised and Expanded, *edited by Enrique Cadenas and Lester Packer*

Flavonoids in Health and Disease: Second Edition, Revised and Expanded, *edited by Catherine A. Rice-Evans and Lester Packer*

Redox–Genome Interactions in Health and Disease, *edited by Jürgen Fuchs, Maurizio Podda, and Lester Packer*

Thiamine: Catalytic Mechanisms in Normal and Disease States, *edited by Frank Jordan and Mulchand S. Patel*

Phytochemicals in Health and Disease, *edited by Yongping Bao and Roger Fenwick*

Carotenoids in Health and Disease, *edited by Norman I. Krinsky, Susan T. Mayne, and Helmut Sies*

Herbal and Traditional Medicine: Molecular Aspects of Health, *edited by Lester Packer, Choon Nam Ong, and Barry Halliwell*

For more information about this series, please visit: https://www.crcpress.com/Oxidative-Stress-and-Disease/book-series/CRCOXISTRDIS

The Dynamic Nature of Mitochondria

From Ultrastructure to Health and Disease

Edited by Andreas S. Reichert

CRC Press
Taylor & Francis Group
Boca Raton London New York

CRC Press is an imprint of the
Taylor & Francis Group, an **informa** business

Designed cover image: Psychedelic art 'Mitochondria texting'. Original data obtained by Arun Kumar Kondadi showing mammalian mitochondria expressing MIC13-SNAP localized to cristae and crista junctions which was visualized by live-cell STED nanoscopy. Psychedelic modification of data image was created by Andreas S. Reichert.

First edition published 2024
by CRC Press
2385 NW Executive Center Drive, Suite 320, Boca Raton FL 33431

and by CRC Press
4 Park Square, Milton Park, Abingdon, Oxon, OX14 4RN

CRC Press is an imprint of Taylor & Francis Group, LLC

ISBN: 9781032117133 (hbk)
ISBN: 9781032117164 (pbk)
ISBN: 9781003221180 (ebk)

DOI: 10.1201/9781003221180

Typeset in Times
by KnowledgeWorks Global Ltd.

Contents

PART IV Mitochondria in Health and Disease: From mtDNA Release to Complex I Assembly

PART V Advanced Methods in Mitochondrial Biology and Metabolism Research

Preface

Mitochondrial research has expanded tremendously over the last ~150 years and plays fundamental roles in many scientific fields, including basic and applied research in biology, biochemistry, cell biology, physiology, toxicology, pharmacology, and medicine. Here we give a general overview on major advances along this time but we focus on a conceptual change in our understanding of mitochondrial biology, namely that mitochondria are extremely dynamic in nature at different levels. One level is that they constantly undergo dynamic cristae membrane remodelling within mitochondria and that lipids and proteins are highly dynamic moving between distinct membrane compartments. Another level is the dynamic reshaping of the mitochondrial network by fusion and fission events between mitochondria. Moreover, we know that mitochondria are in close contact to different organelles such as peroxisomes, lysosomes, and the endoplasmatic reticulum. It is amazing to see how mitochondria are well integrated and connected to nearly all non-mitochondrial processes in a cell and how they control a pleiotropy of functions that have previously been thought to be separated. Mitochondria rule energy conversion and adaptation to changing stress and nutrient conditions and harbour many essential other cellular functions including anabolic and catabolic functions as well as signalling functions. Thus, the dynamic nature of mitochondria which also includes its flexibility to interact with other organelles and pathways is essential for cells to adapt to changing conditions. It is thus not surprising that alterations or an inability to ensure this dynamic behaviour is linked to ageing and human diseases.

How mitochondrial ultrastructure is established is elegantly described in **Chapter 1** by **Patrick Horten** and **Heike Rampelt**, focusing on the intricate interplay of several players such as namely OPA1, the F_1F_O ATP synthase, and the MICOS complex. Another level of regulation of mitochondrial ultrastructure via proteolysis is very well covered in **Chapter 2** by **Gunjan Purohit** and **Oleh Khalimonchuk**. In which way the spatial distribution within the different subcompartments is established and how it can adapt to changing conditions or stresses is well summarized here. A focus is given to the proteolytic control of mitochondrial architecture by the inner membrane protease OMA1. Despite the fact that cristae are highly variable in structure, they were thought to be rather static for a long time. Now different studies using super-resolution nanoscopy in living cells led to an updated view on cristae being highly dynamic instead. These developments and the role of the MICOS complex is covered beautifully in **Chapter 3** by **Ruchika Anand** and **Arun Kondadi**. The role of lipids and its stable and dynamic interactions with protein complexes, in particular with Complex I of the electron transport chain, is very nicely reviewed in **Chapter 4** by **Nanami Senoo** and **Steven Claypool**. How and which lipids determine mitochondrial membrane dynamics and its many physiological functions is one of the underexplored aspects of mitochondrial research. **Chapter 5** by **Jan Riemer** and **Helmut Sies** gives us an excellent overview on the role of mitochondria in redox signalling, whereas **Chapter 6** by **Carla Lopes** and **Nuno Raimundo** give astonishing insights into the complex interplay of mitochondria and lysosomes, and how this is integrated in a diverse set of cellular functions. That mitochondria are signalling hubs raised major attention in recent years. In **Chapter 7** by **Andrea Irazoki** and **David Pla-Martín**, the authors give a wonderful view into the role of mitochondria in regulating the innate immune response. They discuss how mitochondria act as signalling platforms recruiting complexes that initiate signalling to the nucleus. Further, they summarize the different pathways of mtDNA release and how mtDNA acts as a mitochondrial damage-associated molecular pattern initiating inflammatory responses. Released mtDNA in the cytoplasm, inside endosomes, or secreted to the extracellular compartment acts as intra- or intercellular/paracrine signals of mitochondrial dysfunction. How Complex I of the electron transport chain harbouring 45 subunits assembles with the help of assembly factors and how impairments thereof lead to mitochondrial diseases is elaborated excellently in **Chapter 8** by **Luke Formosa**. Last but not least, in **Chapter 9** by

Odette Deen-Rozalen, Marius Nieke, Sara Shumka, Hannes Beyer, and **Matias Zurbriggen**, we get a detailed and amazing overview into state-of-the-art methods in optogenetics to study and manipulate mitochondrial functions and processes linked, including mitophagy and modulation of mitochondrial shape.

Certainly, this book can only cover a minor fraction of the set of mitochondrial functions but it gives important insights on the overall highly dynamic nature of mitochondria and its integration in a larger context. Moreover, it summarizes several technological advances in mitochondrial research leading to these views. In **Chapter 10**, beyond some personal views, I aim to give a synopsis on the field, its development over time, and its possible future. I further propose the '*Dyna*mic-*Tr*ap-and-Fl*ux*' ('DynaTrux') model giving a possible rational for the physiological function of mitochondrial cristae dynamics.

Editing this book was a true challenge and I fully underestimated the work associated with it. Still, it was also inspiring and I thank **Helmut Sies** and **Enrique Cadenas** for giving me the chance to do so. I am deeply grateful to all contributors of this book, who did a fantastic job in sharing their expertise and giving excellent insights into the topic. I hope that all readers find interesting and also unexpected aspects of mitochondrial biology in these chapters, and even more importantly that it inspires their own research and scientific curiosity – enjoy reading.

Andreas S. Reichert

Editor

Andreas S. Reichert has been a Professor and Director of the Institute of Biochemistry and Molecular Biology I at the Heinrich Heine University Düsseldorf, Germany since 2014. He's been a board member of the Biologisch-Medizinisches Forschungszentrum (BMFZ) at Heinrich Heine University Düsseldorf, Germany since 2016, and a speaker of BMFZ since 2021. Prof. Dr. Reichert received his Ph.D. under Prof. Dr. h.c. mult. Svante Pääbo at the Institute of Zoology, Ludwig-Maximilians-Universität München, Germany in 1999, and was a Post-Doc under Prof. Dr. h.c. mult. Svante Pääbo and Prof. Dr. Mario Mörl at the Max Planck Institute for Evolutionary Anthropology, in Leipzig, Germany from 1999–2000. After that he worked as a group leader in the group of Prof. Dr. h.c. mult. Walter Neupert at the Adolf Butenandt Institute for Physiological Chemistry, Ludwig Maximilians-Universität München, Germany from 2000–2007 before he operated as Professor and Head of Mitochondrial Biology at Goethe University Frankfurt am Main, Germany, from 2007 to 2014.

Contributors

Ruchika Anand
Institute of Biochemistry and Molecular Biology I
Medical Faculty and University Hospital Düsseldorf
Heinrich Heine University Düsseldorf
Düsseldorf, Germany

Hannes M. Beyer
Institute of Synthetic Biology
Heinrich Heine University Düsseldorf
Düsseldorf, Germany

Steven M. Claypool
Department of Physiology
Mitochondrial Phospholipid Research Center
Department of Genetic Medicine
Johns Hopkins University School of Medicine
Baltimore, Maryland, USA

Odette Deen-Rozalen
Institute of Synthetic Biology
Heinrich Heine University Düsseldorf
Düsseldorf, Germany

Luke E. Formosa
Department of Biochemistry and Molecular Biology
Monash Biomedicine Discovery Institute
Monash University
Melbourne, Australia

Patrick Horten
Institute of Biochemistry and Molecular Biology
ZBMZ – Zentrum für Biochemie und Molekulare Zellforschung
Faculty of Medicine, CIBSS
Centre for Integrative Biological Signalling Studies
Faculty of Biology
University of Freiburg
Freiburg, Germany

Andrea Irazoki
Department of Biomedical Sciences
University of Copenhagen
Copenhagen, Denmark

Oleh Khalimonchuk
Department of Biochemistry
University of Nebraska–Lincoln
and
Nebraska Redox Biology Center
and
Fred & Pamela Buffett Cancer Center
Omaha, Nebraska, USA

Arun Kumar Kondadi
Institute of Biochemistry and Molecular Biology I
Medical Faculty and University Hospital Düsseldorf
Heinrich Heine University Düsseldorf
Düsseldorf, Germany

Carla Lopes
Multidisciplinary Institute of Ageing
University of Coimbra
Coimbra, Portugal

Marius Nieke
Institute of Synthetic Biology
Heinrich Heine University Düsseldorf
Düsseldorf, Germany

David Pla-Martín
Institute of Biochemistry and Molecular Biology I
Medical Faculty and University Hospital Düsseldorf
Heinrich Heine University Düsseldorf
Düsseldorf, Germany

Gunjan Purohit
Department of Biochemistry
University of Nebraska–Lincoln
Lincoln, Nebraska, USA

Nuno Raimundo
Department of Cellular and Molecular Physiology
and
Penn State Cancer Institute
Penn State College of Medicine
Hershey, Pennsylvania, USA

Heike Rampelt
Institute of Biochemistry and Molecular Biology
ZBMZ – Zentrum für Biochemie und Molekulare Zellforschung
Faculty of Medicine
CIBSS Centre for Integrative Biological Signalling Studies
University of Freiburg
Freiburg, Germany

Andreas S. Reichert
Institute of Biochemistry and Molecular Biology I
Medical Faculty and University Hospital Düsseldorf
Heinrich Heine University Düsseldorf
Düsseldorf, Germany

Jan Riemer
Biochemistry – Redox Metabolism
Institute of Biochemistry
Faculty of Mathematics and Natural Sciences
CECAD
University of Cologne
Cologne, Germany

Nanami Senoo
Department of Physiology
Mitochondrial Phospholipid Research Center
Johns Hopkins University School of Medicine
Baltimore, Maryland, USA

Sara Shumka
Institute of Synthetic Biology
Heinrich Heine University Düsseldorf
Düsseldorf, Germany

Helmut Sies
Institute of Biochemistry and Molecular Biology I
Faculty of Medicine
Heinrich Heine University Düsseldorf
and
Leibniz Research Institute for Environmental Medicine
Düsseldorf, Germany

Matias D. Zurbriggen
Institute of Synthetic Biology
and
CEPLAS – Cluster of Excellence on Plant Sciences
Heinrich Heine University Düsseldorf
Düsseldorf, Germany

Part I

Mitochondrial Ultrastructure

Molecular Mechanisms Shaping the Inner Membrane

1 Roles of the F_1F_O-ATP Synthase and MICOS in Mitochondrial Membrane Organization

*Patrick Horten and Heike Rampelt**

INTRODUCTION

The intricate architecture of the mitochondrial inner membrane is of fundamental importance for many aspects of mitochondrial physiology, including energy metabolism, protein and lipid biogenesis, and mitochondrial inheritance. In contrast to the planar outer membrane, the inner mitochondrial membrane is characterized by large membrane invaginations, the cristae, resulting in topologically and functionally distinct subcompartments of both the inner membrane and the intermembrane space (IMS). The lamellar or tubular cristae are connected to the smooth inner boundary membrane (IBM) by narrow membrane necks called crista junctions (Frey et al., 2002; Mannella, 2006; Zick et al., 2009; Pánek et al., 2020; Klecker & Westermann, 2021). Crista junctions impose restraints on diffusion between the subcompartments of the membrane and the IMS and thereby contribute to their functional specialization. Protein complexes are asymmetrically distributed between the subcompartments: the protein translocation and biogenesis components are localized in the IBM, whereas the oxidative phosphorylation (OXPHOS) complexes are embedded in the cristae membranes (Gilkerson et al., 2003; Vogel et al., 2006; Cogliati et al., 2016; Stoldt et al., 2018). Moreover, the particular architecture of cristae membranes enables optimal respiration and ATP synthesis, and rapidly adapts to metabolic changes and altered availability of substrates (Hackenbrock, 1966; Strauss et al., 2008; Davies et al., 2011; Colina-Tenorio et al., 2020).

Inner membrane topology and compartmentalization is determined by specialized protein machineries in synergy with phospholipids. While many open questions remain, especially regarding the dynamics and variability of the inner membrane, it is clear that the F_1F_O-ATP synthase and the mitochondrial contact site and cristae organizing system (MICOS) are central determinants of inner membrane architecture (Cogliati et al., 2016; Kühlbrandt, 2019; Colina-Tenorio et al., 2020; Eramo et al., 2020; Anand et al., 2021; Mukherjee et al., 2021). These two protein complexes preferentially localize to distinct submitochondrial regions and, among other functions, generate opposite types of membrane curvature that are required to shape cristae in a balanced manner. Dimers and dimer rows of the F_1F_O-ATP synthase are necessary to generate the positive curvature at crista rims, while MICOS stabilizes the negative curvature at crista junctions. Owing to advanced microscopy approaches including super-resolution microscopy, electron tomography and FIB-SEM (focused ion beam scanning electron microscopy) (Wolf et al., 2019; Jakobs et al., 2020; Kondadi et al., 2020; Ohta et al., 2021; Zabeo & Davies, 2022), we have seen considerable progress in the last few years in the study of membrane architecture, as well as of membrane dynamics and remodelling, in unprecedented structural detail. These studies have revealed general principles of cristae morphogenesis. In *Saccharomyces cerevisiae*, there appear to exist at least two parallel pathways for the biogenesis of tubular versus sheet-like cristae (Harner et al., 2016; Kojima et al., 2019; Klecker & Westermann, 2021). Both pathways depend on the function of MICOS as well as the ATP synthase (Harner et al., 2016). The biogenesis of tubular cristae is thought to rely on the

* Corresponding author: heike.rampelt@biochemie.uni-freiburg.de

DOI: 10.1201/9781003221180-2

progressive growth of inner membrane invaginations and, consequently, also depends on phospholipid transport (Kojima et al., 2019). In yeast, the biogenesis of sheet-like cristae additionally requires the action of the dynamin-like GTPase Mgm1 (Harner et al., 2016) which mediates inner membrane fusion and functions as a membrane tether in cristae architecture (Meeusen et al., 2006; Cogliati et al., 2016; Faelber et al., 2019; Gao & Hu, 2021). In contrast, the human Mgm1 ortholog OPA1 is not required for sheet-like cristae (Stephan et al., 2020), but rather stabilizes crista junctions as reported previously (Frezza et al., 2006; Ishihara et al., 2006; Glytsou et al., 2016; Gao & Hu, 2021; Yapa et al., 2021). Instead, it appears that in human cells the Mic10 subcomplex plays an important role in the biogenesis of lamellar cristae (Stephan et al., 2020).

Importantly, it is becoming apparent that cristae membranes are unexpectedly dynamic and heterogeneous. Not only do lamellar cristae disappear within minutes after inactivation of Mgm1 in yeast (Harner et al., 2016), but time-resolved fluorescence imaging has shown that cristae in mammalian mitochondria are highly mobile and appear to even fuse and divide rapidly (Hu et al., 2020; Kondadi et al., 2020). Contrary to the notion that the mitochondrial membrane potential equilibrates across the entire inner membrane, a recent study discovered that individual cristae constitute independent units with differing membrane potentials (Wolf et al., 2019). Moreover, human mitochondria efficiently remodel the existing aberrant inner membranes of MICOS mutants upon re-expression of the missing component (Stephan et al., 2020). In summary, cristae membranes display an astonishing degree of dynamism and malleability beyond the rapidly interconvertible metabolic states of orthodox versus condensed mitochondria that have been known for half a century (Hackenbrock, 1966). The F_1F_O-ATP synthase and MICOS, central determinants of cristae architecture, as well as their functional crosstalk, are key to a mechanistic understanding of mitochondrial cristae morphogenesis and dynamic behaviour.

THE F_1F_O-ATP SYNTHASE

The F_1F_O-ATP synthase, or Complex V of the OXPHOS machinery, synthesizes the vast majority of the cellular ATP, using as its energy source the proton gradient across the inner membrane that is generated by the complexes of the respiratory chain. The F_1F_O-ATP synthase functions like a turbine: translocation of protons across the inner membrane via the membrane-embedded F_O part results in rotation of its c-ring, and the torque is transmitted through the central stalk, driving conformational changes and ATP synthesis by the catalytic F_1 part (Allegretti et al., 2015; Nirody et al., 2020; Almendro-Vedia et al., 2021). However, the mitochondrial F_1F_O-ATP synthase performs an additional crucial function: it induces the strong membrane curvature of cristae rims and edges (Figure 1.1A) (Paumard et al., 2002; Strauss et al., 2008; Davies et al., 2012; Nirody et al., 2020). In contrast to bacterial and chloroplast ATP synthases that are monomeric and evenly distributed, ATP synthases from mitochondria form dimers where the monomers connect at a pronounced angle (almost 90° in the so-called type I dimers of animal and fungal ATP synthase), resulting in membrane bending (Daum et al., 2010; Kühlbrandt, 2019; Nirody et al., 2020). Moreover, *in vivo* dimers associate to oligomeric rows that can reach lengths of more than 1 μm and localize to the strongly curved rims of lamellar cristae (Davies et al., 2011; 2012; Blum et al., 2019; Kühlbrandt, 2019). Oligomerization and membrane bending are widely conserved features of mitochondrial ATP synthases even in organisms with distinct dimer structures or subunit composition such as ciliates, unicellular green algae and others (Blum et al., 2019; Mühleip et al., 2019; 2020; Flygaard et al., 2020; Nirody et al., 2020; Gahura et al., 2022). Mathematical modelling indicated that strongly curved membranes can accommodate a higher charge density, and accordingly cristae are thought to function as proton traps (Strauss et al., 2008). Importantly, the mitochondrial ATP synthase creates its own topological niche within cristae that enables it to function as a proton sink at cristae rims, optimally exploiting the pH gradient across the mitochondrial inner membrane (Strauss et al., 2008; Cogliati et al., 2016; Blum et al., 2019; Toth et al., 2020; Rieger et al., 2021). Moreover, the local membrane potential in cristae was found to be higher than that at the IBM, in a manner directly dependent on ATP synthase activity (Wolf et al., 2019).

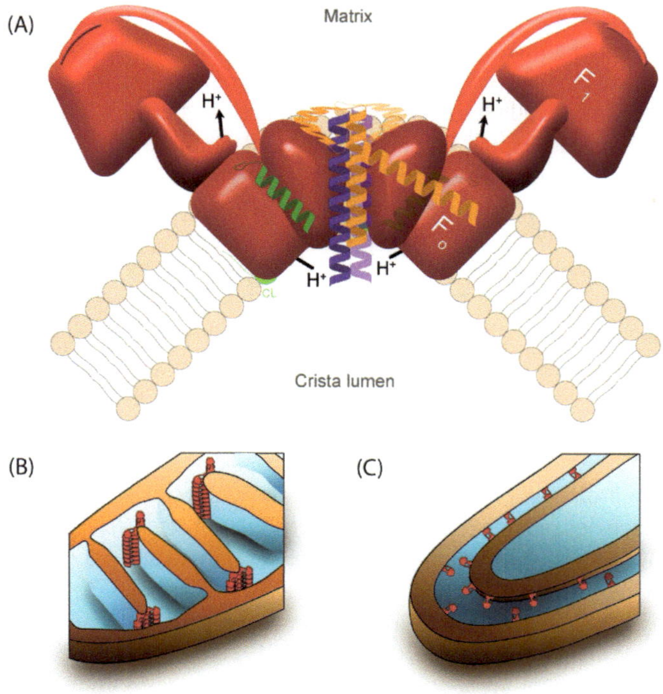

FIGURE 1.1 (A) The F_1F_O-ATP synthase of fungal and animal mitochondria forms dimers at an angle of almost 90° that induce the positive membrane curvature found at the edges and rims of cristae. Dimerization and membrane bending relies on the subunits Atp20 (subunit g, Orange) and Atp21 (subunit e, violet) that form a wedge domain. The subunit Atp19 (subunit k/DAPIT, dark green) additionally stabilizes dimers and dimer rows of the ATP synthase. Dimerization is not required for the catalytic action of the ATP synthase but supports optimal OXPHOS function via its role for cristae architecture. Cardiolipin (CL) stabilizes higher order assemblies of the ATP synthase. (B) In mitochondria from wild-type cells, ATP synthase dimers associate to oligomeric rows that localize to areas of strong inner membrane curvature, in particular to the strongly folded edges of cristae sheets. (C) Loss of ATP synthase dimerization results in aberrant inner membranes that are largely devoid of strong curvature and appear as balloons or onion skins.

Loss of ATP synthase dimerization and, consequently, of oligomerization results in aberrant inner membrane architecture: instead of cristae (Figure 1.1B), the inner membrane forms large balloon- or onion-like structures that can septate the entire mitochondrion and lack the strong membrane curvature seen at cristae rims (Figure 1.1C) (Paumard et al., 2002; Rabl et al., 2009; Davies et al., 2012). Due to the crucial role of ATP synthase higher order assemblies for cristae architecture, their disruption reduces the mitochondrial membrane potential and results in growth defects and human disease (Bornhövd et al., 2006; Habersetzer et al., 2013; Barca et al., 2018; Siegmund et al., 2018; Rampelt et al., 2022). Formation of the membrane bending ATP synthase dimer relies on the subunits Atp20 (subunit g) and Atp21 (subunit e) that form wedge-shaped assemblies bound to Atp4 (subunit b) on both sides of the dimer (Figure 1.1A) (Paumard et al., 2002; Arselin et al., 2004; Hahn et al., 2016; Guo et al., 2017; Pinke et al., 2020; Spikes et al., 2020). Atp20 and Atp21 interact within the wedge via conserved GxxxG motifs in their transmembrane segments (Arselin et al., 2003; Bustos & Velours, 2005; Gahura et al., 2022), but they also connect monomers from adjacent dimers (Arselin et al., 2003; Bustos & Velours, 2005; Pinke et al., 2020; Spikes et al., 2020). Additionally, Atp19 (subunit k, in mammals also known as DAPIT) engages in contacts with subunits from adjacent monomers, thereby stabilizing oligomers; in *S. cerevisiae*, it also stabilizes the oligomerization-competent dimer (Wagner et al., 2010; He et al., 2018; Pinke et al., 2020; Spikes et al., 2020). The monomer-monomer contacts differ between mammalian and yeast mitochondrial ATP synthases, explaining the lower stability of the mammalian dimer (Guo et al., 2017; Pinke et al., 2020; Spikes et al., 2020). Mutation of human DAPIT causes the mitochondriopathy Leigh syndrome due to a pronounced destabilization of ATP synthase dimers and oligomers, resulting in cristae defects (Barca et al., 2018; Siegmund et al., 2018). The membrane-shaping function of the ATP synthase is also relevant in the context of ageing: the inner mitochondrial membrane becomes progressively disordered during ageing in various organisms and ultimately vesiculates (Daum et al., 2013; Brandt et al., 2017). In the ageing model organism *Podospora anserina*, these ultrastructural changes correlate with a loss of oligomeric and dimeric ATP synthase complexes (Brust et al., 2010; Daum et al., 2013).

Higher order assemblies of the ATP synthase are stabilized by the mitochondrial signature phospholipid cardiolipin. Cardiolipin has unique features: it comprises two phosphatidyl groups that harbour two negative charges and four acyl chains, and it has a conical shape in the membrane, resulting in a propensity to segregate to curved membrane regions (van den Brink-van der Laan et al., 2004; Osman et al., 2011; Schlame & Greenberg, 2016; Mårtensson et al., 2017; Almendro-Vedia et al., 2021). Moreover, cardiolipin is uniquely well-suited to alleviate packing stress in the membrane caused by protein complex formation and the extraordinarily high protein concentration of the inner mitochondrial membrane (Ren et al., 2014; Xu et al., 2019; 2021). Consequently, cardiolipin is required for the stability and full function not only of respiratory chain supercomplexes but also of metabolite carriers and protein translocases (Pfeiffer et al., 2003; Horvath & Daum, 2013; Mileykovskaya & Dowhan, 2014; Böttinger et al., 2015; Basu Ball et al., 2017; Mårtensson et al., 2017; Chatzispyrou et al., 2018; Hartley et al., 2019; Rathore et al., 2019). In humans, defects in cardiolipin remodelling that are associated with decreased cardiolipin levels (due to formation of monolyso-cardiolipin) cause Barth syndrome, another mitochondrial disease that presents with cardiomyopathy (Barth et al., 2004; Dudek et al., 2016; Ghosh et al., 2019). Importantly, decrease or lack of cardiolipin results in dissociation of ATP synthase dimers and oligomers (Acehan et al., 2011; Chatzispyrou et al., 2018). Remarkably, cardiolipin molecules specifically bind ATP synthase in different organisms, in particular as part of the Atp20/Atp21 wedge domain (Duncan et al., 2016; Mühleip et al., 2019; Spikes et al., 2020; Gahura et al., 2022).

The oligomeric state of the ATP synthase can also be regulated by protein-protein interactions involving additional factors. Under certain metabolic conditions, the activity of the ATP synthase is subject to regulation by inhibitory factor 1 (IF1) which, dependent on the pH, inserts in between the α and β subunits and is thought to block catalysis of the reverse reaction that would result in ATP hydrolysis (Cabezón et al., 2003; García-Aguilar & Cuezva, 2018; Gatto et al., 2022). Since IF1 forms dimers that link adjacent F_1 particles, binding of IF1 induces tetramerization of the ATP synthase (Gu et al., 2019; Pinke et al., 2020; Mendoza-Hoffmann et al., 2022). It remains unclear whether the structure of the IF1-stabilized ATP synthase tetramer corresponds to the arrangement of ATP synthase in the non-inhibited oligomeric rows. However, forcing ATP synthase tetramerization causes alterations in cristae ultrastructure (Campanella et al., 2008; Weissert et al., 2021), suggesting that IF1 regulates not only enzymatic activity but also the membrane deforming activity of the ATP synthase, with implications for mitochondrial bioenergetics and apoptotic resistance (Campanella et al., 2008; Faccenda et al., 2013; Rieger et al., 2021). In plant mitochondria whose monomeric ATP synthase does not have a membrane-shaping function, catalysis in the dark, which would result in ATP hydrolysis due to an inadequate energy status, is shut down by a different mechanism involving reversible disulfide bond formation in the γ subunit of the central stalk (Nalin & McCarty, 1984; Hahn et al., 2018; Kühlbrandt, 2019).

In a prime example of regulatory crosstalk between membrane-shaping machineries, Mic10, a core component of MICOS (see the next section), interacts with dimers and oligomers of the ATP synthase and stabilizes them (Eydt et al., 2017; Rampelt et al., 2017; Rampelt et al., 2022). Recent work has demonstrated that the interaction is conserved even in the evolutionarily very distant unicellular parasite *Trypanosoma brucei* (Cadena et al., 2021). This regulatory function of Mic10 is important for efficient metabolic adaptation and optimal mitochondrial physiology in yeast mitochondria

(Rampelt et al., 2022). This interaction may support the actions of the two machineries during cristae biogenesis (see section "Antagonism and crosstalk of MICOS and F_1F_O-ATP synthase").

THE MITOCHONDRIAL CONTACT SITE AND CRISTAE ORGANIZING SYSTEM

The cisternal or tubular cristae membranes are separated from the contiguous but topologically distinct inner boundary membrane (IBM) by relatively uniform membrane necks, the crista junctions. Crista junctions are circular or slit-like and typically 15–40 nm in diameter, depending on the species (Zick et al., 2009). These structures display a pronounced membrane curvature and require the widely conserved MICOS for their stability. MICOS is a heterooligomeric protein complex comprising at least six subunits in yeast and seven in human mitochondria that consists of two modules (Figure 1.2A) (Hoppins et al., 2011; von der Malsburg et al., 2011; Harner et al., 2011; Alkhaja et al., 2012; Pfanner et al., 2014; Guarani et al., 2015; Kozjak-Pavlovic, 2017; Colina-Tenorio et al., 2020; Anand et al., 2021; Mukherjee et al., 2021). The subunits are thought to be present in multiple copies, it is unclear however whether MICOS has a defined stoichiometry. Upon disruption of MICOS, crista junctions are lost, and the resulting cristae membranes are aberrant and lack a connection to the IBM (Figure 1.2B). Depending on the cell type and the subunit deletion, the inner membrane in MICOS-deficient mitochondria forms elongated membrane stacks or structures resembling a tube or onion; in some cases, inner membrane invaginations are mostly absent (Rabl et al., 2009; Darshi

(A)

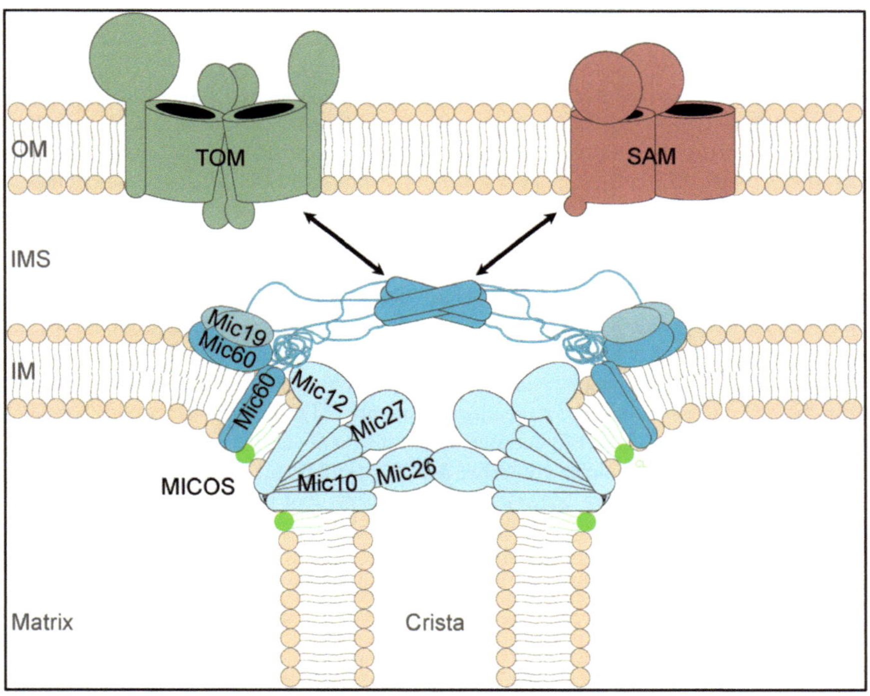

FIGURE 1.2 (A) MICOS (mitochondrial contact site and cristae organizing system) is a multimeric protein complex predominantly localized at crista junctions, the neck-like membrane structures that connect cristae to the inner boundary membrane. The MICOS subcomplex consisting of Mic60 and Mic19 forms contact sites between the inner and outer membrane (IM and OM) by interacting with the mitochondrial protein biogenesis machineries TOM (translocase of the outer mitochondrial membrane) and SAM (sorting and assembly machinery). In addition, the Mic60-Mic19 subcomplex has been suggested to form a dome over the crista junction. The Mic10 subcomplex that also includes Mic12 (QIL1 in mammals), Mic26 and Mic27 induces membrane curvature via oligomerization of Mic10 and influences the phospholipid environment of MICOS by binding cardiolipin (CL).

(Continued)

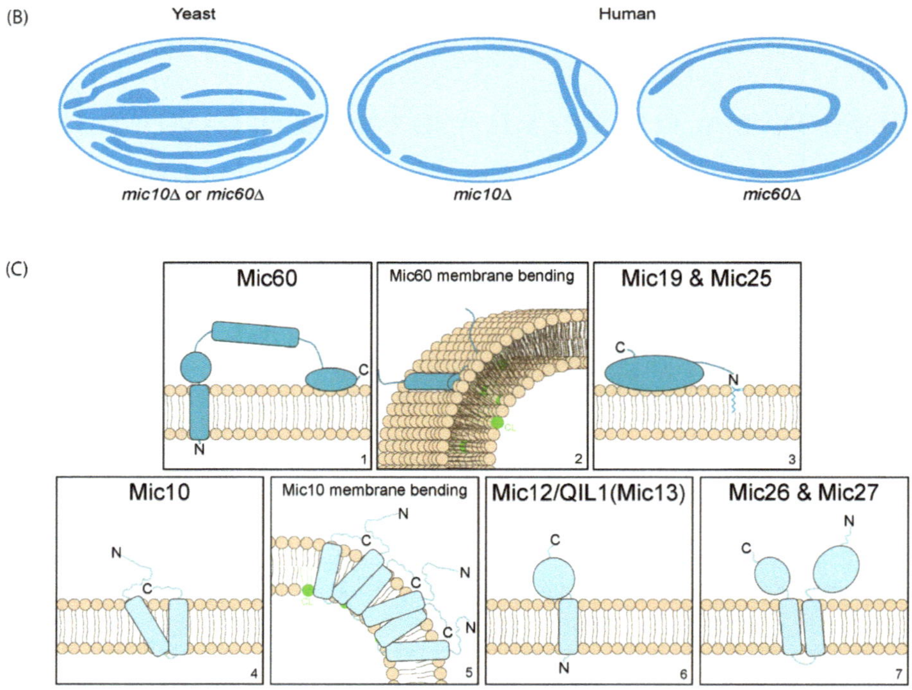

FIGURE 1.2 *(Continued)* (B) Mitochondrial inner membrane architecture in MICOS-deficient cells. The almost complete loss of crista junctions is a common feature for MICOS mutants, but the resulting cristae architecture differs somewhat between yeast and human cells. In yeast (left panel), strong MICOS mutations such as loss of Mic10 or Mic60 result in the accumulation of mostly parallel stacks of cristae membranes. Human mitochondria lacking Mic10 or Mic60 (middle and right panel) display a reduction of cristae membranes as well as aberrant membrane architecture. IMS/crista lumen shown in dark blue, matrix in light blue. (C) Topologies of MICOS subunits at the inner membrane and modes of membrane shaping. All MICOS subunits face the intermembrane space (IMS). The Mic60 subcomplex comprises Mic60 and Mic19, and in vertebrate cells, the Mic19 ortholog Mic25 (panels 1–3). Mic60 (panel 1) has an N-terminal transmembrane segment followed by several α-helical domains that form homomeric interactions or that bind the lipid bilayer (panel 2) or Mic19 (panel 3). Mic60 induces membrane bending by an amphipathic helix that asymmetrically displaces phospholipids from the IMS-facing monolayer of the inner membrane. Mic19 and Mic25 are anchored to the membrane by N-terminal myristoylation (panel 3) and interact with Mic60 via their CHCH (coiled-coil helix coiled-coil helix) domain. The Mic10 subcomplex comprises Mic10, Mic12 (QIL1/Mic13 in mammalian cells), Mic26 and Mic27 (panels 4–7). Mic10 has two transmembrane segments that form a hairpin (panel 4). This wedge-like shape together with its ability to oligomerize (panel 5) makes Mic10 a strong membrane-shaping factor. The phospholipid cardiolipin (CL) is important for Mic10 oligomer stability. Mic12/QIL1 (panel 6) bridges the Mic10 and Mic60 subcomplexes. Mic26 and Mic27 (panel 7) are related proteins that have regulatory functions for the Mic10 subcomplex.

et al., 2011; Harner et al., 2011; Hoppins et al., 2011; von der Malsburg et al., 2011; Alkhaja et al., 2012; Friedman et al., 2015; Guarani et al., 2015; Stephan et al., 2020). MICOS disruption results in a decreased membrane potential, impaired respiratory growth, lower efficiency of specific mitochondrial protein and cofactor biogenesis pathways, defects in mitochondrial dynamics and inheritance and impaired ability to adapt to metabolic changes (Rabl et al., 2009; Harner et al., 2011; Hoppins et al., 2011; von der Malsburg et al., 2011; Darshi et al., 2011; Alkhaja et al., 2012; Bohnert et al., 2012; Itoh et al., 2013; Ott et al., 2015; Guarani et al., 2015; Bohnert et al., 2015; Callegari et al., 2019; Anand et al., 2020; Dietz et al., 2021; Jakubke et al., 2021; Rampelt et al., 2022). Moreover, mutations in MICOS subunits can be causative for human mitochondriopathies: patient mutations in QIL1 (the human ortholog of Mic12) and MIC26 cause severe mitochondrial diseases (see below; Guarani et al., 2016; Zeharia et al., 2016; Gödiker et al., 2018; Benincá et al., 2021; Peifer-Weiß et al., 2023). Mutations in the Mic60 mitochondrial targeting sequence with a dominant effect in reducing Mic60 levels have been identified in patients with Parkinson's disease (Tsai et al., 2018; Van Laar et al., 2019). In addition, alterations in the levels of MICOS subunits, in particular Mic60, have been implicated in the pathology of diabetes, epilepsy and other conditions both in patients and in disease models (Colina-Tenorio et al., 2020; Eramo et al., 2020; Mukherjee et al., 2021).

The two core components of MICOS, Mic60 and Mic10, are the evolutionarily oldest subunits. They exist in all major eukaryotic lineages and their presence correlates with that of inner membrane invaginations (Muñoz-Gómez et al., 2015; 2016; Huynen et al., 2016; Kaurov et al., 2018; Muñoz-Gómez et al., 2023). Mic60 homologs were identified even in a number of α-proteobacteria and thus predate the endosymbiotic event that gave rise to mitochondria (Muñoz-Gómez et al., 2015; Muñoz-Gómez et al., 2023). Since Mic10 and Mic60 both induce membrane curvature (see below), these results show that MICOS is an evolutionarily ancient membrane-shaping complex of the inner membrane. Interestingly, Mic10 and Mic60 form independent subcomplexes (von der Malsburg et al., 2011; Harner et al., 2011; Hoppins et al., 2011; Bohnert et al., 2015; Friedman et al., 2015; Guarani et al., 2015; Anand et al., 2016; Zerbes et al., 2016): the Mic60 subcomplex additionally includes its regulatory partner Mic19 (in metazoa also the closely related Mic25) (Xie et al., 2007; Rabl et al., 2009; Darshi et al., 2011; An et al., 2012; Ott et al., 2012; Ding et al., 2015; Tang et al., 2019). The Mic10 complex comprises Mic12 (yeast) or QIL1/Mic13 (animals) which connects the two subcomplexes, as well as the regulatory subunits Mic26 and Mic27 (Weber et al., 2013; Bohnert et al., 2015; Guarani et al., 2015; Koob et al., 2015; Zerbes et al., 2016; Rampelt et al., 2018; Urbach et al., 2021).

Mic60, also known as mitofilin, performs various functions in mitochondria that are non-redundant with those of other MICOS subunits. It is anchored in the inner membrane by an N-terminal transmembrane helix and exposes its bulk to the IMS, and it forms dimers as well as larger assemblies (Figure 1.2A) (Rabl et al., 2009; Körner et al., 2012; Zerbes et al., 2012; Hessenberger et al., 2017). Mic60, independently or together with its subcomplex partner Mic19, forms contact sites between the inner and outer mitochondrial membranes by interacting with protein translocases such as the translocase of the outer membrane (TOM) and the sorting and assembly machinery (SAM), as well as with other outer membrane proteins (Harner et al., 2011; von der Malsburg et al., 2011; Bohnert et al., 2012; Körner et al., 2012; Ott et al., 2012; Zerbes et al., 2012; Kaurov et al., 2018). In human cells, MICOS and the SAM complex form a stable higher order assembly, the mitochondrial intermembrane space bridging (MIB) complex (Xie et al., 2007; Darshi et al., 2011; Ott et al., 2012; 2015; Huynen et al., 2016; Sastri et al., 2017; Tang et al., 2019). These OM interactions of Mic60 directly facilitate the biogenesis of some mitochondrial proteins and phospholipids (Harner et al., 2011; von der Malsburg et al., 2011; Bohnert et al., 2012; Aaltonen et al., 2016; Kaurov et al., 2018; Eichenberger et al., 2019; Callegari et al., 2019). Since downregulation of mammalian SAM components results in MICOS destabilization as well as in a defective cristae architecture (Ott et al., 2012; 2015; Tang et al., 2019), the contact sites may also anchor crista junctions to the outer membrane and thereby stabilize them.

Interactions of Mic60 with OM proteins show differential dependence on its C-terminal mitofilin domain and the coiled-coil domain that includes a long extended α-helix (Figure 1.2C, panel 1) (Körner et al., 2012; Zerbes et al., 2012; Jumper et al., 2021). In the evolutionarily distant *T. brucei* whose MICOS complex is highly diverged, a mitofilin domain appears to be lacking in Mic60 but may be supplied in *trans* by trypanosome-specific MICOS subunits (Kaurov et al., 2018; Hashimi, 2019; Jumper et al., 2021). The coiled-coil domain mediates formation of a tetramer consisting of two antiparallel dimers (Figure 1.2A) (Bock-Bierbaum et al., 2022). The interactions of Mic60 are modulated by its MICOS subcomplex partner Mic19 via Mic19's CHCH (coiled-coil helix coiled-coil helix) domain, also known as a twin CX_9C domain. This domain, which is widespread among IMS proteins, is stabilized by conserved disulfide bonds, and proteins are imported by the mitochondrial import and assembly (MIA) pathway (Sakowska et al., 2015; Wiedemann & Pfanner, 2017; Habich et al., 2019). Association of Mic19 with the Mic60 mitofilin domain shifts Mic60 from the dimeric to the tetrameric form and also enhances the membrane-deforming activity of Mic60 (Hessenberger et al., 2017; Bock-Bierbaum et al., 2022). Mic19 (along with the orthologous metazoan-specific Mic25, Figure 1.2C, panel 3) is the only MICOS subunit that is not an integral inner membrane protein (Harner et al., 2011; Hoppins et al., 2011; von der Malsburg et al., 2011) and therefore its import does not depend on the mitochondrial membrane potential. Instead, Mic19 and Mic25 bind the inner membrane via an N-terminal myristoyl group that is required for their mitochondrial targeting and import (Figure 1.2C, panel 3) (Darshi et al., 2011; 2012; Ueda et al., 2019).

Importantly, Mic60 induces membrane curvature via an amphipathic helix that is thought to insert partially into the IMS-facing leaflet of the inner membrane (Figure 1.2C, panel 1 + 2) (Hessenberger et al., 2017; Tarasenko et al., 2017), a similar mechanism as that employed by many BAR-domain proteins (Kozlov et al., 2014; Simunovic et al., 2019). This membrane-deforming activity is required for MICOS integrity and function, and it is conserved from prokaryotic Mic60 (Hessenberger et al., 2017; Tarasenko et al., 2017). Interestingly, Mic60-mediated membrane bending is inhibited by binding of a toxic disease-relevant metabolite (Zhou et al., 2022). A recent model based on several partial structures of Mic60 and Mic19 suggests that Mic60 tetramers span the diameter of crista junctions and form a net across the opening (Figure 1.2A) (Bock-Bierbaum et al., 2022), which could explain the uniform diameter of crista junctions as well as a diffusion bottleneck for soluble proteins at the crista junction. Interestingly, there may be species-dependent differences in the question whether it is Mic60 and/or Mic19 that contacts the outer membrane (Darshi et al., 2011; 2012; von der Malsburg et al., 2011; Körner et al., 2012; Zerbes et al., 2012; Tang et al., 2019; Abudu et al., 2021, Muñoz-Gómez et al., 2023). In this context, a study reported that stress-induced cleavage of the Mic19 N-terminus by the protease OMA1 disrupts the MIB complex (Tang et al., 2019).

The Mic60 subcomplex is regulated by phosphorylation as well as by oxidation. Mic60 communicates with PINK1, a kinase that is crucial for mitophagy of damaged mitochondria, with PINK1 mutations causing familial Parkinson's disease (Pickles et al., 2018; Eldeeb et al., 2022). On the one hand, phosphorylation of Mic60 and Mic19 by protein kinase A (PKA) disrupts the interaction of Mic60 with PINK1 and thereby protects mitochondria from PINK1-facilitated clearance (Akabane et al., 2016). On the other hand, PINK1 itself phosphorylates Mic60, enhancing Mic60 oligomerization and simultaneously stabilizing crista junctions, which might be particularly important in energy-demanding tissues or subcellular regions (Tsai et al., 2018). Underlining the contribution of MICOS dysfunction to neurodegeneration, Mic60 mutant variants with a defective biogenesis were described in some patients with Parkinson's disease (Tsai et al., 2018). In *Drosophila*, Mic60 was also identified as part of a redox signalling pathway that is activated during ageing as well as in disease models for Parkinson's and Friedreich's ataxia (Li et al., 2021). This pathway involves Miro proteins, outer membrane GTPases that couple mitochondria to motor adaptors and regulate several aspects of mitochondrial function and behaviour (Eberhardt et al., 2020). The authors found that oxidative stress induces disulfide formation in Mic60, resulting in an enhanced interaction with Miro, and Miro accumulation in the outer membrane (Li et al., 2021). Preventing Mic60 oxidation restored mitochondrial functions and improved the lifespan, possibly via a mechanism involving mitophagy (Eberhardt et al., 2020; Li et al., 2021).

The second core subunit Mic10 is a small protein with two transmembrane domains that, across various organisms, have an asymmetric length and a very short connecting loop, forming a hairpin (Figure 1.2C, panel 4) (Barbot et al., 2015; Bohnert et al., 2015; Jumper et al., 2021). The resulting wedge shape in the membrane is thought to displace more lipids in the IMS leaflet of the membrane, in a manner similar to the membrane-shaping reticulon protein family in the endoplasmic reticulum (Hu et al., 2011; Barbot et al., 2015). Importantly, Mic10 oligomerizes via conserved glycine-rich motifs in both of its TM segments, which significantly boosts its ability to induce membrane curvature (Figure 1.2C, panel 5) (Barbot et al., 2015; Bohnert et al., 2015). Interestingly, MICOS in *T. brucei* includes two Mic10 orthologs whose differing properties are not yet fully understood (Kaurov et al., 2018; Cadena et al., 2021). Oligomerization-deficient Mic10 mutants cannot support crista junctions despite interacting with the other MICOS subunits

(Barbot et al., 2015; Bohnert et al., 2015). Thus, membrane shaping by both Mic60 and Mic10 is required for MICOS function.

The Mic10 subcomplex includes the auxiliary subunits Mic12 (QIL1/Mic13 in mammals), Mic26 and Mic27. Mic12 connects the subcomplex to the Mic60-Mic19(-Mic25) module (Figure 1.2C, panel 6) (Guarani et al., 2015; Anand et al., 2016; Zerbes et al., 2016; Urbach et al., 2021). Interestingly, Mic12/QIL1 contains a highly conserved sequence motif that is required for its function and that may be part of an amphipathic helix (Huynen et al., 2016; Urbach et al., 2021). Pathogenic QIL1 mutations result in loss of the protein and are accompanied by destabilization of other MICOS subunits, ultimately leading to a severe mitochondrial encephalopathy with hepatopathy that can be fatal (Guarani et al., 2016; Zeharia et al., 2016; Gödiker et al., 2018). Mic26 and Mic27 (Figure 1.2C, panel 7) regulate Mic10 oligomerization in yeast and might also influence the phospholipid environment of MICOS (Weber et al., 2013; Koob & Reichert, 2014; Zerbes et al., 2016; Rampelt et al., 2018; Anand et al., 2020). Mic26 and Mic27 were described as part of the apolipoprotein O family, but a new report demonstrated that they do not exist as secreted forms (Lamant et al., 2006; Koob & Reichert, 2014; Koob et al., 2015; Lubeck et al., 2023). Yeast versus vertebrate Mic26 and Mic27 are the result of independent gene duplications and therefore might perform different functions in these organisms (Muñoz-Gómez et al., 2015; Huynen et al., 2016). Two distinct pathogenic mutations have been described for human MIC26. A substitution that alters its behaviour and membrane interaction caused an X-linked mitochondrial disease with various neuromuscular defects (Benincá et al., 2021), whereas a nonsense mutation that strongly destabilizes MIC26 resulted in severe progeria-like symptoms that led to death in infancy (Peifer-Weiß et al., 2023). In yeast, Mic26 and Mic27 antagonistically affect the oligomeric state of Mic10, with Mic26 destabilizing the complex, perhaps to enhance its dynamics, while Mic27 stabilizes Mic10 oligomers (Zerbes et al., 2016; Rampelt et al., 2018). Interestingly, deleting the stabilizing subunit Mic27 in yeast ameliorates the crista junction defect of *mic60Δ* cells and restores their respiratory growth (Eydt et al., 2017). One interpretation of these findings may be that respiratory growth is impaired not merely due to the absence of the contact site anchor Mic60 but also because of excessive, imbalanced assembly of the Mic10 subcomplex. This observation indicates that the Mic60 and Mic10 subcomplexes can act as antagonists. In line with this notion, deletion of Mic26, which further stabilizes the Mic10 subcomplex, aggravates the Mic60 deletion phenotype (Eydt et al., 2017).

The classification of Mic26 and Mic27 as apolipoproteins suggests that they may bind lipids. In fact, human Mic26 was found to bind the mitochondrial signature lipid cardiolipin *in vitro*, and cells lacking both subunits showed reduced overall cardiolipin levels (Weber et al., 2013; Anand et al., 2020). In yeast, cardiolipin is necessary to stabilize Mic10 oligomers, but this does not depend on Mic26 or Mic27 (Rampelt et al., 2018). Rather, the results suggest that the curvature-sensitive cardiolipin binds to the membrane bending factor Mic10. Moreover, cardiolipin is required for focal localization of the Mic10 subcomplex in the absence of Mic60 (Friedman et al., 2015). Thus, the intrinsic membrane-shaping properties of both cardiolipin and Mic10, combined with the stabilizing effect of cardiolipin on Mic10 oligomers, could synergistically facilitate assembly of the Mic10 subcomplex at the curved membranes of crista junctions.

ROLES OF MICOS IN INTEGRATING MITOCHONDRIA IN THEIR CELLULAR CONTEXT

Due to MICOS' crucial functions in stabilizing crista junctions and connecting the outer and inner mitochondrial membranes, this complex may perform a central role in communication of mitochondria with other machineries and organelles, in particular with the ER (van der Laan et al., 2012; Eisenberg-Bord & Schuldiner, 2017; Tamura et al., 2020; Zung & Schuldiner, 2020; Klecker & Westermann, 2021). In fact, MICOS has been found to play important roles in a diverse array of cellular functions.

MICOS shapes the mitochondrial inner membrane not only by interacting with specific phospholipids, it also participates in phospholipid biogenesis and transport processes. MICOS-dependent contact sites enable Psd1, the inner membrane decarboxylase that converts phosphatidylserine (PS) to phosphatidylethanolamine (PE), to act on PS in the outer membrane (Aaltonen et al., 2016). Destabilization of MICOS can also contribute to defects in cardiolipin biogenesis, and MICOS deletions display synthetic genetic defects with cardiolipin biosynthetic enzymes (Hoppins et al., 2011; Harner et al., 2014; Friedman et al., 2015; Anand et al., 2020). Importantly, the synthesis of various phospholipids is shared between mitochondria and the ER and requires interorganellar transport of some lipid species (Horvath & Daum, 2013; Tatsuta & Langer, 2017; Tamura et al., 2020). Mitochondria are closely associated with the endoplasmic reticulum via ER-mitochondria contact sites (ERMCS). Aside from lipid exchange, ERMCSs are involved in a multitude of functions including calcium signalling and ER-assisted mitochondrial division (Murley & Nunnari, 2016; Eisenberg-Bord & Schuldiner, 2017; Lackner, 2019; Tamura et al., 2020). The observation that MICOS clusters are localized nearby ERMCs (Murley & Nunnari, 2016; Qin et al., 2020; Tirrell et al., 2020; Monteiro-Cardoso et al., 2022), together with the involvement of MICOS in lipid biogenesis mentioned earlier, raises the question whether MICOS is part of the infrastructure that facilitates lipid transport between mitochondria and the ER. In fact, a recent study found that the lipid transport proteins ORP5 and ORP8 are localized at ERMCs, in close vicinity to MICOS and SAM. These machineries contribute to the transport of PS from the ER to mitochondria and its subsequent mitochondrial conversion to PE (Monteiro-Cardoso et al., 2022). MICOS in *Arabidopsis* was also found as part of a large protein complex important for lipid trafficking between mitochondria and plastids (Michaud et al., 2016).

Moreover, MICOS is important for the segregation, quality control and inheritance of mitochondrial nucleoids (Itoh et al., 2013; Yang et al., 2015; Li et al., 2016; Qin et al., 2020; Jakubke et al., 2021). Mitochondrial division is assisted by the ER and the actin cytoskeleton and takes place at replicating nucleoids (Boldogh et al., 2003; Frederick et al., 2004; Friedman et al., 2011; Korobova et al., 2013; Lewis et al., 2016). MICOS accordingly may play an important role in mtDNA inheritance and quality control, either indirectly by maintaining cristae architecture and mitochondrial dynamics, or directly via coupling the nucleoid to complexes spanning several membranes (van der Laan et al., 2012; Itoh et al., 2013; Yang et al., 2015; Li et al., 2016; Qin et al., 2020; Jakubke et al., 2021).

ERMCs and the GTPase Miro are also emerging as factors in positioning of MICOS, regulating cristae architecture and, via the Miro interactors TRAK1/2 and Myosin 19, possibly in maintaining membrane architecture during mitochondrial transport (Hoppins et al., 2011; Tsai et al., 2017; López Doménech et al., 2018; Oeding et al., 2018; Modi et al., 2019; Qin et al., 2020; Tirrell et al., 2020; Li et al., 2021; Nahacka et al., 2022; Shi et al., 2022). In several organisms, MICOS forms clusters that are arranged in helical bands (Hoppins et al., 2011; Jans et al., 2013; Stoldt et al., 2019; Kondadi et al., 2020). It will be interesting to investigate if contact sites between mitochondria and the ER have a role in establishing this distribution pattern.

Intriguingly, quality control of mitochondria also involves MICOS: mammalian SAMM50 was recently discovered to function as an adaptor for piecemeal basal mitophagy of SAM as well as MICOS, a pathway that is upregulated during metabolic adaptation to respiratory growth (Le Guerroué et al., 2017; Abudu et al., 2021). This raises the exciting question whether increased degradation of MICOS and SAM during oxidative phosphorylation reflects a requirement for enhanced quality control to remove damaged components, or whether it represents an adaptive regulatory mechanism that biochemically remodels crista junctions by replacing protein components and disrupting their interactions within and outside of mitochondria.

In summary, the close physical and functional connection of MICOS to cristae and respiratory metabolism on the one hand, and various communication processes at the OM on the other hand, makes it a central intermediary and integrator of mitochondrial physiology in its cellular context.

ANTAGONISM AND CROSSTALK OF MICOS AND F_1F_O-ATP SYNTHASE

MICOS and the ATP synthase generate opposing types of curvature in the inner membrane: while dimers and oligomers of the ATP synthase drive formation of positive curvature at cristae rims and tubules, MICOS stabilizes the negatively curved membrane funnel at crista junctions. These characteristics explain why the two machineries act as antagonists in various phenotypic read-outs. Overexpression of the membrane-shaping MICOS subunits Mic60 or Mic10 causes branching of cristae and membrane deformations consistent with increased negative curvature (Rabl et al., 2009; Bohnert et al., 2015), while Mic60 downregulation results in excessive oligomerization of the ATP synthase and an imbalance towards positive membrane curvature (Rabl et al., 2009; Rampelt et al., 2017; Stephan et al., 2020). In fact, the elongated, lamellar cristae stacks observed in MICOS mutants also depend on ATP synthase dimerization (Rabl et al., 2009; Hoppins et al., 2011). Thus, MICOS and ATP synthase are in direct competition for the generation of membrane curvature. Moreover, defects of MICOS mutants in growth, mitochondrial dynamics and in the segregation of nucleoids are ameliorated by additional disruption of ATP synthase dimerization (Rabl et al., 2009; Itoh et al., 2013), indicating that a balance between the two machineries is central for mitochondrial physiology. Interestingly, loss of ATP synthase-mediated positive curvature does not result in enhanced MICOS stability; instead, MICOS is also destabilized in ATP synthase dimerization mutants (Harner et al., 2016). However, the septated and branched inner membranes in these mutants reflect the presence of negative curvature (Rabl et al., 2009). In human cells, the localization of Mic60 clusters is aberrant in mitochondria lacking dimeric ATP synthase (Stephan et al., 2020). In conclusion, the machineries show interdependence that may be mediated by the loss of native cristae architecture or by direct crosstalk between the machineries.

Surprisingly, the MICOS core subunit Mic10 was found to interact with dimeric ATP synthase by a series of robust approaches (Eydt et al., 2017; Rampelt et al., 2017; Rampelt & van der Laan, 2017; Cadena et al., 2021; Rampelt et al., 2022) (Figure 1.3). The interaction is specific for Mic10, while other MICOS subunits are not required (Eydt et al., 2017; Rampelt et al., 2017). The interaction was reported not only from yeast but also from the evolutionarily very distant eukaryote *T. brucei* (Cadena et al., 2021) and may be conserved in human mitochondria (Stephan et al., 2020). Work employing either Mic10 overexpression or Mic10 variants that function selectively at the ATP synthase has shown that the interaction stabilizes oligomeric species of the ATP synthase (Rampelt et al., 2017; 2022). This substoichiometric regulatory function of Mic10 has turned out to be crucial for mitochondrial physiology (Figure 1.3): Mic10 variants that cannot support MICOS function, such as oligomerization-deficient mutants or a fusion of Mic10 with its interaction partner Atp21 at the ATP synthase, nevertheless restore the mitochondrial membrane potential and efficient metabolic adaptation as well as respiratory growth of the cells (Rampelt et al., 2022). This unexpected finding indicates that the regulatory crosstalk between the two machineries is of central importance for metabolism and mitochondrial physiology.

Mic10 could bind the ATP synthase at crista junctions, or alternatively the interaction could involve the Mic10 pool that is present in cristae membranes (Harner et al., 2011; Bohnert et al., 2015; Eydt et al., 2017), since it is independent of an intact MICOS complex (Rampelt et al., 2017). These two scenarios are not mutually exclusive: *in vivo* Mic10 might associate with dimeric ATP synthase at crista junctions and subsequently dissociate from MICOS. Such a mechanism might be relevant during cristae biogenesis when assembly and membrane-shaping actions of MICOS and ATP synthase dimers or oligomers need to be balanced. A regulatory function of Mic10 in this process can potentially explain its role in metabolic adaptation (Rampelt et al., 2022).

In conclusion, recent studies have shed light on the unexpectedly dynamic nature and tight control of the inner membrane, as well as on the molecular communication between membrane-shaping machineries. The emerging picture indicates that the intricate topologies and dynamic behaviours of the inner mitochondrial membrane arise from the crosstalk of individual players. MICOS and the

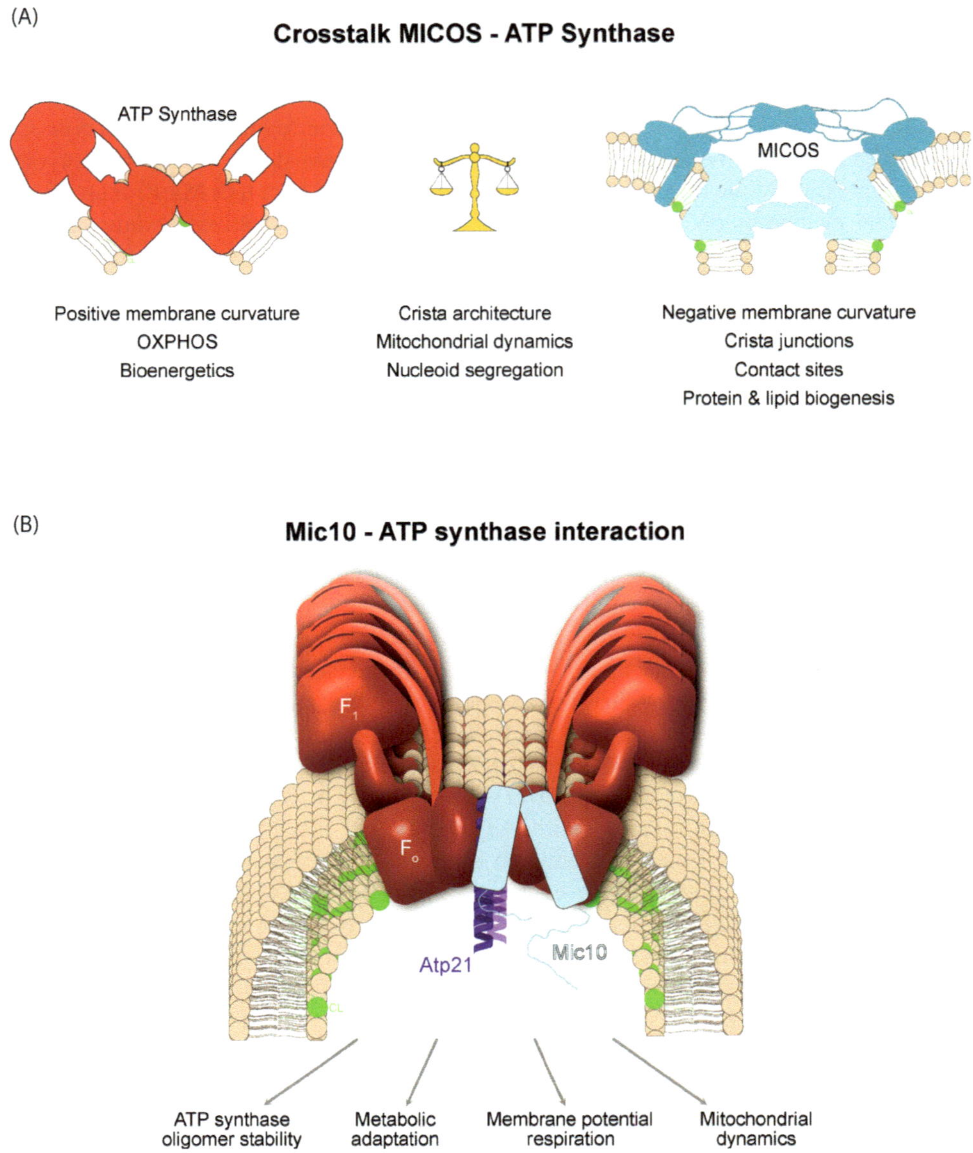

FIGURE 1.3 (A) The ATP synthase and MICOS each shape the inner mitochondrial membrane in distinct ways and thereby, on the one hand, perform independent functions. On the other hand, they share a crosstalk that in some contexts manifests as antagonism. Consequently, a balanced function of both machineries is also required for optimal mitochondrial physiology. (B) The MICOS core component Mic10 interacts with the ATP synthase via its dimerization subunit Atp21 (subunit e), potentially enabling communication between the two major cristae-shaping machineries ATP synthase and MICOS. This novel function of Mic10 results in a stabilization of ATP synthase oligomers and plays a crucial role in supporting multiple facets of mitochondrial physiology, impacting on the ability of cells to undergo metabolic adaptation and perform efficient respiration.

F_1F_O-ATP synthase are a prime example for this paradigm. Additionally, the properties and interactions of protein complexes are subject to regulation or remodelling by post-translational modifications and the phospholipid environment. The exciting progress of recent years has brought us a step closer towards elucidating the mechanistic basis of cristae dynamics operating on short time scales as well as of those governing global reorganization of the inner membrane.

ACKNOWLEDGEMENTS

We thank Nikolaus Pfanner, Jakob D. Busch, Kuo Song and Lilia Colina-Tenorio for helpful comments and discussion. Work included in this study has also been performed in partial fulfilment of the requirements for the doctoral thesis of P.H. at the University of Freiburg. This work was supported by the Deutsche Forschungsgemeinschaft (FOR 2848/P02) and Germany's Excellence Strategy/Initiative CIBSS-EXC-2189 (Project ID 390939984).

REFERENCES

Aaltonen MJ, Friedman JR, Osman C, Salin B, di Rago J-P, Nunnari J, Langer T & Tatsuta T (2016) MICOS and phospholipid transfer by Ups2-Mdm35 organize membrane lipid synthesis in mitochondria. J Cell Biol **213**: 525–534

Abudu YP, Shrestha BK, Zhang W, Palara A, Brenne HB, Larsen KB, Wolfson DL, Dumitriu G, Øie CI, Ahluwalia BS, Levy G, Behrends C, Tooze SA, Mouilleron S, Lamark T & Johansen T (2021) SAMM50 acts with p62 in piecemeal basal- and OXPHOS-induced mitophagy of SAM and MICOS components. J Cell Biol **220**: e202009092

Acehan D, Malhotra A, Xu Y, Ren M, Stokes DL & Schlame M (2011) Cardiolipin affects the supramolecular organization of ATP synthase in mitochondria. Biophys J **100**: 2184–2192

Akabane S, Uno M, Tani N, Shimazaki S, Ebara N, Kato H, Kosako H & Oka T (2016) PKA regulates PINK1 stability and Parkin recruitment to damaged mitochondria through phosphorylation of MIC60. Mol Cell **62**: 371–384

Alkhaja AK, Jans DC, Nikolov M, Vukotic M, Lytovchenko O, Ludewig F, Schliebs W, Riedel D, Urlaub H, Jakobs S & Deckers M (2012) MINOS1 is a conserved component of mitofilin complexes and required for mitochondrial function and cristae organization. Mol Biol Cell **23**: 247–257

Allegretti M, Klusch N, Mills DJ, Vonck J, Kühlbrandt W & Davies KM (2015) Horizontal membrane-intrinsic α-helices in the stator a-subunit of an F-type ATP synthase. Nature **521**: 237–240

Almendro-Vedia V, Natale P, González DV, Lillo MP, Aragones JL & López-Montero I (2021) How rotating ATP synthases can modulate membrane structure. Arch Biochem Biophys **708**: 108939

An J, Shi J, He Q, Lui K, Liu Y, Huang Y & Sheikh MS (2012) CHCM1/CHCHD6, novel mitochondrial protein linked to regulation of mitofilin and mitochondrial cristae morphology. J Biol Chem **287**: 7411–7426

Anand R, Kondadi AK, Meisterknecht J, Golombek M, Nortmann O, Riedel J, Peifer-Weiß L, Brocke-Ahmadinejad N, Schlütermann D, Stork B, Eichmann TO, Wittig I & Reichert AS (2020) MIC26 and MIC27 cooperate to regulate cardiolipin levels and the landscape of OXPHOS complexes. Life Sci Alliance **3**: e202000711–17

Anand R, Reichert AS & Kondadi AK (2021) Emerging roles of the MICOS complex in cristae dynamics and biogenesis. Biology (Basel) **10**:600

Anand R, Strecker V, Urbach J, Wittig I & Reichert AS (2016) Mic13 is essential for formation of crista junctions in mammalian cells. PLOS ONE **11**: e0160258

Arselin G, Giraud M-F, Dautant A, Vaillier J, Brèthes D, Coulary-Salin B, Schaeffer J & Velours J (2003) The GxxxG motif of the transmembrane domain of subunit e is involved in the dimerization/oligomerization of the yeast ATP synthase complex in the mitochondrial membrane. Eur J Biochem **270**: 1875–1884

Arselin G, Vaillier J, Salin B, Schaeffer J, Giraud M-F, Dautant A, Brèthes D & Velours J (2004) The modulation in subunits e and g amounts of yeast ATP synthase modifies mitochondrial cristae morphology. J Biol Chem **279**: 40392–40399

Barbot M, Jans DC, Schulz C, Denkert N, Kroppen B, Hoppert M, Jakobs S & Meinecke M (2015) Mic10 oligomerizes to bend mitochondrial inner membranes at cristae junctions. Cell Metab **21**: 756–763

Barca E, Ganetzky RD, Potluri P, Juanola-Falgarona M, Gai X, Li D, Jalas C, Hirsch Y, Emmanuele V, Tadesse S, Ziosi M, Akman HO, Chung WK, Tanji K, McCormick EM, Place E, Consugar M, Pierce EA, Hakonarson H, Wallace DC, et al. (2018) USMG5 Ashkenazi Jewish founder mutation impairs mitochondrial complex V dimerization and ATP synthesis. Hum Mol Genet **27**: 3305–3312

Barth PG, Valianpour F, Bowen VM, Lam J, Duran M, Vaz FDRM & Wanders RJA (2004) X-linked cardioskeletal myopathy and neutropenia (Barth syndrome): An update. Am J Med Genet **126A**: 349–354

Basu Ball W, Neff JK & Gohil VM (2017) The role of nonbilayer phospholipids in mitochondrial structure and function. FEBS Lett **592**: 1273–1290

Benincá C, Zanette V, Brischigliaro M, Johnson M, Reyes A, Valle DAD, Robinson AJ, Degiorgi A, Yeates A, Telles BA, Prudent J, Baruffini E, Santos MLSF, de Souza RLR, Fernandez-Vizarra E, Whitworth AJ

& Zeviani M (2021) Mutation in the MICOS subunit gene APOO (MIC26) associated with an X-linked recessive mitochondrial myopathy, lactic acidosis, cognitive impairment and autistic features. J Med Genet **58**: 155–167.

Blum TB, Hahn A, Meier T, Davies KM & Kühlbrandt W (2019) Dimers of mitochondrial ATP synthase induce membrane curvature and self-assemble into rows. Proc Natl Acad Sci U.S.A. **116**: 4250–4255

Bock-Bierbaum T, Funck K, Wollweber F, Lisicki E, Malsburg von der K, Malsburg von der A, Laborenz J, Noel JK, Hessenberger M, Jungbluth S, Bernert C, Kunz S, Riedel D, Lilie H, Jakobs S, van der Laan M & Daumke O (2022) Structural insights into crista junction formation by the Mic60-Mic19 complex. Sci Adv **8**: eabo4946

Bohnert M, Wenz L-S, Zerbes RM, Horvath SE, Stroud DA, Malsburg von der K, Müller JM, Oeljeklaus S, Perschil I, Warscheid B, Chacinska A, Veenhuis M, van der Klei IJ, Daum G, Wiedemann N, Becker T, Pfanner N & van der Laan M (2012) Role of mitochondrial inner membrane organizing system in protein biogenesis of the mitochondrial outer membrane. Mol Biol Cell **23**: 3948–3956

Bohnert M, Zerbes RM, Davies KM, Mühleip AW, Rampelt H, Horvath SE, Boenke T, Kram A, Perschil I, Veenhuis M, Kühlbrandt W, van der Klei IJ, Pfanner N & van der Laan M (2015) Central role of Mic10 in the mitochondrial contact site and cristae organizing system. Cell Metab **21**: 747–755

Boldogh IR, Nowakowski DW, Yang H-C, Chung H, Karmon S, Royes P & Pon LA (2003) A protein complex containing Mdm10p, Mdm12p, and Mmm1p links mitochondrial membranes and DNA to the cytoskeleton-based segregation machinery. Mol Biol Cell **14**: 4618–4627

Bornhövd C, Vogel F, Neupert W & Reichert AS (2006) Mitochondrial membrane potential is dependent on the oligomeric state of F_1F_O-ATP synthase supracomplexes. J Biol Chem **281**: 13990–13998

Böttinger L, Ellenrieder L & Becker T (2015) How lipids modulate mitochondrial protein import. J Bioenerg Biomembr **48**: 125–135

Brandt T, Mourier A, Tain LS, Partridge L, Larsson N-G & Kühlbrandt W (2017) Changes of mitochondrial ultrastructure and function during ageing in mice and *Drosophila*. eLife **6**: 3343

Brust D, Daum B, Breunig C, Hamann A, Kühlbrandt W & Osiewacz HD (2010) Cyclophilin D links programmed cell death and organismal aging in *Podospora anserina*. Aging Cell **9**: 761–775

Bustos DM & Velours J (2005) The modification of the conserved GXXXG motif of the membrane-spanning segment of subunit g destabilizes the supramolecular species of yeast ATP synthase. J Biol Chem **280**: 29004–29010

Cabezón E, Montgomery MG, Leslie AGW & Walker JE (2003) The structure of bovine F_1-ATPase in complex with its regulatory protein IF_1. Nat Struct Mol Biol **10**: 744–750

Cadena LR, Gahura O, Panicucci B, Zíková A & Hashimi H (2021) Mitochondrial contact site and cristae organization system and F_1F_O-ATP synthase crosstalk is a fundamental property of mitochondrial cristae. mSphere **6**: e0032721

Callegari S, Müller T, Schulz C, Lenz C, Jans DC, Wissel M, Opazo F, Rizzoli SO, Jakobs S, Urlaub H, Rehling P & Deckers M (2019) A MICOS-TIM22 association promotes carrier import into human mitochondria. J Mol Biol **431**: 2835–2851

Campanella M, Casswell E, Chong S, Farah Z, Wieckowski MR, Abramov AY, Tinker A & Duchen MR (2008) Regulation of mitochondrial structure and function by the F1Fo-ATPase inhibitor protein, IF1. Cell Metab **8**: 13–25

Chatzispyrou IA, Guerrero-Castillo S, Held NM, Ruiter JPN, Denis SW, IJlst L, Wanders RJ, van Weeghel M, Ferdinandusse S, Vaz FM, Brandt U & Houtkooper RH (2018) Barth syndrome cells display widespread remodeling of mitochondrial complexes without affecting metabolic flux distribution. BBA - Mol Basis Dis **1864**: 3650–3658

Cogliati S, Enriquez JA & Scorrano L (2016) Mitochondrial cristae: Where beauty meets functionality. Trends Biochem Sci **41**: 261–273

Colina-Tenorio L, Horten P, Pfanner N & Rampelt H (2020) Shaping the mitochondrial inner membrane in health and disease. J Intern Med **287**: 645–664

Darshi M, Mendiola VL, Mackey MR, Murphy AN, Koller A, Perkins GA, Ellisman MH & Taylor SS (2011) ChChd3, an inner mitochondrial membrane protein, is essential for maintaining crista integrity and mitochondrial function. J Biol Chem **286**: 2918–2932

Darshi M, Trinh KN, Murphy AN & Taylor SS (2012) Targeting and import mechanism of coiled-coil helix coiled-coil helix domain-containing protein 3 (ChChd3) into the mitochondrial intermembrane space. J Biol Chem **287**: 39480–39491

Daum B, Nicastro D, Austin II J, McIntosh JR & Kühlbrandt W (2010) Arrangement of photosystem II and ATP synthase in chloroplast membranes of spinach and pea. Plant Cell **22**: 1299–1312

Daum B, Walter A, Horst A, Osiewacz HD & Kühlbrandt W (2013) Age-dependent dissociation of ATP synthase dimers and loss of inner-membrane cristae in mitochondria. Proc Natl Acad Sci U.S.A. **110**: 15301–15306

Davies KM, Anselmi C, Wittig I, Faraldo-Gómez JD & Kühlbrandt W (2012) Structure of the yeast F1Fo-ATP synthase dimer and its role in shaping the mitochondrial cristae. Proc Natl Acad Sci U.S.A. **109**: 13602–13607

Davies KM, Strauss M, Daum B, Kief JH, Osiewacz HD, Rycovska A, Zickermann V & Kühlbrandt W (2011) Macromolecular organization of ATP synthase and complex I in whole mitochondria. Proc Natl Acad Sci U.S.A. **108**: 14121–14126

Dietz JV, Willoughby MM, Piel RB, Ross TA, Bohovych I, Addis HG, Fox JL, Lanzilotta WN, Dailey HA, Wohlschlegel JA, Reddi AR, Medlock AE & Khalimonchuk O (2021) Mitochondrial contact site and cristae organizing system (MICOS) machinery supports heme biosynthesis by enabling optimal performance of ferrochelatase. Redox Biol **46**: 102125

Ding C, Wu Z, Huang L, Wang Y, Xue J, Chen S, Deng Z, Wang L, Song Z & Chen S (2015) Mitofilin and CHCHD6 physically interact with Sam50 to sustain cristae structure. Sci Rep **5**: 16064

Dudek J, Cheng I-F, Chowdhury A, Wozny K, Balleininger M, Reinhold R, Grunau S, Callegari S, Toischer K, Wanders RJ, Hasenfuß G, Brügger B, Guan K & Rehling P (2016) Cardiac-specific succinate dehydrogenase deficiency in Barth syndrome. EMBO Mol Med **8**: 139–154

Duncan AL, Robinson AJ & Walker JE (2016) Cardiolipin binds selectively but transiently to conserved lysine residues in the rotor of metazoan ATP synthases. Proc Natl Acad Sci U.S.A. **113**: 8687–8692

Eberhardt EL, Ludlam AV, Tan Z & Cianfrocco MA (2020) Miro: A molecular switch at the center of mitochondrial regulation. Protein Sci **29**: 1269–1284

Eichenberger C, Oeljeklaus S, Bruggisser J, Mani J, Haenni B, Kaurov I, Niemann M, Zuber B, Lukeš J, Hashimi H, Warscheid B, Schimanski B & Schneider A (2019) The highly diverged trypanosomal MICOS complex is organized in a nonessential integral membrane and an essential peripheral module. Mol Microbiol **112**: 1731–1743

Eisenberg-Bord M & Schuldiner M (2017) Mitochatting – If only we could be a fly on the cell wall. BBA - Mol Cell Res **1864**: 1469–1480

Eldeeb MA, Thomas RA, Ragheb MA, Fallahi A & Fon EA (2022) Mitochondrial quality control in health and in Parkinson's disease. Physiol Rev **102**: 1721–1755

Eramo MJ, Lisnyak V, Formosa LE & Ryan MT (2020) The 'mitochondrial contact site and cristae organising system' (MICOS) in health and human disease. J Biochem **167**: 243–255

Eydt K, Davies KM, Behrendt C, Wittig I & Reichert AS (2017) Cristae architecture is determined by an interplay of the MICOS complex and the F_1F_O ATP synthase via Mic27 and Mic10. Microb Cell **4**: 259–272

Faccenda D, Tan CH, Seraphim A, Duchen MR & Campanella M (2013) IF1 limits the apoptotic-signalling cascade by preventing mitochondrial remodelling. Cell Death Differ **20**: 686–697

Faelber K, Dietrich L, Noel JK, Wollweber F, Pfitzner A-K, Mühleip AW, Sánchez R, Kudryashev M, Chiaruttini N, Lilie H, Schlegel J, Rosenbaum E, Hessenberger M, Matthaeus C, Kunz S, Malsburg von der A, Noé F, Roux A, van der Laan M, Kühlbrandt W, et al. (2019) Structure and assembly of the mitochondrial membrane remodelling GTPase Mgm1. Nature **571**: 429–433

Flygaard RK, Mühleip AW, Tobiasson V & Amunts A (2020) Type III ATP synthase is a symmetry-deviated dimer that induces membrane curvature through tetramerization. Nat Commun **11**: 5342

Frederick RL, McCaffery JM, Cunningham KW, Okamoto K & Shaw JM (2004) Yeast Miro GTPase, Gem1p, regulates mitochondrial morphology via a novel pathway. J Cell Biol **167**: 87–98

Frey TG, Renken CW & Perkins GA (2002) Insight into mitochondrial structure and function from electron tomography. Biochim Biophys Acta **1555**: 196–203

Frezza C, Cipolat S, Martins de Brito O, Micaroni M, Beznoussenko GV, Rudka T, Bartoli D, Polishuck RS, Danial NN, De Strooper B & Scorrano L (2006) OPA1 controls apoptotic cristae remodeling independently from mitochondrial fusion. Cell **126**: 177–189

Friedman JR, Lackner LL, West M, DiBenedetto JR, Nunnari J & Voeltz GK (2011) ER tubules mark sites of mitochondrial division. Science **334**: 358–362

Friedman JR, Mourier A, Yamada J, McCaffery JM & Nunnari J (2015) MICOS coordinates with respiratory complexes and lipids to establish mitochondrial inner membrane architecture. eLife **4**: e07739

Gahura O, Mühleip AW, Hierro-Yap C, Panicucci B, Jain M, Hollaus D, Slapničková M, Zíková A & Amunts A (2022) An ancestral interaction module promotes oligomerization in divergent mitochondrial ATP synthases. Nat Commun **13**: 5989

Gao S & Hu J (2021) Mitochondrial fusion: The machineries in and out. Trends Cell Biol **31**: 62–74

García-Aguilar A & Cuezva JM (2018) A review of the inhibition of the mitochondrial ATP synthase by IF1 *in vivo*: Reprogramming energy metabolism and inducing mitohormesis. Front Physiol **9**: 1322

Gatto C, Grandi M, Solaini G, Baracca A & Giorgio V (2022) The F1Fo-ATPase inhibitor protein IF1 in pathophysiology. Front Physiol **13**: 917203

Ghosh S, Iadarola DM, Basu Ball W & Gohil VM (2019) Mitochondrial dysfunctions in Barth syndrome. IUBMB Life **71**: 791–801

Gilkerson RW, Selker JML & Capaldi RA (2003) The cristal membrane of mitochondria is the principal site of oxidative phosphorylation. FEBS Lett **546**: 355–358

Glytsou C, Calvo E, Cogliati S, Mehrotra A, Anastasia I, Rigoni G, Raimondi A, Shintani N, Loureiro M, Vazquez J, Pellegrini L, Enríquez JA, Scorrano L & Soriano ME (2016) Optic atrophy 1 is epistatic to the core MICOS component MIC60 in mitochondrial cristae shape control. Cell Rep **17**: 3024–3034

Gödiker J, Grüneberg M, DuChesne I, Reunert J, Rust S, Westermann C, Wada Y, Classen G, Langhans CD, Schlingmann KP, Rodenburg RJ, Pohlmann R & Marquardt T (2018) QIL1-dependent assembly of MICOS complex–lethal mutation in C19ORF70 resulting in liver disease and severe neurological retardation. J Hum Genet **63**: 707–716

Gu J, Zhang L, Zong S, Guo R, Liu T, Yi J, Wang P, Zhuo W & Yang M (2019) Cryo-EM structure of the mammalian ATP synthase tetramer bound with inhibitory protein IF1. Science **364**: 1068–1075

Guarani V, Jardel C, Chrétien D, Lombès A, Bénit P, Labasse C, Lacène E, Bourillon A, Imbard A, Benoist J-F, Dorboz I, Gilleron M, Goetzman ES, Gaignard P, Slama A, Elmaleh-Bergès M, Romero NB, Rustin P, Ogier de Baulny H, Paulo JA, et al. (2016) QIL1 mutation causes MICOS disassembly and early onset fatal mitochondrial encephalopathy with liver disease. eLife **5**: e17163

Guarani V, McNeill EM, Paulo JA, Huttlin EL, Fröhlich F, Gygi SP, Van Vactor D & Harper JW (2015) QIL1 is a novel mitochondrial protein required for MICOS complex stability and cristae morphology. eLife **4**: e06265

Guo H, Bueler SA & Rubinstein JL (2017) Atomic model for the dimeric FO region of mitochondrial ATP synthase. Science **358**: 936–940

Habersetzer J, Larrieu I, Priault M, Salin B, Rossignol R, Brèthes D & Paumard P (2013) Human F1F0 ATP synthase, mitochondrial ultrastructure and OXPHOS impairment: A (super-)complex matter? PLOS ONE **8**: e75429

Habich M, Salscheider SL, Murschall LM, Hoehne MN, Fischer M, Schorn F, Petrungaro C, Ali M, Erdogan AJ, Abou-Eid S, Kashkar H, Dengjel J & Riemer J (2019) Vectorial import via a metastable disulfide-linked complex allows for a quality control step and import by the mitochondrial disulfide relay. Cell Rep **26**: 759–774.e5

Hackenbrock CR (1966) Ultrastructural bases for metabolically linked mechanical activity in mitochondria: I. Reversible ultrastructural changes with change in metabolic steady state in isolated liver mitochondria. J Cell Biol **30**: 269–297

Hahn A, Parey K, Bublitz M, Mills DJ, Zickermann V, Vonck J, Kühlbrandt W & Meier T (2016) Structure of a complete ATP synthase dimer reveals the molecular basis of inner mitochondrial membrane morphology. Mol Cell **63**: 445–456

Hahn A, Vonck J, Mills DJ, Meier T & Kühlbrandt W (2018) Structure, mechanism, and regulation of the chloroplast ATP synthase. Science **360**: eaat4318

Harner ME, Körner C, Walther D, Mokranjac D, Kaesmacher J, Welsch U, Griffith J, Mann M, Reggiori F & Neupert W (2011) The mitochondrial contact site complex, a determinant of mitochondrial architecture. EMBO J **30**: 4356–4370

Harner ME, Unger A-K, Geerts WJ, Mari M, Izawa T, Stenger M, Geimer S, Reggiori F, Westermann B & Neupert W (2016) An evidence based hypothesis on the existence of two pathways of mitochondrial crista formation. eLife **5**: e18853

Harner ME, Unger A-K, Izawa T, Walther DM, Ozbalci C, Geimer S, Reggiori F, Brügger B, Mann M, Westermann B & Neupert W (2014) Aim24 and MICOS modulate respiratory function, tafazzin-related cardiolipin modification and mitochondrial architecture. eLife 3: e01684

Hartley AM, Lukoyanova N, Zhang Y, Cabrera-Orefice A, Arnold S, Meunier B, Pinotsis N & Maréchal A (2019) Structure of yeast cytochrome c oxidase in a supercomplex with cytochrome bc1. Nat Struct Mol Biol 26: 78–83

Hashimi H (2019) A parasite's take on the evolutionary cell biology of MICOS. PLOS Pathog **15**: e1008166

He J, Ford HC, Carroll J, Douglas C, Gonzales E, Ding S, Fearnley IM & Walker JE (2018) Assembly of the membrane domain of ATP synthase in human mitochondria. Proc Natl Acad Sci U.S.A. **115**: 2988–2993

Hessenberger M, Zerbes RM, Rampelt H, Kunz S, Xavier AH, Purfürst B, Lilie H, Pfanner N, van der Laan M & Daumke O (2017) Regulated membrane remodeling by Mic60 controls formation of mitochondrial crista junctions. Nat Commun **8**: 15258

Hoppins SC, Collins SR, Cassidy-Stone A, Hummel E, DeVay RM, Lackner LL, Westermann B, Schuldiner M, Weissman JS & Nunnari J (2011) A mitochondrial-focused genetic interaction map reveals a scaffold-like complex required for inner membrane organization in mitochondria. J Cell Biol **195**: 323–340

Horvath SE & Daum G (2013) Lipids of mitochondria. Prog Lipid Res **52**: 590–614

Hu C, Shu L, Huang X, Yu J, Li L, Gong L, Yang M, Wu Z, Gao Z, Zhao Y, Chen L & Song Z (2020) OPA1 and MICOS Regulate mitochondrial crista dynamics and formation. Cell Death Dis **11**: 940

Hu J, Prinz WA & Rapoport TA (2011) Weaving the web of ER tubules. Cell **147**: 1226–1231

Huynen MA, Mühlmeister M, Gotthardt K, Guerrero-Castillo S & Brandt U (2016) Evolution and structural organization of the mitochondrial contact site (MICOS) complex and the mitochondrial intermembrane space bridging (MIB) complex. Biochim Biophys Acta **1863**: 91–101

Ishihara N, Fujita Y, Oka T & Mihara K (2006) Regulation of mitochondrial morphology through proteolytic cleavage of OPA1. EMBO J **25**: 2966–2977

Itoh K, Tamura Y, Iijima M & Sesaki H (2013) Effects of Fcj1-Mos1 and mitochondrial division on aggregation of mitochondrial DNA nucleoids and organelle morphology. Mol Biol Cell **24**: 1842–1851

Jakobs S, Stephan T, Ilgen P & Brüser C (2020) Light microscopy of mitochondria at the nanoscale. Annu Rev Biophys **49**: 289–308

Jakubke C, Roussou R, Maiser A, Schug C, Thoma F, Bunk D, Hörl D, Leonhardt H, Walter P, Klecker T & Osman C (2021) Cristae-dependent quality control of the mitochondrial genome. Sci Adv **7**: eabi8886

Jans DC, Wurm CA, Riedel D, Wenzel D, Stagge F, Deckers M, Rehling P & Jakobs S (2013) STED super-resolution microscopy reveals an array of MINOS clusters along human mitochondria. Proc Natl Acad Sci U.S.A. **110**: 8936–8941

Jumper J, Evans R, Pritzel A, Green T, Figurnov M, Ronneberger O, Tunyasuvunakool K, Bates R, Žídek A, Potapenko A, Bridgland A, Meyer C, Kohl SAA, Ballard AJ, Cowie A, Romera-Paredes B, Nikolov S, Jain R, Adler J, Back T, et al. (2021) Highly accurate protein structure prediction with AlphaFold. Nature **596**: 583–589

Kaurov I, Vancová M, Schimanski B, Cadena LR, Heller J, Bílý T, Potěšil D, Eichenberger C, Bruce H, Oeljeklaus S, Warscheid B, Zdráhal Z, Schneider A, Lukeš J & Hashimi H (2018) The diverged trypanosome MICOS complex as a hub for mitochondrial cristae shaping and protein import. Curr Biol **28**: 3393–3407.e5

Klecker T & Westermann B (2021) Pathways shaping the mitochondrial inner membrane. Open Biol **11**: 210238

Kojima R, Kakimoto Y, Furuta S, Itoh K, Sesaki H, Endo T & Tamura Y (2019) Maintenance of cardiolipin and crista structure requires cooperative functions of mitochondrial dynamics and phospholipid transport. Cell Rep **26**: 518–528.e6

Kondadi AK, Anand R, Hänsch S, Urbach J, Zobel T, Wolf DM, Segawa M, Liesa M, Shirihai OS, Weidtkamp Peters S & Reichert AS (2020) Cristae undergo continuous cycles of membrane remodelling in a MICOS-dependent manner. EMBO Rep **21**: e49776

Koob S & Reichert AS (2014) Novel intracellular functions of apolipoproteins: The ApoO protein family as constituents of the mitofilin/MINOS complex determines cristae morphology in mitochondria. Biol Chem **395**: 285–296

Koob S, Barrera M, Anand R & Reichert AS (2015) The non-glycosylated isoform of MIC26 is a constituent of the mammalian MICOS complex and promotes formation of crista junctions. Biochim Biophys Acta **1853**: 1551–1563

Körner C, Barrera M, Dukanovic J, Eydt K, Harner ME, Rabl R, Vogel F, Rapaport D, Neupert W & Reichert AS (2012) The C-terminal domain of Fcj1 is required for formation of crista junctions and interacts with the TOB/SAM complex in mitochondria. Mol Biol Cell **23**: 2143–2155

Korobova F, Ramabhadran V & Higgs HN (2013) An actin-dependent step in mitochondrial fission mediated by the ER-associated formin INF2. Science **339**: 464–467

Kozjak-Pavlovic V (2017) The MICOS complex of human mitochondria. Cell Tissue Res **367**: 83–93

Kozlov MM, Campelo F, Liska N, Chernomordik LV, Marrink S-J & McMahon HT (2014) Mechanisms shaping cell membranes. Curr Opin Cell Biol **29**: 53–60

Kühlbrandt W (2019) Structure and mechanisms of F-type ATP synthases. Annu Rev Biochem **88**: 515–549

Lackner LL (2019) The expanding and unexpected functions of mitochondria contact sites. Trends Cell Biol **29**: 580–590

Lamant M, Smih F, Harmancey R, Philip-Couderc P, Pathak A, Roncalli J, Galinier M, Collet X, Massabuau P, Senard J-M & Rouet P (2006) ApoO, a novel apolipoprotein, is an original glycoprotein up-regulated by diabetes in human heart. J Biol Chem **281**: 36289–36302

Le Guerroué F, Eck F, Jung J, Starzetz T, Mittelbronn M, Kaulich M & Behrends C (2017) Autophagosomal content profiling reveals an LC3C- dependent piecemeal mitophagy pathway. Mol Cell **68**: 786–796.e6

Lewis SC, Uchiyama LF & Nunnari J (2016) ER-mitochondria contacts couple mtDNA synthesis with mitochondrial division in human cells. Science **353**: aaf5549

Li H, Ruan Y, Zhang K, Jian F, Hu C, Miao L, Gong L, Sun L, Zhang X, Chen S, Chen H, Liu D & Song Z (2016) Mic60/mitofilin determines MICOS assembly essential for mitochondrial dynamics and mtDNA nucleoid organization. Cell Death Differ **23**: 380–392

Li L, Conradson DM, Bharat V, Kim MJ, Hsieh C-H, Minhas PS, Papakyrikos AM, Durairaj AS, Ludlam A, Andreasson KI, Partridge L, Cianfrocco MA & Wang X (2021) A mitochondrial membrane-bridging machinery mediates signal transduction of intramitochondrial oxidation. Nat Metab **3**: 1242–1258

López Doménech G, Covill Cooke C, Ivankovic D, Halff EF, Sheehan DF, Norkett R, Birsa N & Kittler JT (2018) Miro proteins coordinate microtubule- and actin-dependent mitochondrial transport and distribution. EMBO J **37**: 321–336

Lubeck M, Derkum NH, Naha R, Strohm R, Driessen MD, Belgardt B-F, Roden M, Stühler K, Anand R, Reichert AS & Kondadi AK (2023) MIC26 and MIC27 are bona fide subunits of the MICOS complex in mitochondria and do not exist as glycosylated apolipoproteins. PLOS ONE **18**: e0286756

Mannella CA (2006) Structure and dynamics of the mitochondrial inner membrane cristae. Biochim Biophys Acta **1763**: 542–548

Mårtensson CU, Doan KN & Becker T (2017) Effects of lipids on mitochondrial functions. Biochim Biophys Acta **1862**: 102–113

Meeusen SL, DeVay RM, Block J, Cassidy-Stone A, Wayson S, McCaffery JM & Nunnari J (2006) Mitochondrial inner-membrane fusion and crista maintenance requires the dynamin-related GTPase Mgm1. Cell **127**: 383–395

Mendoza-Hoffmann F, Zarco-Zavala M, Ortega R, Celis-Sandoval H, Torres-Larios A & García-Trejo JJ (2022) Evolution of the inhibitory and non-inhibitory ε, ζ, and IF1 subunits of the F_1F_O-ATPase as related to the endosymbiotic origin of mitochondria. Microorganisms **10**: 1372

Michaud M, Gros V, Tardif M, Brugière S, Ferro M, Prinz WA, Toulmay A, Mathur J, Wozny M, Falconet D, Maréchal E, Block MA & Jouhet J (2016) AtMic60 is involved in plant mitochondria lipid trafficking and is part of a large complex. Curr Biol **26**: 627–639

Mileykovskaya E & Dowhan W (2014) Cardiolipin-dependent formation of mitochondrial respiratory supercomplexes. Chem Phys Lipids **179**: 42–48

Modi S, López Doménech G, Halff EF, Covill Cooke C, Ivankovic D, Melandri D, Arancibia-Cárcamo IL, Burden JJ, Lowe AR & Kittler JT (2019) Miro clusters regulate ER-mitochondria contact sites and link cristae organization to the mitochondrial transport machinery. Nat Commun **10**: 4399

Monteiro-Cardoso VF, Rochin L, Arora A, Houcine A, Jääskeläinen E, Kivelä AM, Sauvanet C, Le Bars R, Marien E, Dehairs J, Neveu J, Khallouki El N, Santonico E, Swinnen JV, Tareste D, Olkkonen VM & Giordano F (2022) ORP5/8 and MIB/MICOS link ER-mitochondria and intra-mitochondrial contacts for non-vesicular transport of phosphatidylserine. Cell Rep **40**: 111364

Mühleip AW, Flygaard RK, Ovciarikova J, Lacombe A, Fernandes P, Sheiner L & Amunts A (2020) ATP synthase hexamer assemblies shape cristae of Toxoplasma mitochondria. Nat Commun **12**: 120

Mühleip AW, McComas SE & Amunts A (2019) Structure of a mitochondrial ATP synthase with bound native cardiolipin. eLife **8**: e51179

Mukherjee I, Ghosh M & Meinecke M (2021) MICOS and the mitochondrial inner membrane morphology – when things get out of shape. FEBS Lett **595**: 1159–1183

Muñoz-Gómez SA, Cadena LR, Gardiner AT, Leger MM, Sheikh S, Connell LB, Bily T, Kopejtka K, Beatty JT, Koblizek M, Roger AJ, Slamovits CH, Lukes J & Hashimi H (2023) Intracytoplasmic-membrane development in alphaproteobacteria involves the homolog of the mitochondrial crista-developing protein Mic60. Curr Biol **33**: 1–13

Muñoz-Gómez SA, Slamovits CH, Dacks JB & Wideman JG (2016) The evolution of MICOS: Ancestral and derived functions and interactions. Commun Integr Biol **8**: e1094593

Muñoz-Gómez SA, Slamovits CH, Dacks JB, Baier KA, Spencer KD & Wideman JG (2015) Ancient homology of the mitochondrial contact site and cristae organizing system points to an endosymbiotic origin of mitochondrial cristae. Curr Biol **25**: 1489–1495

Murley A & Nunnari J (2016) The emerging network of mitochondria-organelle contacts. Mol Cell **61**: 648–653

Nahacka Z, Novak J, Zobalova R & Neuzil J (2022) Miro proteins and their role in mitochondrial transfer in cancer and beyond. Front Cell Dev Biol **10**: 937753

Nalin CM & McCarty RE (1984) Role of a disulfide bond in the gamma subunit in activation of the ATPase of chloroplast coupling factor 1. J Biol Chem **259**: 7275–7280

Nirody JA, Budin I & Rangamani P (2020) ATP synthase: Evolution, energetics, and membrane interactions. J Gen Physiol **152**: e201912475

Oeding SJ, Majstrowicz K, Hu X-P, Schwarz V, Freitag A, Honnert U, Nikolaus P & Bähler M (2018) Identification of Miro1 and Miro2 as mitochondrial receptors for myosin XIX. J Cell Sci **131**: jcs219469

Ohta K, Hirashima S, Miyazono Y, Togo A & Nakamura K-I (2021) Correlation of organelle dynamics between light microscopic live imaging and electron microscopic 3D architecture using FIB-SEM. Microscopy (Oxf) **70**: 161–170

Osman C, Voelker DR & Langer T (2011) Making heads or tails of phospholipids in mitochondria. J Cell Biol **192**: 7–16

Ott C, Dorsch E, Fraunholz M, Straub SP & Kozjak-Pavlovic V (2015) Detailed analysis of the human mitochondrial contact site complex indicate a hierarchy of subunits. PLOS ONE **10**: e0120213

Ott C, Ross K, Straub SP, Thiede B, Götz M, Goosmann C, Krischke M, Mueller MJ, Krohne G, Rudel T & Kozjak-Pavlovic V (2012) Sam50 functions in mitochondrial intermembrane space bridging and biogenesis of respiratory complexes. Mol Cell Biol **32**: 1173–1188

Pánek T, Eliáš M, Vancová M, Lukeš J & Hashimi H (2020) Returning to the fold for lessons in mitochondrial crista diversity and evolution. Curr Biol **30**: R575–R588

Paumard P, Vaillier J, Coulary B, Schaeffer J, Soubannier V, Mueller DM, Brèthes D, di Rago J-P & Velours J (2002) The ATP synthase is involved in generating mitochondrial cristae morphology. EMBO J **21**: 221–230

Peifer-Weiß L, Kurban M, David C, Lubeck M, Kondadi AK, Nemer G, Reichert AS & Anand R (2023) A X-linked nonsense APOO/MIC26 variant causes a lethal mitochondrial disease with progeria-like phenotypes. Clin Genet **104**: 659–668.

Pfanner N, van der Laan M, Amati P, Capaldi RA, Caudy AA, Chacinska A, Darshi M, Deckers M, Hoppins SC, Icho T, Jakobs S, Ji J, Kozjak-Pavlovic V, Meisinger C, Odgren PR, Park SK, Rehling P, Reichert AS, Sheikh MS, Taylor SS, et al. (2014) Uniform nomenclature for the mitochondrial contact site and cristae organizing system. J Cell Biol **204**: 1083–1086

Pfeiffer K, Gohil VM, Stuart RA, Hunte C, Brandt U, Greenberg ML & Schägger H (2003) Cardiolipin stabilizes respiratory chain supercomplexes. J Biol Chem **278**: 52873–52880

Pickles S, Vigié P & Youle RJ (2018) Mitophagy and quality control mechanisms in mitochondrial maintenance. Curr Biol **28**: R170–R185

Pinke G, Zhou L & Sazanov LA (2020) Cryo-EM structure of the entire mammalian F-type ATP synthase. Nat Struct Mol Biol **27**: 1077–1085

Qin J, Guo Y, Xue B, Shi P, Chen Y, Su QP, Hao H, Zhao S, Wu C, Yu L, Li D & Sun Y (2020) ER-mitochondria contacts promote mtDNA nucleoids active transportation via mitochondrial dynamic tubulation. Nat Commun **11**: 4471

Rabl R, Soubannier V, Scholz R, Vogel F, Mendl N, Vasiljev-Neumeyer A, Körner C, Jagasia R, Keil T, Baumeister W, Cyrklaff M, Neupert W & Reichert AS (2009) Formation of cristae and crista junctions in mitochondria depends on antagonism between Fcj1 and Su e/g. J Cell Biol **185**: 1047–1063

Rampelt H & van der Laan M (2017) The yin & yang of mitochondrial architecture — Interplay of MICOS and F1Fo-ATP synthase in cristae formation. Microb Cell **4**: 236–239

Rampelt H, Bohnert M, Zerbes RM, Horvath SE, Warscheid B, Pfanner N & van der Laan M (2017) Mic10, a core subunit of the mitochondrial contact site and cristae organizing system, interacts with the dimeric F1Fo-ATP synthase. J Mol Biol **429**: 1162–1170

Rampelt H, Wollweber F, Gerke C, de Boer R, van der Klei IJ, Bohnert M, Pfanner N & van der Laan M (2018) Assembly of the mitochondrial cristae organizer Mic10 is regulated by Mic26-Mic27 antagonism and cardiolipin. J Mol Biol **430**: 1883–1890

Rampelt H, Wollweber F, Licheva M, de Boer R, Perschil I, Steidle L, Becker T, Bohnert M, van der Klei I, Kraft C, van der Laan M & Pfanner N (2022) Dual role of Mic10 in mitochondrial cristae organization and ATP synthase-linked metabolic adaptation and respiratory growth. Cell Rep **38**: 110290

Rathore S, Berndtsson J, Marin-Buera L, Conrad J, Carroni M, Brzezinski P & Ott M (2019) Cryo-EM structure of the yeast respiratory supercomplex. Nat Struct Mol Biol **26**: 50–57

Ren M, Phoon CKL & Schlame M (2014) Metabolism and function of mitochondrial cardiolipin. Prog Lipid Res **55**: 1–16

Rieger B, Arroum T, Borowski M-T, Villalta J & Busch KB (2021) Mitochondrial F1 FO ATP synthase determines the local proton motive force at cristae rims. EMBO Rep **22**: e52727

Sakowska P, Jans DC, Mohanraj K, Riedel D, Jakobs S & Chacinska A (2015) The oxidation status of Mic19 regulates MICOS assembly. Mol Cell Biol **35**: 4222–4237

Sastri M, Darshi M, Mackey M, Ramachandra R, Ju S, Phan S, Adams S, Stein K, Douglas CR, Kim JJ, Ellisman MH, Taylor SS & Perkins GA (2017) Sub-mitochondrial localization of the genetic-tagged mitochondrial intermembrane space-bridging components Mic19, Mic60 and Sam50. J Cell Sci **130**: 3248–3260

Schlame M & Greenberg ML (2016) Biosynthesis, remodeling and turnover of mitochondrial cardiolipin. Biochim Biophys Acta - Mol Cell Biol Lipids **1862**: 3–7

Shi P, Ren X, Meng J, Kang C, Wu Y, Rong Y, Zhao S, Jiang Z, Liang L, He W, Yin Y, Li X, Liu Y, Huang X, Sun Y, Li B & Wu C (2022) Mechanical instability generated by Myosin 19 contributes to mitochondria cristae architecture and OXPHOS. Nat Commun **13**: 2673–14

Siegmund SE, Grassucci R, Carter SD, Barca E, Farino ZJ, Juanola-Falgarona M, Zhang P, Tanji K, Hirano M, Schon EA, Frank J & Freyberg Z (2018) Three-dimensional analysis of mitochondrial crista ultrastructure in a patient with Leigh syndrome by *in situ* cryoelectron tomography. iScience **6**: 83–91

Simunovic M, Evergren E, Callan-Jones A & Bassereau P (2019) Curving cells inside and out: Roles of BAR domain proteins in membrane shaping and its cellular implications. Annu Rev Cell Dev Biol **35**: 111–129

Spikes TE, Montgomery MG & Walker JE (2020) Structure of the dimeric ATP synthase from bovine mitochondria. Proc Natl Acad Sci U.S.A. **117**: 23519–23526

Stephan T, Brüser C, Deckers M, Steyer AM, Balzarotti F, Barbot M, Behr TS, Heim G, Hübner W, Ilgen P, Lange F, Pacheu-Grau D, Pape JK, Stoldt S, Huser T, Hell SW, Möbius W, Rehling P, Riedel D & Jakobs S (2020) MICOS assembly controls mitochondrial inner membrane remodeling and crista junction redistribution to mediate cristae formation. EMBO J **39**: 217–224

Stoldt S, Stephan T, Jans DC, Brüser C, Lange F, Keller-Findeisen J, Riedel D, Hell SW & Jakobs S (2019) Mic60 exhibits a coordinated clustered distribution along and across yeast and mammalian mitochondria. Proc Natl Acad Sci U.S.A. **5**: 201820364

Stoldt S, Wenzel D, Kehrein K, Riedel D, Ott M & Jakobs S (2018) Spatial orchestration of mitochondrial translation and OXPHOS complex assembly. Nat Cell Biol **20**: 528–534

Strauss M, Hofhaus G, Schröder RR & Kühlbrandt W (2008) Dimer ribbons of ATP synthase shape the inner mitochondrial membrane. EMBO J **27**: 1154–1160

Tamura Y, Kawano S & Endo T (2020) Lipid homeostasis in mitochondria. Biol Chem **401**: 821–833

Tang J, Zhang K, Dong J, Yan C, Hu C, Ji H, Chen L, Chen S, Zhao H & Song Z (2019) Sam50-Mic19-Mic60 axis determines mitochondrial cristae architecture by mediating mitochondrial outer and inner membrane contact. Cell Death Differ **1763**: 542

Tarasenko D, Barbot M, Jans DC, Kroppen B, Sadowski B, Heim G, Möbius W, Jakobs S & Meinecke M (2017) The MICOS component Mic60 displays a conserved membrane-bending activity that is necessary for normal cristae morphology. J Cell Biol **216**: 889–899

Tatsuta T & Langer T (2017) Intramitochondrial phospholipid trafficking. BBA - Mol Cell Biol Lipids **1862**: 81–89

Tirrell PS, Nguyen KN, Luby-Phelps K & Friedman JR (2020) MICOS subcomplexes assemble independently on the mitochondrial inner membrane in proximity to ER contact sites. J Cell Biol **219**: e202003024

Toth A, Meyrat A, Stoldt S, Santiago R, Wenzel D, Jakobs S, Ballmoos von C & Ott M (2020) Kinetic coupling of the respiratory chain with ATP synthase, but not proton gradients, drives ATP production in cristae membranes. Proc Natl Acad Sci U.S.A. **117**: 2412–2421

Tsai P-I, Lin C-H, Hsieh C-H, Papakyrikos AM, Kim MJ, Napolioni V, Schoor C, Couthouis J, Wu R-M, Wszolek ZK, Winter D, Greicius MD, Ross OA & Wang X (2018) PINK1 phosphorylates MIC60/Mmitofilin to control structural plasticity of mitochondrial crista junctions. Mol Cell **69**: 744–756.e6

Tsai P-I, Papakyrikos AM, Hsieh C-H & Wang X (2017) *Drosophila* MIC60/mitofilin conducts dual roles in mitochondrial motility and crista structure. Mol Biol Cell **28**: 3471–3479

Ueda E, Tamura Y, Sakaue H, Kawano S, Kakuta C, Matsumoto S & Endo T (2019) Myristoyl group-aided protein import into the mitochondrial intermembrane space. Sci Rep **9**: 1185–12

Urbach J, Kondadi AK, David C, Naha R, Deinert K, Reichert AS & Anand R (2021) Conserved GxxxG and WN motifs of MIC13 are essential for bridging two MICOS subcomplexes. BBA - Biomembr **1863**: 183683

van den Brink-van der Laan E, Antoinette Killian J & de Kruijff B (2004) Nonbilayer lipids affect peripheral and integral membrane proteins via changes in the lateral pressure profile. BBA - Biomembr **1666**: 275–288

van der Laan M, Bohnert M, Wiedemann N & Pfanner N (2012) Role of MINOS in mitochondrial membrane architecture and biogenesis. Trends Cell Biol **22**: 185–192

Van Laar VS, Otero PA, Hastings TG & Berman SB (2019) Potential role of Mic60/mitofilin in Parkinson's disease. Front Neurosci **12**: 898

Vogel F, Bornhövd C, Neupert W & Reichert AS (2006) Dynamic subcompartmentalization of the mitochondrial inner membrane. J Cell Biol **175**: 237–247

von der Malsburg K, Müller JM, Bohnert M, Oeljeklaus S, Kwiatkowska P, Becker T, Loniewska-Lwowska A, Wiese S, Rao S, Milenkovic D, Hutu DP, Zerbes RM, Schulze-Specking A, Meyer HE, Martinou J-C, Rospert S, Rehling P, Meisinger C, Veenhuis M, Warscheid B, et al. (2011) Dual role of mitofilin in mitochondrial membrane organization and protein biogenesis. Dev Cell **21**: 694–707

Wagner K, Perschil I, Fichter CD & van der Laan M (2010) Stepwise assembly of dimeric F_1F_O-ATP synthase in mitochondria involves the small F_o-subunits k and i. Mol Biol Cell **21**: 1494–1504

Weber TA, Koob S, Heide H, Wittig I, Head B, van der Bliek A, Brandt U, Mittelbronn M & Reichert AS (2013) APOOL is a cardiolipin-binding constituent of the mitofilin/MINOS protein complex determining cristae morphology in mammalian mitochondria. PLOS ONE **8**: e63683

Weissert V, Rieger B, Morris S, Arroum T, Psathaki OE, Zobel T, Perkins G & Busch KB (2021) Inhibition of the mitochondrial ATPase function by IF1 changes the spatiotemporal organization of ATP synthase. BBA - Bioenerg **1862**: 148322

Wiedemann N & Pfanner N (2017) Mitochondrial machineries for protein import and assembly. Annu Rev Biochem **86**: 685–714

Wolf DM, Segawa M, Kondadi AK, Anand R, Bailey ST, Reichert AS, Bliek AM, Shackelford DB, Liesa M & Shirihai OS (2019) Individual cristae within the same mitochondrion display different membrane potentials and are functionally independent. EMBO J **38**: 1–21

Xie J, Marusich MF, Souda P, Whitelegge J & Capaldi RA (2007) The mitochondrial inner membrane protein Mitofilin exists as a complex with SAM50, metaxins 1 and 2, coiled-coil-helix coiled-coil-helix domain-containing protein 3 and 6 and DnaJC11. FEBS Lett **581**: 3545–3549

Xu Y, Anjaneyulu M, Donelian A, Yu W, Greenberg ML, Ren M, Owusu-Ansah E & Schlame M (2019) Assembly of the complexes of oxidative phosphorylation triggers the remodeling of cardiolipin. Proc Natl Acad Sci U.S.A. **116**: 11235–11240

Xu Y, Erdjument-Bromage H, Phoon CKL, Neubert TA, Ren M & Schlame M (2021) Cardiolipin remodeling enables protein crowding in the inner mitochondrial membrane. EMBO J **40**: e108428

Yang R-F, Sun L-H, Zhang R, Zhang Y, Luo Y-X, Zheng W, Zhang Z-Q, Chen H-Z & Liu D-P (2015) Suppression of Mic60 compromises mitochondrial transcription and oxidative phosphorylation. Sci Rep **5**: 7990

Yapa NMB, Lisnyak V, Reljic B & Ryan MT (2021) Mitochondrial dynamics in health and disease. FEBS Lett **595**: 1184–1204

Zabeo D & Davies KM (2022) Studying membrane modulation mechanisms by electron cryo-tomography. Curr Opin Struct Biol **77**: 102464

Zeharia A, Friedman JR, Tobar A, Saada A, Konen O, Fellig Y, Shaag A, Nunnari J & Elpeleg O (2016) Mitochondrial hepato-encephalopathy due to deficiency of QIL1/MIC13 (C19orf70), a MICOS complex subunit. Eur J Hum Genet **24**: 1778–1782

Zerbes RM, Bohnert M, Stroud DA, Malsburg von der K, Kram A, Oeljeklaus S, Warscheid B, Becker T, Wiedemann N, Veenhuis M, van der Klei IJ, Pfanner N & van der Laan M (2012) Role of MINOS in mitochondrial membrane architecture: Cristae morphology and outer membrane interactions differentially depend on mitofilin domains. J Mol Biol **422**: 183–191

Zerbes RM, Höß P, Pfanner N, van der Laan M & Bohnert M (2016) Distinct roles of Mic12 and Mic27 in the mitochondrial contact site and cristae organizing system. J Mol Biol **428**: 1485–1492

Zhou J, Duan M, Wang X, Zhang F, Zhou H, Ma T, Yin Q, Zhang J, Tian F, Wang G & Yang C (2022) A feedback loop engaging propionate catabolism intermediates controls mitochondrial morphology. Nat Cell Biol **24**: 526–537

Zick M, Rabl R & Reichert AS (2009) Cristae formation-linking ultrastructure and function of mitochondria. Biochim Biophys Acta **1793**: 5–19

Zung N & Schuldiner M (2020) New horizons in mitochondrial contact site research. Biol Chem **401**: 793–809

2 Proteolytic Regulation of Mitochondrial Architecture

Gunjan Purohit and Oleh Khalimonchuk

INTRODUCTION

Mitochondria are double membrane-restricted organelles with a distinct set of sub-compartments. The outer mitochondrial membrane (OMM) is generally uniform and relatively permeable to small molecules and ions, while the inner mitochondrial membrane (IMM) is topologically heterogenous and comprises two distinct domains: the inner boundary membrane (IBM) and the inner cristae membrane (ICM). The IBM is a planar membrane closely positioned with the OMM, whereas the ICM is the region that extends into the matrix forming large invaginations commonly referred to as cristae [1–4]. Such ultrastructural organization creates a major basis for mitochondrial functions. In many cells, mitochondria form highly interconnected and dynamic networks extending across the cell. In addition to their well-studied energy conversion function, mitochondria are also linked to a plethora of other vital functions, including the maintenance of ion homeostasis, cellular signaling, and inflammatory responses. Mitochondrial architecture is highly dynamic and tightly linked to cellular physiology. To date, several machineries have been identified that regulate mitochondrial structure and function. Mounting evidence indicates that mitochondria sense both stress and physiological cues and respond by changing their shape and form, and, consequently, functional outputs. Among the factors that control said processes are several proteases that have been identified as key factors involved in the maintenance of mitochondrial shape and function [5]. Mitochondria undergo fusion and fission cycles in response to various cellular stimuli. Loss of membrane potential, ATP, and redox potential result in ultrastructural remodeling and rapid fragmentation of mitochondrial network, whereas proliferation and starvation lead to mitochondrial fusion. Likewise, mitochondria fuse and form reticulated networks to support energy requirement of cells. Mitochondria fission protects healthy pool of mitochondria under conditions of stress, wherein prolonged stress cues cause rearrangement of mitochondrial architecture and removal of segregated non-functional mitochondria by mitophagy or activate apoptotic program. Interestingly, mounting evidence suggests that mitochondrial ultrastructure, which is mediated by several factors including the mitochondrial contact site and cristae junction organizing system (MICOS) machinery, is maintained largely independent of the mitochondrial network dynamics despite sharing some of the proteins known to regulate each process.

In this chapter, we primarily focus on the role of conserved ATP-independent mitochondrial protease overlapping with matrix AAA protease 1 (OMA1) in regulation of mitochondrial network morphology and MICOS. These and other recently uncovered relevant mechanisms provide important new insights into the dynamic nature of mitochondrial architecture and deepen fundamental understanding of mitochondrial biology. Equally important, they offer new prospective targets for therapeutic interventions in a plethora of early-to-late-onset human diseases.

OMA1 METALLOPEPTIDASE

Overview of Mitochondrial Proteases

Mitochondria are central organelles for cellular metabolism and signaling, and the cellular and mitochondrial homeostatic mechanisms are tightly intertwined. Mitochondrial proteome comprises about 1,300 proteins that regulate its structure, function, and biogenesis [6]. A subset of these proteins

 DOI: 10.1201/9781003221180-3

is dedicated to the quality control processes within the organelle. Several mitochondrial quality control mechanisms have been identified, such as mitochondrial unfolded protein response (UPRmt), mitochondria-specific autophagy (commonly referred to as mitophagy), mitochondria-derived vesicles, and others [5, 7, 8]. Remarkably, mitochondrial proteases (from here on referred to as mitoproteases) – a diverse group of conserved proteolytic enzymes – are critical to maintain quality control owing to their versatile protease activity and one way or the other are involved in said mechanisms. The broad function of mitoproteases is to inhibit excess protein accumulation, degrade misfolded or orphaned proteins, process proteins for import and maturation, and regulate protein activity by targeted cleavage or degradation. The mitochondrion harbors over 40 proteolytic enzymes, and a number of outstanding reviews on mitoproteases and their roles in mitochondrial function is available [9]. Herein, we will mainly focus on the M48-type metallopeptidase OMA1, which has now emerged as a critical regulator of a variety of mitochondrial features and functions including mitochondrial architecture.

OMA1 Topology and Organization

In humans, OMA1 gene is located at the short arm of chromosome 1. It contains a total of nine exons, eight of which are protein-coding ones. The OMA1 transcript translates into a 524 amino acid-long polypeptide with a predicted molecular weight of 60.1 kDa. The structure of the OMA1 protein has not been solved yet, but the computational modeling of protein structure by sequence similarity suggests the presence of at least one bona fide transmembrane helical domain and an unstructured region at the N-terminus of the protein [10] (Figure 2.1). The translated OMA1 protein contains an N-terminal mitochondrial targeting sequence. This presequence appears to be removed by a currently unknown protease during OMA1 import into the mitochondria. Such proteolytic processing yields a 381 amino acid-long mature polypeptide. The protease is an integral membrane protein of the inner mitochondrial membrane and contains at least one membrane-spanning helical domain. OMA1 belongs to the metzincins family of proteases, which are ATP-independent M48-type zinc metalloproteases. The N-terminal portion of the mature OMA1 is located to the matrix region, whereas the C-terminal moiety of the enzyme contains the M48 proteolytic domain facing the intermembrane space (IMS) [11]. The proteolytic activity of OMA1 is conferred by the residues 144-163 of said domain. The N-terminal region of the mammalian enzyme appears to harbor a charged sensor domain, which is important for OMA1 stress activation. Of note, upon its activation the mammalian OMA1 appears to self-proteolyze in a sequence-independent manner [11–13]. Studies

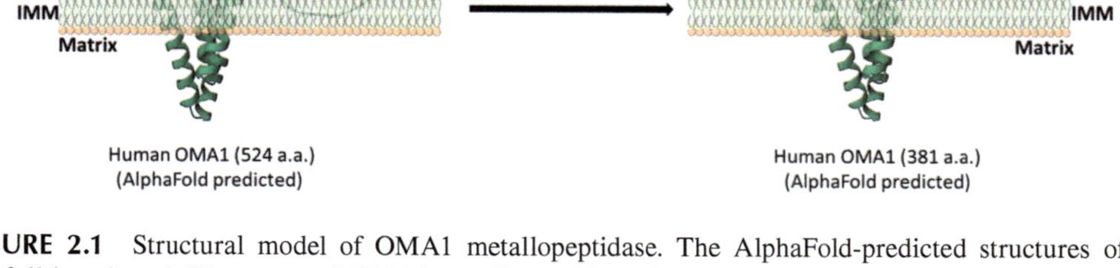

FIGURE 2.1 Structural model of OMA1 metallopeptidase. The AlphaFold-predicted structures of the (A) full-length and (B) processed OMA1 metallopeptidase. OMA1 is a conserved mitochondrial protease equipped with an M48 metalloendopeptidase domain facing the intermembrane space. The protein appears to be anchored within the IMM through a predicted transmembrane domain. The N-terminal of the enzyme harbors an unstructured region, which appears to be cleaved during OMA1 biogenesis.

in yeast established that the enzyme exists as ~300 kDa, presumably homo-oligomeric, complex with several copies of OMA1 present within the oligomer [11, 14]. In line with this notion, the co-expression of the two catalytically impaired mutants of OMA1, E324Q, and Δ144-163 partially restores the enzyme's activity [12], further indicating the formation of a homo-multimeric complex.

Under conditions of mitochondrial stress, the enzyme appears to undergo conformational rearrangements within the oligomer, leading to its activation [14]. The oligomerization is independent of OMA1's catalytic domain and likely involves its conserved C-terminal portion. One notable difference between the yeast and mammalian enzymes is its post-activation fate: while stress-triggered yeast Oma1 remains stable [14], the activated human OMA1 is eventually removed – either via autoproteolysis [12, 15] or by YME1L protease [16]. The latter two events were proposed to constitute a feedback loop mechanism ensuring that the activated enzyme is kept at bay and not causing any undesired proteolysis down the line.

Another regulatory mechanism was recently suggested by studies in yeast that identified a set of highly conserved cysteine residues in Oma1 [17]. These intermembrane space (IMS)-exposed residues, Cys-272 and Cys-332, were shown to dynamically tune Oma1 oligomerization and activity in a redox-dependent manner, and proposed to act as a redox-tunable switch to optimize Oma1 proteolytic function in response to specific homeostatic challenges [17].

The disulfide bond between Cys-272 and Cys-332 is required to maintain the semi-oxidized state of Oma1. Any mutation of these residues leads to destabilization of the Oma1 high-mass complex or yields a metastable protein in the case of Cys-332 [17]. Notably, the mammalian OMA1 is also equipped with these residues and the critical role of said cysteines has been conformed very recently [18]. Nonetheless, further studies to resolve molecular determinants for stress activation, substrate recognition, and redox tuning of OMA1 are warranted.

ROLE OF OMA1 IN THE INNER MEMBRANE HOMEOSTASIS

OMA1 was originally discovered in yeast as a backup protease for the matrix-AAA (m-AAA) metalloprotease, facilitating the cleavage of misfolded mitochondrial membrane protein insertase Oxa1 [11]. Subsequent work on yeast established that Oma1 specifically degrades misassembled Cox1 subunit of cytochrome *c* oxidase (CcO) protein in a subset of CcO assembly mutants, most notably cells lacking yeast-specific assembly factor Coa2. Coa2-deleted cells are respiratory deficient, but this phenotype can be rescued by genetic deletion of Oma1 [13, 19]. Another context-specific substrate of yeast Oma1 is a mutant variant of phosphatidylserine decarboxylase Psd1 [20]. In general, the substrate repertoire of the yeast peptidase remains underexplored and awaits further systematic analysis. Importantly, studies in yeast established that Oma1 genetically interacts with other mitochondrial quality control proteases, thereby reflecting partial functional redundancy of these enzymes [14].

As to the mammalian OMA1, several relatively well-defined targets have been reported. The mammalian enzyme has been found to target other proteins – alone or in concert with other proteases – such as respiratory Complex III assembly factor UQCC3 [21], mitochondrial phosphatase PGAM5 [22, 23], mutant variants of the ubiquitin kinase PINK1 [24], DAP3-binding cell death enhancer protein DELE1 [25, 26], and the IMM GTPase OPA1 [27, 28]. Under certain conditions, OMA1 was also found to proteolytically degrade mtDNA-encoded proteins MTCO1, CYTB, and ND1 proteins [29]. Conversely, the deletion of OMA1 leads to increased steady-state levels of these proteins.

Under specific circumstances, OMA1 can also cleave some mutant variants of PINK1 kinase. PINK1 is a ubiquitin kinase and an effector protein of mitophagy pathway and certain mutations in this protein have been linked to early-onset Parkinson's disease. Under normal conditions, PINK1 is imported into the mitochondria and then rapidly degraded by rhomboid protease PARL. However, when mitochondrial depolarization persists, PINK1 import is inhibited and the protein accumulates at the OMM, thereby initiating the mitophagy process.

TOMM7 protein of the translocase of the outer mitochondrial membrane (TOM) complex is required for accumulation of PINK1 at mitochondria after CCCP treatment. Depletion of TOMM7 or mutation of few negatively charged amino acids at the C-terminal allows PINK1's entry in depolarized mitochondria, wherein it gets processed by OMA1 protease [24]. Of note, OMA1 was also found to cleave serine-threonine phosphatase PGAM5 in coordination with the mitochondrial rhomboid protease PARL upon mitochondrial depolarization. PGAM5 is a substrate for PARL, which is a part of SPY scaffolding complex encompassing stomatin-like protein SLP2 and two IMM proteases – PARL and YME1L [22]. OMA1 has been postulated to transiently associate with the SPY complex, but this model is based on OMA1 overexpression experiments and thus – just like the role of the SPY complex in the IMM organization/remodeling – awaits further investigation. Nonetheless, the activity of both PARL and OMA1 in the cleavage of PGAM5 appears to be mediated by SLP2 [20].

Mitochondrial retrograde responses are now recognized as a facet of the pan-cellular integrated stress response (ISR) in mammalian cells [30]. The ISR is triggered by perturbations to mitochondrial homeostasis and leads to the phosphorylation of translation initiation factor eIF2α, thereby resulting in the suppression of global cytosolic translation while activating specific ISR effector translation factors ATF4 and CHOP [25, 26]. This pathway commences with the stress-activated OMA1 cleaving mitochondria-targeted DELE1, to generate a short form of the protein – sDELE1, which then slips back into the cytosol where it interacts with heme-regulated eIF2α kinase HRI. The interaction of sDELE1 with HRI in the cytosol promotes eIF2α phosphorylation, thereby triggering the induction of its downstream effectors, ATF4 and CHOP [25].

Perhaps the best-characterized target of OMA1 is optic atrophy 1 (OPA1) protein, which is a dynamin-like GTPase located in the inner mitochondrial membrane and is essential for maintaining mitochondrial fusion and ultrastructure. OPA1 has a complex expression pattern, with eight isoforms – each with tissue-specific variations in abundance – generated by alternate splicing at exons 4, 4b, and 5b of the *OPA1* gene [31–33] (Figure 2.2). These isoforms exist as precursor forms, which are cleaved by the mitochondrial processing peptidase to generate long isoforms seen as two

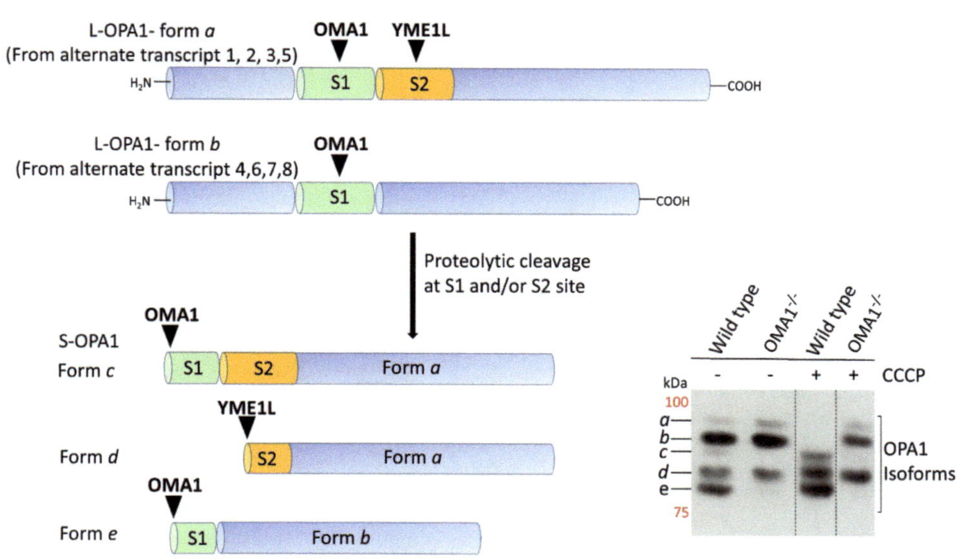

FIGURE 2.2 Molecular organization of OPA1 GTPase. The gene encoding OPA1 produces up to eight different alternative transcript variants. These transcripts are translated into two long forms – '*a*' and a slightly shorter form '*b*' – of the enzyme commonly known as L-OPA1. Both forms contain a proteolytic site, S1, recognized by the OMA1 protease, whereas only the '*b*' form contains a YME1L protease-specific target site designated S2. Under normal physiological conditions, the YME1L cleaves relevant L-OPA1 variants to generate three shorter forms of OPA1 (S-OPA1). The OMA1 is largely dormant and only engages in L-OPA1 processing at S1 site upon homeostatic insults. A representative OPA1 immunoblot analysis demonstrates the appearance of the L-OPA1 variants as five distinct bands in mitochondria isolated from the indicated mouse embryonic fibroblasts. Additional details can be found in the text.

bands (*a* and *b* of L-OPA1) on a polyacrylamide gel. The isoforms of L-OPA1 are further cleaved by mitochondrial protease YME1L, resulting in three shorter bands (*c*, *d*, and *e* of S-OPA1). All OPA1 isoforms contain a common cleavage site, S1 [34], with additional cleavage site S2 present in isoforms 4, 6, 7, and 8 [34]. Stress-activated OMA1 specifically cleaves L-OPA1 variants at S1 site, whereas YME1L processes L-OPA1 at S2 site under basal conditions [27, 28, 33]. Recently, yet another YME1L-dependent cleavage site, S3, was identified in L-OPA1 [35]. OMA1-mediated cleavage of L-OPA1 generates two shorter forms, *c* and *e*, and YME1L cleavage at S2 site generates an extra form, *d*. Under basal conditions, YME1L is constitutively active and OMA1 is largely dormant; therefore, two L-OPA1 forms and three S-OPA1 forms can be observed. These forms exist roughly in an equilibrium and are believed to be responsible for balanced fusion of the mitochondrial network. However, under various stress conditions such as loss of mitochondrial potential, ATP depletion, oxidative stress, inhibition of ETC complex proteins, and rapid changes in cellular redox potential, OMA1 is stress-activated and converts all available L-OPA1 species into the soluble S-OPA1 form, thereby shifting the equilibrium and triggering rapid network fragmentation [12, 27, 36]. Likewise, the oligomerization of pro-apoptotic BAX and BAK proteins, which is a critical step in the intrinsic apoptosis pathway, activates OMA1, thus causing further cleavage of L-OPA1 and triggering cytochrome *c* release [37]. The long OPA1 variants that localize to the IBM/ICM boundaries are believed to oligomerize, thereby locking cristae, thus preventing cytochrome *c* release, and consequently inhibiting the initiation of canonical apoptosis [38, 39] (Figure 2.3).

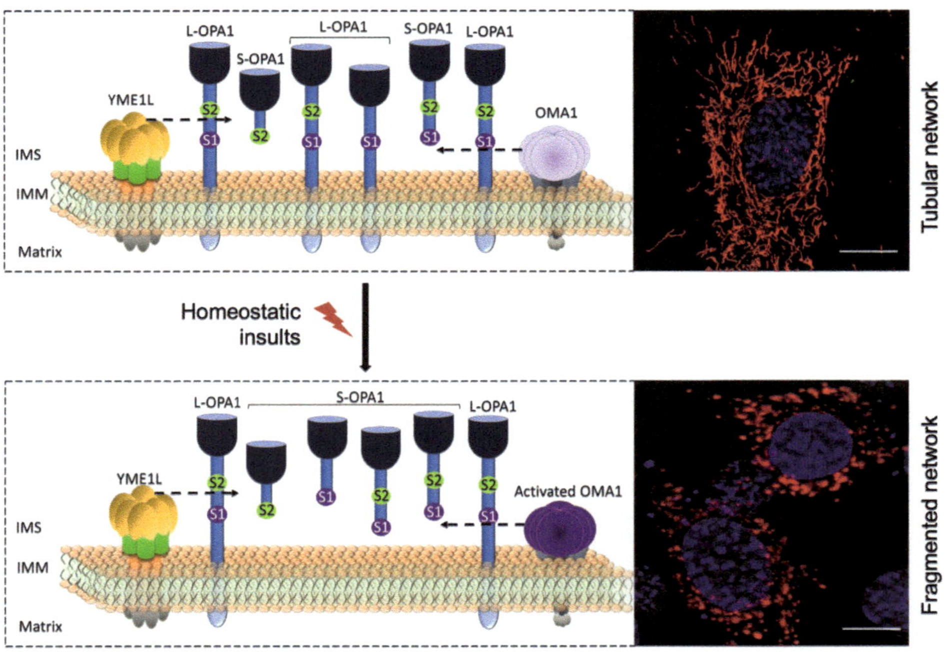

FIGURE 2.3 Functional cooperation between proteases in regulation of IMM dynamics. Schematic depiction of functional cooperation between the OMA1 and YME1L metalloproteases in processing the long forms of OPA1 GTPase. OPA1 is a key IMM remodeling factor engaged in both regulation of the mitochondrial fusion and cristae dynamics. The i-AAA protease YME1L primarily processes a subset of L-OPA1 ('*b*' form) variants harboring the protease-specific S2 cleavage under normal physiological conditions, yielding nearly equimolar amounts of L-OPA1 ('*a*' form) and S-OPA1, thus balancing IMM fusion and fission events. The L-OPA1 is also important for stabilization of cristae junctions. While OMA1 has little to no involvement in L-OPA1 processing under normal conditions, it is rapidly activated in response to physiological challenges or homeostatic insults and executes proteolytic processing of all the S1 site-equipped L-OPA1 forms, thereby triggering mitochondrial network fragmentation and cristae junctions remodeling. Further details can be found in the text.

Indeed, overexpression of L-OPA1 or its variants lacking the S1 cleavage site has been shown to render cells more stress-tolerant and exert tissue-protective effects [40]. Of note, OMA1-deleted cells also exhibit protection against apoptosis-mediated cell death, albeit other aspects of stress tolerance are not improved [41], suggesting more complex role of the enzyme in mitochondrial physiology. In YME1L-depleted cells, L-OPA1 is not processed at the S2 site, leading to increased mitochondrial fragmentation, susceptibility to apoptosis, and deformed cristae morphology [42, 43]. Of note, depletion of OMA1 in YME1L-devoid cells can alleviate some but not all cellular stress phenotypes, further highlighting OMA1's complex role in the IMM homeostasis [36]. Interestingly, in neuronal cells, YME1L is partially degraded by OMA1 [44]. Conversely, YME1L degrades OMA1 under membrane-depolarized conditions in an ATP concentration-dependent manner [16]. While such reciprocity is quite intriguing, it has not been observed in several other types of cells and may be specific to cell context [12, 16, 36] and needs to be investigated further. A more common theme in stress activation of the mammalian OMA1 appears to be its autoproteolysis that occurs upon the enzyme activation, although it is delayed in time when compared to L-OPA1 processing [12, 45] and likely represents a feedback loop mechanism to ensure the reversibility of L-OPA1 cleavage and eliminates hyperactive protease and thus minimizes potential collateral damage [15].

As indicated earlier, the short form of OPA1 promotes mitochondrial fragmentation, whereas stabilization or enhance expression of L-OPA1 isoform protects cristae morphology and inhibit apoptosis-mediated cell death [40]. Conversely, OMA1-executed cleavage of L-OPA1 inhibits mitochondrial fusion and induces partitioning of mitochondrial network, which is believed to aid in removing the damaged/defective mitochondria by mitophagy [46]. It is noteworthy, however, that the above mechanism does not represent the sole way to remove certain mitochondrial populations as other mitophagy and mitophagy-like pathways have been reported for which mitochondrial network fragmentation is either dispensable or not required [47–49]. The importance of OMA1 catalytic function in the IMM shaping and remodeling is underscored by the fact that cells expressing proteolytically inactive H331A and E324Q variants of the protease are unable to process L-OPA1 and largely mimic OMA1-devoid cells [11, 27, 37].

Several other factors have been reported to influence OMA1 stability and/or activity. For instance, OMA1 turnover is regulated by prohibitins PHB1 and PHB2, and phospholipid cardiolipin [50]. The prohibitins stabilize cardiolipin, which in turn enhances OMA1 turnover. Interestingly, such regulation appears to be independent of membrane potential or YME1L [50]. Related to this notion, prohibitins were shown to prevent hypoxia-induced apoptosis by reducing the release of cytochrome *c* [51] through a mechanism that is likely to involve OMA1. Likewise, an optic atrophy patient mutation G337E in the AFG3L2 subunit of the m-AAA protease also results in hyperactivation of OMA1 and depletion of the L-OPA1 isoform [23].

Another factor, the mitochondrial matrix chaperone p32/C1QBP, reportedly modulates mitochondrial ultrastructure and metabolism through the regulation of OMA1-mediated cleavage of L-OPA1 [26]. Consistently, cells lacking p32/C1QBP show a shift toward glycolysis and an increase in apoptosis. Another example is estrogen receptor α that binds to the promoter region and controls the expression of OMA1 in MIN6 pancreatic β cells [27]. An overexpression of OMA1 in MIN6 cells increases the levels of PERK and eIF2α phosphorylation and the expression of CHOP protein [52]. These events are likely related to OMA1-DELE1-HRI signaling axis of the integrated stress response. Leptins have also been reported to regulate the activation/degradation of OMA1 through phosphorylated GSK3 kinase and ubiquitin, thus leading to stabilization of L-OPA1 [28]. Such stabilization protects human mesenchymal stem cells from both apoptosis and ischemic injury when leptin treatment is applied.

Recently, a new OMA1-dependent stress response mechanism has been described to prevent the extreme fusion of mitochondrial network [53]. In DRP1-deficient cells, extreme mitochondrial fusion is reportedly inhibited by OMA1-mediated L-OPA1 processing and mitofusin MFN2 downregulation. Briefly, in DRP1-deleted cells, mitochondrial membrane potential decreases in short pulses, referred to as "flickering", thereby activating OMA1 proteolytic activity. Given its role in protecting cells from bioenergetic deficit stemming from the extreme mitochondrial fusion, the mechanism was named "mitochondrial safeguard" or "Mitosafe". Related to this notion, a novel microprotein Mco8/Mdi1 was reported recently in yeast. Akin to the Mitosafe mechanism, it protects mitochondria from hyperfusion [54]. Mdi1 is an IMS-localized protein that inhibits excessive mitochondrial fusion presumably by distortion of IMM. It will be interesting to know if similar or related proteins exist in mammalian mitochondria and whether it has any role in the Mitosafe mechanism.

THE MICOS MACHINERY

MICOS Complex

The mitochondria contact site and cristae organizing system – commonly referred to as MICOS – is a large, evolutionarily conserved multiprotein complex residing at the interface of the IBM and CM known as cristae junctions. Along with other factors, the complex is responsible for shaping and maintaining the structure of cristae and cristae junctions as well as connecting the IMM and OMM, thereby forming intermembrane contacts (Figure 2.4). The essential components of the MICOS machinery are summarized in Table 2.1. Several other proteins interact with the complex to ensure its structural integrity and functioning. Biochemical analyses established that MICOS comprises

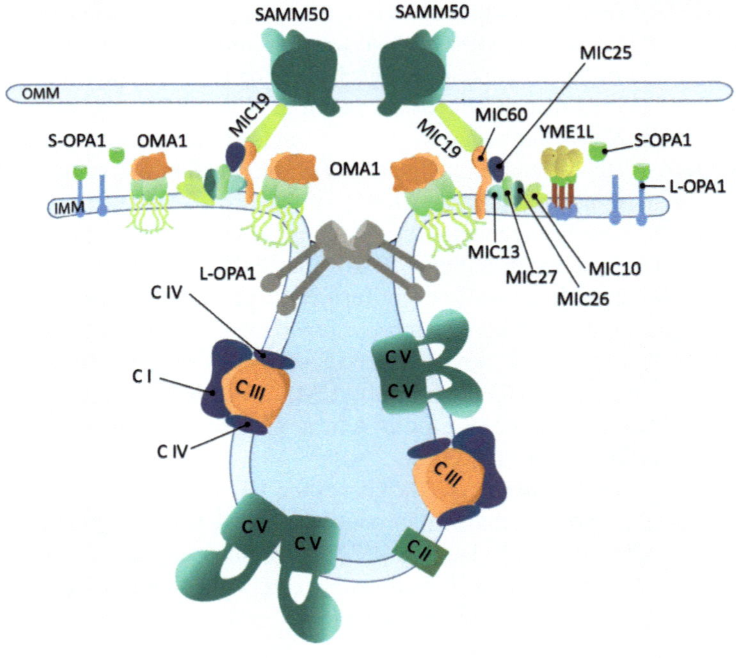

FIGURE 2.4 Molecular organization of the MICOS machinery. The schematic shows the core MICOS machinery. The MICOS complex proteins are located at the cristae junction sites of the inner membrane demarcating the boundaries between the inner boundary membrane and intercristae domains of the IMM. The ICM domain (i.e., cristae) harbor the oxidation phosphorylation system complexes, which play a role in shaping these membrane structures, whereas the MICOS machinery supports cristae formation at the junction sites. Other proteins such as metallopeptidase OMA1 and GTPase OPA1 are associated with the MICOS machinery with the latter enzyme regulating cristae diameter. The MICOS-mediated intermembrane contacts sites, often referred to as an MIB complex are formed through the interactions with OMM proteins such as SAMM50, thus shaping up the organelles' membrane architecture. The OMA1-MICOS axis appears to stabilize these structures through a currently unknown mechanism. See the text for additional details.

TABLE 2.1
Core Components of the Mitochondrial Contact Site and Cristae Organizing System

Common Name	Mammals	Yeast	Estimated MW (kDa)	Observed MW (kDa)
MIC60	Mitofilin	Mic60	90	~80
MIC19	CHCHD3	Mic19	26	26
MIC25	CHCHD6	Mic25	26	~29
MIC10	MINOS1	Mic10	9	9
MIC26	APOO	Mic26	22	22
MIC27	APOOL	Mic27	29	29
MIC13	QIL1	Mic13	13	13

two subcomplexes: one subcomplex includes MIC60, MIC19, and MIC25 subunits and bridges the IMM and OMM; the other subcomplex encompasses MIC27, MIC26, MIC13, and MIC10 subunits and regulates crisate junctions. Of note, the MIC26 and MIC27 subunits belong to the apolipoprotein family and are involved in later stages of MICOS assembly but are dispensable for the stability of the complex per se. Both MIC19 and MIC25 subunits contain coiled-coil-helix-coiled-coil-helix domains and interact directly with MIC60. The loss of either MIC19 or MIC25 subunit leads to a reduction of the other protein in this tandem [55, 56]. Moreover, the overexpression of MIC25 rescues mitochondrial morphology in a MIC19 knockout background [57]. Furthermore, MIC19 interacts with the OMM protein SAMM50, thereby bridging two membranes. Consequently, depletion of MIC19 results in the loss of both MIC60 and SAMM50 [57,58]. Additionally, MICOS complex was shown to associate with various other proteins – both within the IMM (MIA40, DNAJC1, CHCHD2, and CHCHD10) and the OMM (MIRO1/2, SLC25A46, VDAC). The nature of these associations remains to be investigated but, as a common denominator, depletion of said proteins results in aberrant cristae structure [59–61].

MICOS Functions in Mitochondria

The MICOS complex is one of the key factors that define mitochondrial membrane architecture. It is crucial for maintaining the structure of cristae and cristae junctions. Depleting virtually any core component of the complex results in profound ultrastructural abnormalities. For example, deletion of MIC60 results in the complete loss of cristae junctions, loss of the IMM-OMM contacts, and ultrastructural abnormalities that manifest as concentric rings or tubular stacks of the IMM [62]. Similarly, deletion of MIC10 also leads to abnormal mitochondrial membrane architecture [62]. MIC26 and MIC27 regulate their protein levels antagonistically, i.e., the deletion of one protein results in a moderate cristae morphology defect, whereas a double deletion of both proteins causes severe ultrastructural defects and affects the stability of respiratory chain supercomplexes [63]. By contrast, the effect of MIC25 deletion is severe and leads to complete disruption of cristae morphology with cristae-less hollow mitochondria and defects in cellular growth and metabolic defects [56]. Finally, the deletion of MIC19 has been reported to result in either complete loss of cristae or abnormal tubular cristae [55]. Such severe phenotypes underscore the key cooperative role of the MICOS machinery with other IMM proteins and lipids in altering physical state and dynamically shaping the IMM at cristae junctions, thereby defining cristae shape and positions.

MICOS functions are not limited to cristae shaping. For example, the MIC10 subcomplex has been shown to cooperate with the TIM22 and TIM23 IMM translocases to facilitate the import of carrier proteins [64]. Of note, in human pathogen *Trypanosoma brucei*, the TbMIC20 subunit of the trypanosomal MICOS complex was found to regulate protein import [65]. Furthermore, as we have already pointed out above, MICOS is also associated with the SAMM50 protein – a core component of the OMM-residing SAM complex – and may be involved in regulation of the sorting an assembly of β-barrel proteins such as VDAC in the OMM [57, 58]. Additionally, MICOS and SAMM50 form a higher order mitochondrial intermembrane space bridging (MIB) complex, which has been shown to be critical for mitochondrial architecture and IMM remodeling in response to physiological challenges and other crucial functions such as transport of lipid biosynthetic intermediates or intermediate porphyrins to supply the process of heme biosynthesis [57, 66, 67]. In a related fashion, MICOS has been implicated in the regulation of mitophagy. This is achieved through the interaction of MICOS with the ubiquitin kinase PINK1. Under normal physiological conditions, PINK1 was shown to phosphorylate MIC60 to maintain its oligomerization state, which is required for the formation and maintenance of cristae junctions. PINK1 deficit in the IMS due to its arrested import under mitochondrial stress conditions leads to the significantly reduced number of cristae junctions, subsequently causing cristae to morph into a "fuzzy onion" or vacuolar shape [65].

Remarkably, the overexpression of MIC60 mimics the phosphorylation phenotype, stabilizing its oligomeric state and rescuing cristae junctions and overall mitochondrial ultrastructure [65]. Another IMM protein, immunity-related GTPase M (IRGM), has been implicated in the regulation of MIC60 stability and the mitophagy process. In SH-SY5Y neuronal cells, depletion of IRGM promotes resistance to PINK1-Parkin-mediated mitophagy [68]. IRGM promotes the degradation of MIC60 and stabilizes PINK1, leading to its accumulation on the surface of depolarized mitochondria. Finally, the MICOS machinery appears to have a role in mtDNA transcription. The MIC60 subunit has been shown to associate with mitochondrial transcription factors TFAM, TFB2M, and TFB1M located in the mitochondrial matrix [69]. A decrease in MIC60 levels results in a reduction in mitochondrial transcription due to a decrease in TFAM binding to the mtDNA. Such a reduction results in decreased expression of genes encoding core OXPHOS subunits and, consequently, reduced respiratory activity [69]. The molecular basis of the MICOS association with TFAM proteins needs to be clarified, however. It remains to be investigated whether this association is direct or mediated through other factors, such as AAA protein ATAD3A, which is known to associate with both MICOS machinery and mtDNA [70, 71].

PROTEOLYTIC CONTROL OF THE MICOS MACHINERY

MICOS and OMA1 Protease

In this section, we primarily focus on recently identified functional association between the OMA1 protease and the MICOS machinery. Until recently, the main role of OMA1 was believed to be the stress-triggered cleavage of L-OPA1 species to induce fragmentation of the mitochondrial network. However, several recent studies have reported a specific interaction between OMA1 and the MIC60 and MIC19 subunits of the MICOS/MIB complex [57]. Importantly, this interaction is independent of OPA1, which has also been found to associate with the MICOS machinery [72]. Although the molecular underpinnings of said association remain to be characterized in more detail, the MIC60 proteins does not appear to be a substrate of OMA1 but rather its interacting partner [72]. The situation is even less clear in the case of MIC19 subunit – while one study reported that OMA1 cleaves MIC19 at its N-terminal domain under stress conditions [57], the other independent reports were unable to corroborate this claim [72, 73]. Therefore, the functional relationship between OMA1 and MIC19 warrants further investigation. Regardless, these studies revealed that OMA1-dependent support of the MICOS/MIB machinery and IM-OMM contacts is crucial for overall mitochondrial architecture and bioenergetics under both basal and stress conditions [57, 72]. Indeed, OMA1 appears to be critical for stability of the MIB complex and the OMA1-MICOS/MIB axis plays an important role in the IMM and cristae junctions remodeling [18]. The deletion of OMA1 results in labile MICOS/MIB complex due to changes in the MIC60 subcomplex and reduced steady state levels of the MIC19, MIC13, and MIC10 subunits. This manifests in the increased number of mitochondria exhibiting aberrant ultrastructure, even though L-OPA1 variants remain preserved. The population of mitochondria with vesicular or "fuzzy onion"-style cristae is further expanded in the stressed cells [72]. In line with this notion is the finding that expression of a synthetic IMM-OMM protein bridge, mitoT, can remedy most of the defects arising from impaired intermembrane contacts seen in OMA1-deficient cells, including bioenergetics deficit and enhanced apoptotic resistance stemming from impeded cytochrome *c* release [72].

Other Factors

In addition to the abovementioned factors, other intermembrane space-facing proteases could contribute to the MICOS complex stability and mitochondrial architecture overall. For instance, the i-AAA protease YME1L has been shown to regulate mitochondrial cristae dynamics [33, 36, 42, 43]. As indicated earlier, in the absence of YME1L – and independently to changes in OPA1 processing – mitochondrial fragmentation and abnormal cristae morphology were observed [43]. Moreover, YME1L was suggested to mediate proteolytic degradation of MIC60 in the MIC19-depleted cells [74]. Considering YME1L's role in L-OPA1 processing, it will be interesting to interrogate the mechanism by which YME1L mediates mitochondrial ultrastructure regulation. It should be noted, however, that YME1L is a multifaceted enzyme, and its loss has been associated with a plethora of molecular and physiological consequences, such as dysregulated phospholipid homeostasis in the IMM. Therefore, correlation versus causation issue should be carefully considered when investigating this issue.

Another candidate molecule to influence mitochondrial architecture is a β-lactamase family protein LACTB. This unusual and conserved serine protease resides in the IMS facing the intermembrane space. It forms long filamentous homopolymer structures that are thought to regulate the enzyme's catalytic activity and – potentially – the intermembrane distance and IMM compartmentalization [75–79]. It has been hypothesized that contrary to the MICOS complex that appears to be bringing the OMM and IMM closer, LACTB filaments might increase the spatial distance between the mitochondrial membranes [75]. Such a spatial separation could therefore limit the transfer of metabolites/precursors from the mitochondrial surface (i.e., OMM) and the IMM compartments. At present, no protein substrates are known for LACTB and the role of this protease remains elusive. It will be interesting to investigate its role in mitochondrial architecture regulation.

REGULATION OF MITOCHONDRIAL ARCHITECTURE IN THE CONTEXT OF DISEASE

MICOS in Sickness and Health

Altered mitochondrial architecture is a hallmark of many pathological states [80, 81] but has long been a subject of a "cause versus consequence" debate. Recent studies on MICOS offer some new insights to this question. Considering the significance of MICOS machinery, it is not surprising that a variety of clinical manifestations have been lined to mutations or altered expression of the MICOS-related proteins. To date, pathologic mutations have been reported in the MIC26/APOO component of MICOS [82]. These include the Ile117Thr substitution that has a complex neuromuscular and metabolic manifestation including progressive developmental delay, lactic acidosis, muscle weakness, hypotonia, weight loss, gastrointestinal and body temperature dysautonomia, repetitive infections, cognitive impairment, and autistic behavior [82]. Subsequent studies in yeast and fruit fly genetic models have validated the mutation's impact on MICOS and mitochondrial functions [82]. Another patient with a remarkably similar set of clinical manifestations was reported to contain a 275 kb-deletion impacting five annotated genes including MIC26, among which the latter appears to be the only disease-associated one [83].

More pathologies have been linked to altered expression of MICOS subunits. For instance, in dopamine quinone–treated neurons mimicking the conditions of neurodegenerative Parkinson's disease (PD), the levels of MIC60 were found to be significantly decreased [84]. Moreover, in dopamine-treated dopaminergic cell lines, the deletion of MIC60 exacerbates cell death, while its overexpression was shown to provide protection [85]. Similarly, and as discussed earlier, the ubiquitin kinase PINK1 can phosphorylate MIC60 and enhance its oligomerization in the fruit fly *Drosophila melanogaster* genetic model of PD [68]. Mutation of the MIC60 phosphorylation site leads to abnormal cristae morphology. Reciprocally, depletion of both PINK1 and Parkin also causes cristae abnormalities. The overexpression of DmMIC60 rescues the cristae defects caused by PINK1 depletion but not those caused by Parkin depletion [65]. Furthermore, PINK1 overexpression in a DmMIC60 knockout background cannot rescue the aberrant cristae phenotype, indicating that DmMIC60 functions downstream of PINK1 in maintaining mitochondrial architecture. Furthermore, MIC60 is also a substrate of protein kinase A (PKA), which phosphorylates it at the Ser-528 residue [72]. The deletion of MIC60 decreases the level of PINK1 in depolarized mitochondria and hinders the recruitment of Parkin. Mutation of the Ser-528 leads to the localization of PINK1 to mitochondrial surface and the recruitment of Parkin to induce mitophagy. Similarly, MIC19 has also been found to be a target of PKA and functions similarly to MIC60 in Parkin recruitment [86].

The alteration of MICOS complex levels have been also reported in various cancers [87]. For example, anticancer drugs that induce DNA damage have been shown to downregulate the levels of MIC25/CHCHD6, which appears to make breast and colon cancer cells more susceptible to further therapeutic interventions [56]. By contrast, induction of CHCHD6 expression leads to drug resistance [56]. Elevated levels of MIC60 have been observed in pancreatic, prostate, hepatic, breast, gastric, and pediatric ependymoma cancers, whereas MIC13 was found to be upregulated in inflammatory breast cancer [88–92]. However, it remains to be established whether these changes in MICOS complex levels are the cause or an effect.

Finally, alterations in MICOS machinery have been linked to obesity and obesity-associated complications [93–95]. Studies have shown that high-fat diets lead to a reduction in MIC60 expression in mouse models, which translates to changes in mitochondrial metabolism and a decrease in cristae number [93, 94]. On the other hand, overexpression of MIC60 in diabetic patients has been shown to improve cardiac function, whereas overexpression of MIC25 had the opposite effect [93]. Additionally, a decrease in the association between the MIC60-SAM50-MIC19 axis of the MICOS/MIB complex has been observed in patients with diabetes-induced cardiomyopathy [95]. Furthermore, MIC19 levels were found to be significantly reduced in both obese and diabetic patients [96].

OMA1 in Health and Disease

Although OMA1 appears to play a major and multifaceted role in cellular physiology, the current understanding of its role in human disease remains somewhat limited. Conserved heterozygous mutations in OMA1 such as His69Tyr and Glu272Gly have been reported in several patients with sporadic amyotrophic lateral sclerosis (ALS) [97]. Considering the protease's ability to form homo-oligomers [11, 14], it is plausible that the presence of mutant OMA1 variants in oligomers may interfere with its stability or function. However, further studies are warranted to interrogate this scenario. Mutations in MICOS-associated factors CHCHD2 and CHCHD10 have been linked to both PD and ALS [98]. Interestingly, depleting CHCHD2 and CHCHD10 in mouse models leads to mitochondrial fragmentation through OMA1-mediated cleavage of L-OPA1 [99].

Many insights into OMA1's role in disease have been mostly gleaned from studies in the rodent models. While deleting OMA1 in the mouse model does not cause any developmental abnormalities, it does lead to physiological and metabolic impediments [41]. When fed a high-fat diet, these animals exhibit a notable difference in body weight and white adipose tissue, with higher levels of triglycerides and a tendency toward liver steatosis and impaired glucose tolerance [41]. Unlike their wild-type counterparts, OMA1-deleted mice are unable to maintain their core body temperature, exhibiting a lower temperature during nighttime and under cold stress conditions [41]. These findings underscore OMA1's role in regulation of mitochondrial and cellular metabolic functions and are in line with a recent report demonstrating impaired metabolic flux of pyruvate in OMA1-deleted cells [100]. The impaired cristae remodeling, mitochondrial fragmentation, and cytochrome *c* release, and consequential apoptotic resistance in OMA1-deleted cells can be physiologically beneficial in certain tissue context. For example, the proper balance between OMA1 and YME1L in processing of L-OPA1 isoforms is crucial for maintaining the function of adult cardiac tissue. The absence of YME1L in mice results in embryonic death in whole-body knockout animals and a cardiac hypertrophy characterized by impaired mitochondrial morphology, when the gene is specifically deleted in adult hearts [101]. Quite remarkably, this issue can be resolved through simultaneous deletion of OMA1, which restores both the mitochondrial shape and cardiac functionality [101]. Likewise, in some, but not all, mouse model of heart failure, OMA1 was found to play a critical role [100, 102]. The excessive production of reactive oxygen species in heart failure mice activates OMA1, leading to cleavage of L-OPA1 and disruption of the mitochondrial structure, ultimately causing cell death. Again, the disease phenotype can be remedied by OMA1 depletion [103]. Related to this notion, a variant allele, rs17117699T>G, of OMA1 was identified to increase the risk of heart failure [104]. It should be highlighted, however, that loss of OMA1 does not always provide physiological benefit as described above. Indeed, while stabilizing the L-OPA1 forms can have certain neuroprotective effect, not only depletion of OMA1 in YME1L-deleted neurons provides appreciable rescue effect but, in some cases, exacerbates physiological penalties of said deletion [105]. These observations illustrate the complexity of OMA1's role in mitochondrial physiology and underscore its unique functions in certain tissues.

Finally, a string of recent studies implicated OMA1 in the progression and development of several types of cancer. For instance, the high OMA1 levels have been associated with poor prognosis and clinical outcomes in gastric, colorectal, squamous cell lung carcinoma, and sarcoma [18, 98, 106]. By contrast, in other cancers such as lung adenocarcinoma and breast carcinoma, OMA1 acts as a tumor suppressor, preventing the development and progressions of these types of cancer [106, 107]. Of note, a recent study discovered recurrent in-frame fusions of TACSTD2-OMA1 in breast cancer patients [108]. Although the molecular bases of these effects remain to be clarified, in the latter case, depletion of OMA1 in metastatic breast adenocarcinoma cells has been shown to promote metastasis through epithelial-mesenchymal transition and induce metabolic changes as well as altered mitochondrial retrograde responses [107]. Similarly, in colorectal cancer, the OMA1-OPA1 axis leads to metabolic reprogramming, causing a shift in metabolism from OXPHOS to glycolysis [109]. Interestingly, these effects resemble those described in the previous paragraph and underscore nuanced and context-specific involvement of the protease in these tumors, which remains to be understood in a greater detail.

Finally, OMA1 levels appear to be playing a role in anticancer therapies. For example, cisplatin – a widely used chemotherapy drug – has been shown to elevate the OMA1 levels only in chemosensitive cells and not in chemoresistant cervical tumor cells, leading to increased cleavage of L-OPA1 and resulting in enhanced mitochondrial fragmentation in the former cells [110]. Such behavior was proposed to be due to the dissociation of the prohibitin1-OPA1 complex by cisplatin treatment, which causes prohibitin to bind to phosphorylated p53 within the mitochondria. Thus, silencing p53 may reduce the chemoresistant effects [110] and suggests potential benefits of combinatorial therapies targeting proteins in question.

OUTLOOK

While many aspects of mitochondrial biology have come to light in recent years, our fundamental understanding of the mechanisms that orchestrate and adjust mitochondrial functions in response to physiological cues remains limited. One major gap in knowledge pertains to the dynamic spatiotemporal organization of various functional machineries such as MICOS/MIB and intermembrane contacts that play critical roles in the choreography of mitochondrial membranes – a long-standing enigma that is central to a variety of mitochondrial processes like cofactor synthesis and trafficking. Similarly, precisely how cristae remodeling influences metabolic and signaling outputs of mitochondria awaits further clarification. The functional interplay between the mechanisms underpinning various IMM remodeling pathways is another question to be addressed. Recent studies reporting a new role of cristae in creating and controlling dynamic "gradients" of mitochondrial membrane potential throughout the IMM [111, 112] and other mechanisms at the interface of mitochondrial network dynamics and cristae remodeling such as Mitosafe [53] or a very recent identification of non-MICOS architecture-mediating factors such as Cqd1/ADCK2 [113] underscore the necessity of such studies. Addressing these issues may provide answers as to how cristae remodeling is finely orchestrated and sometimes even reversed during mitochondrial membrane permeability [114] or intracellular pathogen insults [115] and what other factors contribute to these events.

A number of exciting questions pertaining to OMA1-MICOS axis await to be addressed as well. For example, one exciting direction regards recently identified redox regulation of both OMA1 and MICOS [17, 18]. It will be interesting to test whether concerted action of these factors could be organized in a redox-dependent manner. New insights gleaned from the abovementioned studies will undoubtedly open novel exciting research avenues and may aid in targeting said processes in relevant human disorders.

ACKNOWLEDGEMENTS

We apologize to those authors whose work we were unable to cite due to space constraints. We wish to thank Carey Goddard for her help with the illustrations. We also wish to acknowledge the grant support from the National Institutes of Health National Institute of General Medical Sciences, GM131701-01 to O.K.

Conflict of Interest

The authors declare that they have no conflicts of interest with the contents of this text.

REFERENCES

1. J. R. Friedman and J. Nunnari, "Mitochondrial form and function," *Nature*, vol. 505, no. 7483, pp. 335–343, 2014, doi: 10.1038/nature12985.
2. T. G. Frey and C. A. Mannella, "The internal structure of mitochondria," *Trends Biochem. Sci.*, vol. 25, no. 7, pp. 319–324, Jul. 2000, doi: 10.1016/s0968-0004(00)01609-1.
3. M. Zick, R. Rabl, and A. S. Reichert, "Cristae formation-linking ultrastructure and function of mitochondria," *Biochim. Biophys. Acta — Mol. Cell Res.*, vol. 1793, no. 1, pp. 5–19, 2009, doi: 10.1016/j.bbamcr.2008.06.013.
4. C. A. Mannella, "The relevance of mitochondrial membrane topology to mitochondrial function," *Biochim. Biophys. Acta — Mol. Basis Dis.*, vol. 1762, no. 2, pp. 140–147, 2006, doi: 10.1016/j.bbadis.2005.07.001.
5. J. Song, J. M. Herrmann, and T. Becker, "Quality control of the mitochondrial proteome," *Nat. Rev. Mol. Cell Biol.*, vol. 22, no. 1, pp. 54–70, 2021, doi: 10.1038/s41580-020-00300-2.
6. S. Rath *et al.*, "MitoCarta3.0: An updated mitochondrial proteome now with sub-organelle localization and pathway annotations," *Nucleic Acids Res.*, vol. 49, no. D1, pp. D1541–D1547, 2021, doi: 10.1093/nar/gkaa1011.
7. M. Y. W. Ng, T. Wai, and A. Simonsen, "Quality control of the mitochondrion," *Dev. Cell*, vol. 56, no. 7, pp. 881–905, 2021, doi: 10.1016/j.devcel.2021.02.009.
8. O. Khalimonchuk and D. F. Becker, "Molecular determinants of mitochondrial shape and function and their role in glaucoma," *Antioxid. Redox Signal.*, vol. 38, pp. 896–919, 2023, doi: 10.1089/ars.2022.0124.
9. S. Deshwal, K. U. Fiedler, T. Langer, T. Langer, and T. Langer, "Mitochondrial proteases: Multifaceted regulators of mitochondrial plasticity," *Annu. Rev. Biochem.*, vol. 89, pp. 501–528, 2020, doi: 10.1146/annurev-biochem-062917-012739.
10. M. V Alavi, "OMA1 — An integral membrane protease?," *Biochim. Biophys. Acta — Proteins Proteom*, vol. 1869, no. 2, p. 140558, 2021, [Online].doi:10.1016/j.bbapap.2020.140558.

11. M. Käser, M. Kambacheld, B. Kisters-Woike, and T. Langer, "Oma1, a novel membrane-bound metallopeptidase in mitochondria with activities overlapping with the m-AAA protease," *J. Biol. Chem.*, vol. 278, no. 47, pp. 46414–46423, 2003, doi: 10.1074/jbc.M305584200.
12. M. J. Baker *et al.*, "Stress-induced OMA1 activation and autocatalytic turnover regulate OPA1-dependent mitochondrial dynamics," *EMBO J.*, vol. 33, no. 6, pp. 578–593, 2014, doi: 10.1002/embj.201386474.
13. O. Khalimonchuk, M. Y. Jeong, T. Watts, E. Ferris, and D. R. Winge, "Selective Oma1 protease-mediated proteolysis of Cox1 subunit of cytochrome oxidase in assembly mutants," *J. Biol. Chem.*, vol. 287, no. 10, pp. 7289–7300, 2012, doi: 10.1074/jbc.M111.313148.
14. I. Bohovych, G. Donaldson, S. Christianson, N. Zahayko, and O. Khalimonchuk, "Stress-triggered activation of the metalloprotease Oma1 involves its C-terminal region and is important for mitochondrial stress protection in yeast," *J. Biol. Chem.*, vol. 289, no. 19, pp. 13259–13272, 2014, doi: 10.1074/jbc.M113.542910.
15. K. Zhang, H. Li, and Z. Song, "Membrane depolarization activates the mitochondrial protease OMA1 by stimulating self-cleavage," *EMBO Rep.*, vol. 15, no. 5, pp. 576–585, 2014, doi: 10.1002/embr.201338240.
16. T. K. Rainbolt, J. Lebeau, C. Puchades, and R. L. Wiseman, "Reciprocal degradation of YME1L and OMA1 adapts mitochondrial proteolytic activity during stress," *Cell Rep.*, vol. 14, no. 9, pp. 2014–2049, 2016, doi: 10.1016/j.celrep.2016.02.011.
17. I. Bohovych, J. V. Dietz, S. Swenson, N. Zahayko, and O. Khalimonchuk, "Redox regulation of the mitochondrial quality control protease Oma1," *Antioxid. Redox Signal.*, vol. 31, no. 6, pp. 429–443, 2019, doi: 10.1089/ars.2018.7642.
18. R. Miallot *et al.*, "An OMA1 redox site controls mitochondrial homeostasis, sarcoma growth, and immunogenicity," *Life Sci. alliance*, vol. 6, no. 6, pp. 1–17, 2023, doi: 10.26508/lsa.202201767.
19. M. Bestwick, O. Khalimonchuk, F. Pierrel, and D. R. Winge, "The role of Coa2 in hemylation of yeast Cox1 revealed by its genetic interaction with Cox10," *Mol. Cell. Biol.*, vol. 30, no. 1, pp. 172–185, 2010, doi: 10.1128/mcb.00869-09.
20. O. B. Ogunbona, O. Onguka, E. Calzada, and S. M. Claypool, "Multitiered and cooperative surveillance of mitochondrial phosphatidylserine decarboxylase 1," *Mol. Cell. Biol.*, vol. 37, no. 17, pp. 1–24, 2017, doi: 10.1128/mcb.00049-17.
21. M. Desmurs *et al.*, "C11orf83, a mitochondrial cardiolipin-binding protein involved in *bc*1 complex assembly and supercomplex stabilization," *Mol. Cell. Biol.*, vol. 35, no. 7, pp. 1139–1156, 2015, doi: 10.1128/mcb.01047-14.
22. T. Wai *et al.*, "The membrane scaffold SLP2 anchors a proteolytic hub in mitochondria containing PARL and the *i*-AAA protease YME1L," *EMBO Rep.*, vol. 17, no. 12, pp. 1844–1856, 2016, doi: 10.15252/embr.201642698.
23. E. P. Sokolov *et al.*, "Effects of hypoxia-reoxygenation stress on mitochondrial proteome and bioenergetics of the hypoxia-tolerant marine bivalve Crassostrea gigas," *J. Proteomics*, vol. 194, no. October 2018, pp. 99–111, 2019, doi: 10.1016/j.jprot.2018.12.009.
24. S. Sekine, C. Wang, D. P. Sideris, E. Bunker, Z. Zhang, and R. J. Youle, "Reciprocal roles of Tom7 and OMA1 during mitochondrial import and activation of PINK1," *Mol. Cell*, vol. 73, no. 5, pp. 1028–1043.e5, 2019, doi: 10.1016/j.molcel.2019.01.002.
25. E. Fessler *et al.*, "A pathway coordinated by DELE1 relays mitochondrial stress to the cytosol," *Nature*, vol. 579, no. 7799, pp. 433–437, 2020, doi: 10.1038/s41586-020-2076-4.
26. X. Guo *et al.*, "Mitochondrial stress is relayed to the cytosol by an OMA1–DELE1–HRI pathway," *Nature*, vol. 579, no. 7799, pp. 427–432, 2020, doi: 10.1038/s41586-020-2078-2.
27. B. Head, L. Griparic, M. Amiri, S. Gandre-Babbe, and A. M. Van Der Bliek, "Inducible proteolytic inactivation of OPA1 mediated by the OMA1 protease in mammalian cells," *J. Cell Biol.*, vol. 187, no. 7, pp. 959–966, 2009, doi: 10.1083/jcb.200906083.
28. S. Ehses *et al.*, "Regulation of OPA1 processing and mitochondrial fusion by m-AAA protease isoenzymes and OMA1," *J. Cell Biol.*, vol. 187, no. 7, pp. 1023–1036, 2009, doi: 10.1083/jcb.200906084.
29. U. Richter, T. Lahtinen, P. Marttinen, F. Suomi, and B. J. Battersby, "Quality control of mitochondrial protein synthesis is required for membrane integrity and cell fitness," *J. Cell Biol.*, vol. 211, no. 2, pp. 373–389, 2015, doi: 10.1083/jcb.201504062.
30. M. Costa-Mattioli and P. Walter, "The integrated stress response: From mechanism to disease," *Science*, vol. 368, no. 6489, 2020, doi: 10.1126/science.aat5314.
31. C. Delettre *et al.*, "Mutation spectrum and splicing variants in the OPA1 gene," *Hum. Genet.*, vol. 109, no. 6, pp. 584–591, 2001, doi: 10.1007/s00439-001-0633-y.
32. C. Delettre *et al.*, "Nuclear gene OPA1, encoding a mitochondrial dynamin-related protein, is mutated in dominant optic atrophy," *Nat. Genet.*, vol. 26, no. 2, pp. 207–210, 2000, doi: 10.1038/79936.

33. Z. Song, H. Chen, M. Fiket, C. Alexander, and D. C. Chan, "OPA1 processing controls mitochondrial fusion and is regulated by mRNA splicing, membrane potential, and Yme1L," *J. Cell Biol.*, vol. 178, no. 5, pp. 749–755, 2007, doi: 10.1083/jcb.200704110.
34. N. Ishihara, Y. Fujita, T. Oka, and K. Mihara, "Regulation of mitochondrial morphology through proteolytic cleavage of OPA1," *EMBO J.*, vol. 25, no. 13, pp. 2966–2977, 2006, doi: 10.1038/sj.emboj.7601184.
35. R. Wang, P. Mishra, S. D. Garbis, A. Moradian, M. J. Sweredoski, and D. C. Chan, "Identification of new OPA1 cleavage site reveals that short isoforms regulate mitochondrial fusion," *Mol. Biol. Cell*, vol. 32, no. 2, pp. 157–168, 2021, doi: 10.1091/MBC.E20-09-0605.
36. R. Anand *et al.*, "The i-AAA protease YME1L and OMA1 cleave OPA1 to balance mitochondrial fusion and fission," *J. Cell Biol.*, vol. 204, no. 6, pp. 919–929, 2014, doi: 10.1083/jcb.201308006.
37. X. Jiang, H. Jiang, Z. Shen, and X. Wang, "Activation of mitochondrial protease OMA1 by Bax and Bak promotes cytochrome c release during apoptosis," *Proc. Natl. Acad. Sci. U.S.A.*, vol. 111, no. 41, pp. 14782–14787, 2014, doi: 10.1073/pnas.1417253111.
38. C. Frezza *et al.*, "OPA1 controls apoptotic cristae remodeling independently from mitochondrial fusion," *Cell*, vol. 126, no. 1, pp. 177–189, 2006, doi: 10.1016/j.cell.2006.06.025.
39. D. A. Patten *et al.*, "OPA1-dependent cristae modulation is essential for cellular adaptation to metabolic demand," *EMBO J.*, vol. 33, no. 22, pp. 2676–2691, 2014, doi: 10.15252/embj.201488349.
40. H. Lee, S. B. Smith, S. S. Sheu, and Y. Yoon, "The short variant of optic atrophy 1 (OPA1) improves cell survival under oxidative stress," *J. Biol. Chem.*, vol. 295, no. 19, pp. 6543–6560, 2020, doi: 10.1074/jbc.RA119.010983.
41. P. M. Quirós *et al.*, "Loss of mitochondrial protease OMA1 alters processing of the GTPase OPA1 and causes obesity and defective thermogenesis in mice," *EMBO J.*, vol. 31, no. 9, pp. 2117–2133, 2012, doi: 10.1038/emboj.2012.70.
42. J. Cesnekova, M. Rodinova, H. Hansikova, J. Zeman, and L. Stiburek, "Loss of mitochondrial AAA proteases AFG3L2 and YME1L impairs mitochondrial structure and respiratory chain biogenesis," *Int. J. Mol. Sci.*, vol. 19, no. 12, 2018, doi: 10.3390/ijms19123930.
43. Y. Ruan, H. Li, K. Zhang, F. Jian, J. Tang, and Z. Song, "Loss of Yme1L perturbates mitochondrial dynamics," *Cell Death Dis.*, vol. 4, no. 10, pp. 1–12, 2013, doi: 10.1038/cddis.2013.414.
44. T. K. Rainbolt, J. M. Saunders, and R. L. Wiseman, "YME 1L degradation reduces mitochondrial proteolytic capacity during oxidative stress," *EMBO Rep.*, vol. 16, no. 1, pp. 97–106, 2015, doi: 10.15252/embr.201438976.
45. M. V. Alavi, "Recent advances in, and challenges of, designing OMA1 drug screens," *Pharmacol. Res.*, vol. 176, p. 106063, 2022, doi: 10.1016/j.phrs.2022.106063.
46. T. D. B. MacVicar and J. D. Lane, "Impaired OMA1-dependent cleavage of OPA1 and reduced DRP1 fission activity combine to prevent mitophagy in cells that are dependent on oxidative phosphorylation," *J. Cell Sci.*, vol. 127, no. 10, pp. 2313–2325, 2014, doi: 10.1242/jcs.144337.
47. D. Bernhardt, M. Müller, A. S. Reichert, and H. D. Osiewacz, "Simultaneous impairment of mitochondrial fission and fusion reduces mitophagy and shortens replicative lifespan," *Sci. Rep.*, vol. 5, pp. 1–9, 2015, doi: 10.1038/srep07885.
48. N. Mendl, A. Occhipinti, M. Müller, P. Wild, I. Dikic, and A. S. Reichert, "Mitophagy in yeast is independent of mitochondrial fission and requires the stress response gene WHI2," *J. Cell Sci.*, vol. 124, no. 8, pp. 1339–1350, 2011, doi: 10.1242/jcs.076406.
49. A. R. Anzell *et al.*, "Mitochondrial fission and mitophagy are independent mechanisms regulating ischemia/reperfusion injury in primary neurons," *Cell Death Dis.*, vol. 12, no. 5, 2021, doi: 10.1038/s41419-021-03752-2.
50. C. J. Anderson *et al.*, "Prohibitin levels regulate OMA1 activity and turnover in neurons," *Cell Death Differ.*, vol. 27, no. 6, pp. 1896–1906, 2020, doi: 10.1038/s41418-019-0469-4.
51. T. Muraguchi, A. Kawawa, and S. Kubota, "Prohibitin protects against hypoxia-induced H9c2 cardiomyocyte cell death," *Biomed. Res.*, vol. 31, no. 2, pp. 113–122, 2010, doi: 10.2220/biomedres.31.113.
52. Z. Zhou *et al.*, "Estrogen receptor α protects pancreatic β-cells from apoptosis by preserving mitochondrial function and suppressing endoplasmic reticulum stress," *J. Biol. Chem.*, vol. 293, no. 13, pp. 4735–4751, 2018, doi: 10.1074/jbc.M117.805069.
53. D. Murata *et al.*, "Mitochondrial Safeguard: a stress response that offsets extreme fusion and protects respiratory function via flickering-induced Oma1 activation," *EMBO J.*, vol. 39, no. 24, pp. 1–17, 2020, doi: 10.15252/embj.2020105074.
54. O. M. Connor, S. K. Matta, and J. R. Friedman, "An intermembrane space protein facilitates completion of mitochondrial division in yeast," *BioRxiv*, 2023, [Online]. Available: https://www.ncbi.nlm.nih.gov/pmc/articles/PMC10081322/pdf/nihpp-2023.03.31.535139v1.pdf.

55. M. Darshi *et al.*, "ChChd3, an inner mitochondrial membrane protein, is essential for maintaining Crista integrity and mitochondrial function," *J. Biol. Chem.*, vol. 286, no. 4, pp. 2918–2932, 2011, doi: 10.1074/jbc.M110.171975.
56. J. An *et al.*, "CHCM1/CHCHD6, novel mitochondrial protein linked to regulation of mitofilin and mitochondrial cristae morphology," *J. Biol. Chem.*, vol. 287, no. 10, pp. 7411–7426, 2012, doi: 10.1074/jbc.M111.277103.
57. J. Tang *et al.*, "Sam50–Mic19–Mic60 axis determines mitochondrial cristae architecture by mediating mitochondrial outer and inner membrane contact," *Cell Death Differ.*, vol. 27, no. 1, pp. 146–160, 2020, doi: 10.1038/s41418-019-0345-2.
58. C. Ding *et al.*, "Mitofilin and CHCHD6 physically interact with Sam50 to sustain cristae structure," *Sci. Rep.*, vol. 5, pp. 1–11, 2015, doi: 10.1038/srep16064.
59. S. A. Munoz-Gomez, C. H. Slamovits, J. B. Dacks, and J. G. Wideman, "The evolution of MICOS: Ancestral and derived functions and interactions," *Commun. Integr. Biol.*, vol. 8, no. 6, pp. 1–5, 2015.
60. M. van der Laan, S. E. Horvath, and N. Pfanner, "Mitochondrial contact site and cristae organizing system," *Curr. Opin. Cell Biol.*, vol. 41, pp. 33–42, 2016, doi: 10.1016/j.ceb.2016.03.013.
61. H. Rampelt, R. M. Zerbes, M. van der Laan, and N. Pfanner, "Role of the mitochondrial contact site and cristae organizing system in membrane architecture and dynamics," *Biochim. Biophys. Acta.—Mol. Cell Res.*, vol. 1864, no. 4, pp. 737–746, 2017, doi: 10.1016/j.bbamcr.2016.05.020.
62. T. Stephan *et al.*, "MICOS assembly controls mitochondrial inner membrane remodeling and crista junction redistribution to mediate cristae formation," *EMBO J.*, vol. 39, no. 14, pp. 1–24, 2020, doi: 10.15252/embj.2019104105.
63. S. Koob, M. Barrera, R. Anand, and A. S. Reichert, "The non-glycosylated isoform of MIC26 is a constituent of the mammalian MICOS complex and promotes formation of crista junctions," *Biochim. Biophys. Acta.—Mol. Cell Res.*, vol. 1853, no. 7, pp. 1551–1563, 2015, doi: 10.1016/j.bbamcr.2015.03.004.
64. S. Callegari *et al.*, "A MICOS–TIM22 association promotes carrier import into human mitochondria," *J. Mol. Biol.*, vol. 431, no. 15, pp. 2835–2851, 2019, doi: 10.1016/j.jmb.2019.05.015.
65. P. I. Tsai *et al.*, "PINK1 phosphorylates MIC60/Mitofilin to control structural plasticity of mitochondrial crista junctions," *Mol. Cell*, vol. 69, no. 5, pp. 744–756.e6, 2018, doi: 10.1016/j.molcel.2018.01.026.
66. M. A. Huynen, M. Mühlmeister, K. Gotthardt, S. Guerrero-Castillo, and U. Brandt, "Evolution and structural organization of the mitochondrial contact site (MICOS) complex and the mitochondrial intermembrane space bridging (MIB) complex," *Biochim. Biophys. Acta.*, vol. 1863, no. 1, pp. 91–101, 2016, doi: 10.1016/j.bbamcr.2015.10.009.
67. V. Kozjak-Pavlovic, "The MICOS complex of human mitochondria," *Cell Tissue Res.*, vol. 367, no. 1, pp. 83–93, 2017, doi: 10.1007/s00441-016-2433-7.
68. X. Guo *et al.*, "IRGM promotes the PINK1-mediated mitophagy through the degradation of Mitofilin in SH-SY5Y cells," *FASEB J.*, vol. 34, no. 11, pp. 14768–14779, 2020, doi: 10.1096/fj.202000943RR.
69. R. F. Yang *et al.*, "Suppression of Mic60 compromises mitochondrial transcription and oxidative phosphorylation," *Sci. Rep.*, vol. 5, pp. 1–8, 2015, doi: 10.1038/srep07990.
70. T. Arguello *et al.*, "ATAD3A has a scaffolding role regulating mitochondria inner membrane structure and protein assembly," *Cell Rep.*, vol. 37, no. 12, p. 110139, 2021, doi: 10.1016/j.celrep.2021.110139.
71. J. He *et al.*, "The AAA+ protein ATAD3 has displacement loop binding properties and is involved in mitochondrial nucleoid organization," *J. Cell Biol.*, vol. 176, no. 2, pp. 141–146, 2007, doi: 10.1083/jcb.200609158.
72. M. P. Viana, R. M. Levytskyy, R. Anand, A. S. Reichert, and O. Khalimonchuk, "Protease OMA1 modulates mitochondrial bioenergetics and ultrastructure through dynamic association with MICOS complex," *iScience*, vol. 24, no. 2, p. 102119, 2021, doi: 10.1016/j.isci.2021.102119.
73. M. K. Shammas *et al.*, "OMA1 mediates local and global stress responses against protein misfolding in CHCHD10 mitochondrial myopathy," *J. Clin. Invest.*, vol. 132, no. 14, pp. 20–24, 2022, doi: 10.1172/JCI157504.
74. H. Li *et al.*, "Mic60/Mitofilin determines MICOS assembly essential for mitochondrial dynamics and mtDNA nucleoid organization," *Cell Death Differ.*, vol. 23, no. 3, pp. 380–392, 2016, doi: 10.1038/cdd.2015.102.
75. Z. Keckesova *et al.*, "LACTB is a tumour suppressor that modulates lipid metabolism and cell state," *Nature*, vol. 543, no. 7647, pp. 681–686, 2017, doi: 10.1038/nature21408.
76. T. S. Smith, C. Southan, K. Ellington, D. Campbell, D. G. Tew, and C. Debouck, "Identification, genomic organization, and mRNA expression of LACTB, encoding a serine β-lactamase-like protein with an amino-terminal transmembrane domain," *Genomics*, vol. 78, no. 1–2, pp. 12–14, 2001, doi: 10.1006/geno.2001.6643.

77. Z. Polianskyte *et al.*, "LACTB is a filament-forming protein localized in mitochondria," *Proc. Natl. Acad. Sci. U.S.A.*, vol. 106, no. 45, pp. 18960–18965, 2009, doi: 10.1073/pnas.0906734106.
78. J. A. Bennett, L. R. Steward, J. Rudolph, A. P. Voss, and H. Aydin, "The structure of the human LACTB filament reveals the mechanisms of assembly and membrane binding," *PLOS Biol.*, vol. 20, no. 12, pp. 1–30, 2022, doi: 10.1371/journal.pbio.3001899.
79. M. Zhang *et al.*, "Structural basis for the catalytic activity of filamentous human serine beta-lactamase-like protein LACTB," *Structure*, vol. 30, no. 5, pp. 685–696.e5, 2022, doi: 10.1016/j.str.2022.02.007.
80. P. Mishra and D. C. Chan, "Mitochondrial dynamics and inheritance during cell division, development and disease," *Nat. Rev. Mol. Cell Biol.*, vol. 15, no. 10, pp. 634–646, 2014, doi: 10.1038/nrm3877.
81. T. Navaratnarajah, R. Anand, A. S. Reichert, and F. Distelmaier, "The relevance of mitochondrial morphology for human disease," *Int. J. Biochem. Cell Biol.*, vol. 134, 2021, doi: 10.1016/j.biocel.2021.105951.
82. C. Benincá *et al.*, "Mutation in the MICOS subunit gene APOO (MIC26) associated with an X-linked recessive mitochondrial myopathy, lactic acidosis, cognitive impairment and autistic features," *J. Med. Genet.*, vol. 58, no. 3, pp. 155–167, 2021, doi: 10.1136/jmedgenet-2020-106861.
83. A. B. Kumarie Latchman, *Evidence for a mitochondrial disease phenotype due to APOO deletion.* 2021, https://jmg.bmj.com/content/58/3/155.responses.
84. V. S. Van Laar, A. A. Dukes, M. Cascio, and T. G. Hastings, "Proteomic analysis of rat brain mitochondria following exposure to dopamine quinone: Implications for Parkinson disease," *Neurobiol. Dis.*, vol. 29, no. 3, pp. 477–489, 2008, doi: 10.1016/j.nbd.2007.11.007.
85. V. S. Van Laar, S. B. Berman, and T. G. Hastings, "Mic60/Mitofilin overexpression alters mitochondrial dynamics and attenuates vulnerability of dopaminergic cells to dopamine and rotenone," *Neurobiol. Dis.*, vol. 91, no. July, pp. 247–261, 2016, doi: 10.1016/j.nbd.2016.03.015.Mic60.
86. S. Akabane *et al.*, "PKA regulates PINK1 stability and Parkin recruitment to damaged mitochondria through phosphorylation of MIC60," *Mol. Cell*, vol. 62, no. 3, pp. 371–384, 2016, doi: 10.1016/j.molcel.2016.03.037.
87. M. J. Eramo, V. Lisnyak, L. E. Formosa, and M. T. Ryan, "The 'mitochondrial contact site and cristae organising system' (MICOS) in health and human disease," *J. Biochem.*, vol. 167, no. 3, pp. 243–255, 2020, doi: 10.1093/jb/mvz111.
88. J. C. Ghosh *et al.*, "Ghost mitochondria drive metastasis through adaptive GCN2/Akt therapeutic vulnerability," *Proc. Natl. Acad. Sci. U.S.A.*, vol. 119, no. 8, 2022, doi: 10.1073/pnas.2115624119.
89. J. V Philley *et al.*, "Complex-I alteration and enhanced mitochondrial fusion are associated with prostate cancer progression," *J. Cell Physiol.*, vol. 231, no. 6, pp. 1364–1374, 2016, doi: https://doi.org/10.1002/jcp.25240.
90. F. Sotgia and M. P. Lisanti, "Mitochondrial biomarkers predict tumor progression and poor overall survival in gastric cancers: Companion diagnostics for personalized medicine," *Oncotarget*, vol. 8, no. 40, pp. 67117–67128, 2017, doi: 10.18632/oncotarget.19962.
91. M. Pérez-Ramírez *et al.*, "Genomics and epigenetics: A study of ependymomas in pediatric patients," *Clin. Neurol. Neurosurg.*, vol. 144, pp. 53–58, 2016, doi: 10.1016/j.clineuro.2016.02.041.
92. I. J. Suárez-Arroyo *et al.*, "The proteome signature of the inflammatory breast cancer plasma membrane identifies novel molecular markers of disease," *Am. J. Cancer Res.*, vol. 6, no. 8, pp. 1720–1740, 2016.
93. D. Thapa, C. E. Nichols, S. E. Lewis, D. L. Shepherd, R. Jagannathan, and J. M. Hollander, "Transgenic overexpression of mitofilin attenuates diabetes mellitus-associated cardiac and mitochondria dysfunction," *J. Mol. Cell Cardiol.*, vol. 79, no. 2, pp. 212–223, 2015, doi: 10.1016/j.yjmcc.2014.11.008.
94. W. A. Baseler *et al.*, "Proteomic alterations of distinct mitochondrial subpopulations in the type 1 diabetic heart: Contribution of protein import dysfunction," *Am. J. Physiol. Regul. Integr. Comp. Physiol.*, vol. 300, no. 2, pp. 186–200, 2011, doi: 10.1152/ajpregu.00423.2010.
95. R. Q. Xue *et al.*, "Pyridostigmine alleviates cardiac dysfunction via improving mitochondrial cristae shape in a mouse model of metabolic syndrome," *Free Radic. Biol. Med.*, vol. 134, pp. 119–132, 2019, doi: 10.1016/j.freeradbiomed.2019.01.011.
96. H. Hwang *et al.*, "Proteomics analysis of human skeletal muscle reveals novel abnormalities in obesity and type 2 diabetes," *Diabetes*, vol. 59, no. 1, pp. 33–42, 2010, doi: 10.2337/db09-0214.
97. H. Daoud *et al.*, "Resequencing of 29 candidate genes in patients with familial and sporadic amyotrophic lateral sclerosis," *Arch. Neurol.*, vol. 68, no. 5, pp. 587–593, 2011, doi: 10.1001/archneurol.2010.351.
98. T. Jiang, Y. Wang, X. Wang, and J. Xu, "CHCHD2 and CHCHD10: Future therapeutic targets in cognitive disorder and motor neuron disorder," *Front. Neurosci.*, vol. 16, no. August, pp. 1–9, 2022, doi: 10.3389/fnins.2022.988265.

99. Y. T. Liu *et al.*, "Loss of CHCHD2 and CHCHD10 activates OMA1 peptidase to disrupt mitochondrial cristae phenocopying patient mutations," *Hum. Mol. Genet.*, vol. 29, no. 9, pp. 1547–1567, 2020, doi: 10.1093/hmg/ddaa077.
100. P. Rivera-Mejías *et al.*, "The mitochondrial protease OMA1 acts as a metabolic safeguard upon nuclear DNA damage," *Cell Rep.*, vol. 42, no. 4, 2023, doi: 10.1016/j.celrep.2023.112332.
101. T. Wai *et al.*, "Imbalanced OPA1 processing and mitochondrial fragmentation cause heart failure in mice," *Science*, vol. 350, no. 6265, 2015, doi: 10.1126/science.aad0116.
102. S. Ahola *et al.*, "OMA1-mediated integrated stress response protects against ferroptosis in mitochondrial cardiomyopathy," *Cell Metab.*, vol. 34, no. 11, pp. 1875–1891.e7, 2022, doi: 10.1016/j.cmet.2022.08.017.
103. R. Acin-Perez *et al.*, "Ablation of the stress protease OMA1 protects against heart failure in mice," *Sci. Transl. Med.*, vol. 10, no. 434, 2018, doi: 10.1126/scitranslmed.aan4935.
104. D. Hu *et al.*, "A common missense variant in OMA1 associated with the prognosis of heart failure," *Cardiovasc. Drugs Ther.*, vol. 34, no. 3, pp. 345–356, 2020, doi: 10.1007/s10557-020-06960-8.
105. H. Sprenger *et al.*, "Loss of the mitochondrial *i*-AAA protease YME1L leads to ocular dysfunction and spinal axonopathy," *EMBO Mol. Med.*, vol. 11, no. 1, pp. 1–20, 2019, doi: 10.15252/emmm.201809288.
106. M. V. Alavi, "Targeted OMA1 therapies for cancer," *Int. J. Cancer*, vol. 145, no. 9, pp. 2330–2341, 2019, doi: 10.1002/ijc.32177.
107. A. Daverey *et al.*, "Depletion of mitochondrial protease OMA1 alters proliferative properties and promotes metastatic growth of breast cancer cells," *Sci. Rep.*, vol. 9, no. 1, pp. 1–15, 2019, doi: 10.1038/s41598-019-49327-2.
108. J. Kim, S. Kim, S. Ko, Y. In, H.-G. Moon, and W. Han, "Recurrent fusion transcripts detected by whole-transcriptome sequencing of 120 primary breast cancer samples," *Genes. Chromosomes Cancer*, vol. 54, pp. 681–691, 2015.
109. Z. Wu *et al.*, "OMA1 reprograms metabolism under hypoxia to promote colorectal cancer development," *EMBO Rep.*, vol. 22, no. 1, pp. 1–18, 2021, doi: 10.15252/embr.202050827.
110. B. Kong, Q. Wang, E. Fung, K. Xue, and B. K. Tsang, "P53 is required for cisplatin-induced processing of the mitochondrial fusion protein L-Opa1 that is mediated by the mitochondrial metallopeptidase Oma1 in gynecologic cancers," *J. Biol. Chem.*, vol. 289, no. 39, pp. 27134–27145, 2014, doi: 10.1074/jbc.M114.594812.
111. B. Gottschalk *et al.*, "MICU1 controls spatial membrane potential gradients and guides Ca^{2+} fluxes within mitochondrial substructures," *Commun. Biol.*, vol. 5, no. 1, 2022, doi: 10.1038/s42003-022-03606-3.
112. D. M. Wolf *et al.*, "Individual cristae within the same mitochondrion display different membrane potentials and are functionally independent," *EMBO J.*, vol. 38, no. 22, pp. 1–21, 2019, doi: 10.15252/embj.2018101056.
113. S. Vázquez-Fonseca *et al.*, "ADCK2 haploinsufficiency reduces mitochondrial lipid oxidation and causes myopathy associated with CoQ deficiency," *J. Clin. Med.*, vol. 8, no. 9, 2019, doi: 10.3390/jcm8091374.
114. H. M. Tang and H. L. Tang, "Anastasis: Recovery from the brink of cell death," *R. Soc. Open Sci.*, vol. 5, no. 9, 2018, doi: 10.1098/rsos.180442.
115. X. Li *et al.*, "Mitochondria shed their outer membrane in response to infection-induced stress," *Science*, vol. 375, no. 6577, 2022, doi: 10.1126/science.abi4343.

Part II

Mitochondrial Cristae and Lipid Dynamics

From Super-Resolution Microscopy to Lipid-OXPHOS Interplay

3 Changes in Cristae Architecture and Function

From Static to Dynamic

Ruchika Anand and Arun Kumar Kondadi*

INTRODUCTION

Mitochondria, nicknamed 'powerhouse', are the major energy source for the cell. Though energy conversion is an important function of mitochondria, they perform many other important cellular functions including calcium signalling, iron-sulfur cluster synthesis, lipid metabolism, ROS production, amino acid metabolism and apoptosis (Monzel et al., 2023). These ancient organelles originated from prokaryotes via endosymbiosis. Over the course of evolution, mitochondria have integrated well into their host and most of the mitochondrial proteins are encoded by the nuclear genome. Mitochondria still contain its own genome that encodes few yet essential proteins and accessory factors involved in oxidative phosphorylation (OXPHOS). Therefore, it is not surprising that any disturbance in mitochondrial structure or function could manifest into human diseases. Genetic disorders in nuclear or mitochondrial encoded genes or regions are collectively known as mitochondrial diseases. They contribute to a large class of inborn errors of metabolism and can occur at the rate of 1.6 in 5000 (Stenton and Prokisch, 2020). Mitochondrial diseases can present multisystemic phenotype such as neurodegeneration, seizures, cardiomyopathy, skeletal muscle myopathy, blindness, deafness, ataxia, liver dysfunction and kidney disease (Vafai and Mootha, 2012). Additionally, several human aliments are associated with faulty mitochondria including diabetes, cancer, neurodegeneration and ageing. Mitochondrial inner membrane (IM) structure is very intriguing and plays an important role in regulating many of its functions. Mitochondria with abnormal ultrastructure are found in pathological conditions (Vincent et al., 2016). However, whether this contributes to the disease progression or occurs due to the disease in a cause-and-effect manner is not understood. Mitochondrial IM structure is known to undergo remodelling under various physiological or metabolic conditions. Still, for several decades, mitochondrial ultrastructure was largely considered static mainly due to the fact that electron microscopy (EM) and electron tomography (ET) were the main techniques employed. These techniques were instrumental in defining the mitochondrial IM architecture at nanoscopic details but in fixed samples, and therefore not suitable for live cell imaging. As scientific discoveries and technological advancements go hand in hand, recent advances in live cell super-resolution (SR) microscopy allowed to image mitochondrial IM in living cells (Hu et al., 2020; Huang et al., 2018; Kondadi et al., 2020a; Kondadi et al., 2020b; Stephan et al., 2019; Wang et al., 2019a). Mitochondrial IM structures like cristae and crista junctions (CJs) could be resolved in a spatial and time-dependent manner at a resolution of 50 nm. These results show fast dynamics of cristae and CJs at the timescale of seconds, dispelling the notion of slow or static mitochondrial internal structures. Here, we review how mitochondrial IM structures are organized and how this arrangement helps mitochondria to modulate its functional requirements depending on external cues like metabolic or energetic demands. The main focus would be on fast cristae dynamics and what could be their potential significance for mitochondrial function.

* Corresponding author: anand@hhu.de

DOI: 10.1201/9781003221180-5

MODULAR DESIGN OF MITOCHONDRIAL INTERNAL STRUCTURES AND HOW IT CAN OPTIMIZE MITOCHONDRIAL FUNCTIONS

Mitochondria are double membrane-bound organelles where the outer membrane (OM) is smooth and encloses the organelle. The IM is more diverse as it comprises characteristic inward folds of the membrane called cristae (Daems and Wisse, 1966; Harmon et al., 1974). The ET techniques used in the 1990s helped in better understanding of three-dimensional (3D) mitochondrial membrane topology and led to proposal of the 'crista junction model' for mitochondrial IM structure (Frey and Mannella, 2000; Perkins et al., 1997) (Figure 3.1). CJs are membrane openings at the neck of cristae with a high inward-directed curvature towards mitochondrial matrix (Perkins et al., 1997). They are proposed as diffusion barriers having diameter between 12 and 40 nm (Frey et al., 2002; Mannella et al., 2013). Presence of CJs divides the IM into two distinct membrane system; cristae membrane (CM) encloses the intracristal space (ICS) and inner boundary membrane (IBM) is the rest of the IM that runs parallel to the OM, both forming the borders of the intermembrane space (IMS). Indeed, the CM and IBM differ in their composition and function (Vogel et al., 2006; Wurm and Jakobs, 2006), CMs mainly host the respiratory chain complexes (RCs) and act as a site for oxidative phosphorylation and ATP production (Gilkerson et al., 2003). The IBM contains

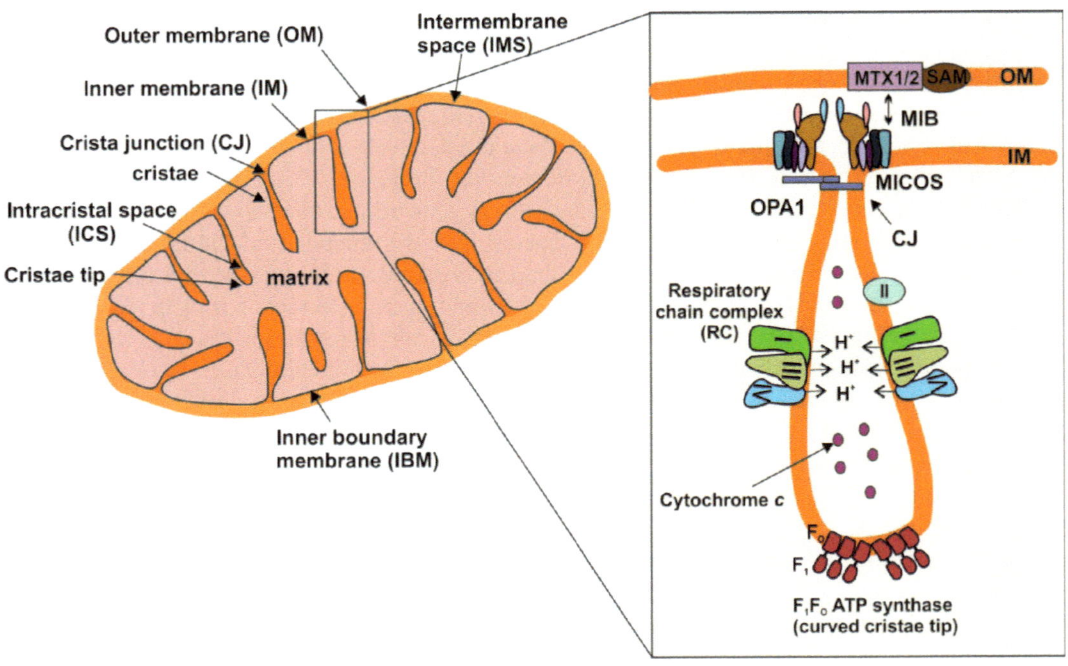

FIGURE 3.1 Modular design of mitochondrial inner membrane architecture. Mitochondria contain two membranes, outer membrane (OM) and inner membrane (IM). The infolds of IM are termed cristae, whereas the IM that is parallel to OM is called inner boundary membrane (IBM). The narrow opening of cristae is called crista junction (CJ). Due to a diffusion barrier, CJs divide mitochondria into various subcompartments. The matrix is the space enclosed by IM which contain machinery for biosynthetic pathways. The space enclosed by the cristae membrane is called intracristal space (ICS) which traps soluble OXPHOS factors such as cytochrome *c*. The space between IM and OM is called intermembrane space (IMS). The inset highlights the molecular players that modulate cristae shape. Cristae shape is organized by an interplay between MICOS, OPA1 and F_1F_O-ATP synthase complex and phospholipid remodelling (not depicted). OXPHOS machinery is located at the flat stem region of cristae, whereas F_1F_O-ATP synthase is present at the cristae tips. The oligomerization of F_1F_O-ATP synthase complex drives the bending of membrane at cristae tips.

the proteins required for mitochondrial import, fusion or fission processes (Vogel et al., 2006; Wurm and Jakobs, 2006). The CJs also divide a mitochondrion into various distinct functional subcompartments (Figure 3.1). The IMS is a site for protein import into mitochondria while the ICS enclosed by cristae can sequester soluble OXPHOS factors such as cytochrome *c*. Mitochondrial matrix enclosed by IM is home to many metabolic biosynthetic processes including the Krebs cycle (Frezza, 2017; Spinelli and Haigis, 2018). While there is a strong negative curvature at the CJs, the relatively flat region of cristae contain RCs (Gilkerson et al., 2003). The positive curvature at the cristae tips are formed by the dimers of F_1F_O-ATP synthase (Brandt et al., 2017; Davies et al., 2012; Salewskij et al., 2020; Strauss et al., 2008). The shape and size of the cristae and other membrane modules can be tailored to optimize the metabolic or energetic demands of the cell (Baker et al., 2019; Cogliati et al., 2016; Mannella, 2006b; Mannella et al., 2013). Mitochondria with higher energy conversion rate prioritize on cristae volume, whereas those with greater metabolite biosynthesis and signalling prioritize on matrix volume (Glancy et al., 2020). Lamellar shape of cristae has flatter surface to accommodate more electron transport chain complexes, and are more prevalent in heart or muscle tissue compared to liver tissue (Brandt et al., 2017; Phillips et al., 2012). The ratio of the curvature to that of flat cristae regions can regulate the proportion between the F_1F_O-ATP synthase and RCs. For example, brown adipose tissue with high energy-dissipating requirement contains lamellar cristae with less curvature even with similar cristae densities compared to heart mitochondria (Kang et al., 2009). The mitochondria in liver have very dense matrix and cristae are less regular and not parallelly arranged to optimize for its metabolic and biosynthetic functions (Brandt et al., 2017; Shami et al., 2021). Hence, mitochondrial IM structure is modulated according to the energetic or metabolic need of each cell type.

MOLECULAR MACHINERY THAT REGULATES CRISTAE STRUCTURE FORMATION

The intricate architecture of mitochondria is formed and maintained by coordination between at least three main protein complexes, namely MICOS, Optic Atrophy Type 1 (OPA1) and F_1F_O-ATP synthase, and lipid remodelling pathways (Klecker and Westermann, 2021). In the following section, we describe how these complexes and their partnership with each other and lipid remodelling components coordinate the formation and remodelling of mitochondrial IM structure.

The MICOS Complex

The mitochondrial contact site and cristae organization system (MICOS) is a large multi-subunit complex required for the formation and maintenance of CJs and contact sites (Harner et al., 2011; Hoppins et al., 2011; von der Malsburg et al., 2011) (Figure 3.1). It is mainly composed of at least seven subunits in mammalian cells, which are termed as MIC'X' with 'X' defined as a number close to their molecular weight in kDa. The MICOS subunits consist of MIC60, MIC27, MIC26, MIC25, MIC19, MIC13 and MIC10. Loss of individual MICOS subunits show disturbed cristae morphology with loss of CJs and accumulation of cristae stacks in the matrix to a variable degree (Kondadi et al., 2020a; Stephan et al., 2020). MICOS subunits are further arranged into two smaller subcomplexes named MIC60-subcomplex (MIC60-MIC19-MIC25) and MIC10-subcomplex (MIC10-MIC13-MIC26-MIC27). Individual subunits of MICOS work in unison to coordinate the formation and maintenance of curved CJs. The main task of sculpturing the membrane is performed by MIC60 and MIC10 as both of them are shown to have membrane-bending activity (Barbot et al., 2015; Bohnert et al., 2015; Hessenberger et al., 2017; Tarasenko et al., 2017). Recent 3D crystal structure of MIC60 showed that tetramers of MIC60 resembling a bow-tie shape encompasses CJ width and provide the stability to CJs (Bock-Bierbaum et al., 2022). MIC19 is a redox regulator and promotes tetramerization of MIC60 and assembly of MIC60-subcomplex (Sakowska et al., 2015).

MIC25 is a homologue of MIC19 which is proposed to perform similar functions as MIC19, albeit loss of MIC25 shows milder defects compared to MIC19 loss (An et al., 2012). MIC13 is involved in assembly of MIC10-subcomplex and acts as a bridge between MIC60- and MIC10-subcomplex (Anand et al., 2016; Guarani et al., 2015) using its conserved GxxxG and WN motifs (Urbach et al., 2021). MIC10 can form large oligomers by interaction between its GxGxG motifs which are proposed to form wedge-like structures of MIC10 and promote membrane sculpturing (Barbot et al., 2015; Bohnert et al., 2015). MIC26 and MIC27 are orthologous proteins belonging to the family of apolipoproteins. The binding of MIC27 to cardiolipin (CL) (Weber et al., 2013) and reduced levels of CL upon combined deletion of *MIC26* and *MIC27* (Anand et al., 2020) indicate their role in phospholipid remodelling within MICOS. Strategically residing at CJs, MICOS can interact with many proteins in different subcompartments of mitochondria to regulate important functions. The MICOS proteins interact with OM proteins SAMM50 (Ott et al., 2012), metaxin1, metaxin2 and DNAJC11 (Ioakeimidis et al., 2014) to form a larger mitochondrial intermembrane space bridging (MIB) complex that can span through IM and OM. The MIB complex regulates cristae morphology (Ott et al., 2012; Ott et al., 2015) and promotes transfer and biosynthesis of phospholipids across ER-mitochondrial contact sites. MICOS is involved in many other mitochondrial functions including mitophagy, mtDNA organization, apoptosis, inflammation and protein import (Anand et al., 2021; Klecker and Westermann, 2021; Kondadi et al., 2020b). Mutations in MICOS subunits are associated with several diseases. A *MIC13* mutation causes mitochondrial encephalopathy with liver dysfunction (Guarani et al., 2016), whereas a *MIC26* mutation was associated with cognitive impairment, mitochondrial myopathy, lactic acidosis and autistics features (Beninca et al., 2021). Another variant of *MIC26* caused lethal mitochondrial disease with progeria-like phenotypes (Peifer-Weiß et. al., 2023). *MIC60* mutations were found in patients with Parkinson's disease (Tsai et al., 2018) and mitochondrial developmental encephalopathy with bilateral optic neuropathy (Marco-Hernández et al., 2022). *Metaxin 2* mutations cause severe laminopathy-like mandibuloacral dysplasia (Elouej et al., 2020).

OPA1

OPA1 is a large dynamin-related GTPase that plays a dual role in regulating mitochondrial morphology and cristae formation (Figure 3.1). Deletion of *OPA1* causes mitochondrial fragmentation, loss of cristae or widening of the lumen and CJs of the residual cristae (Meeusen et al., 2006; Olichon et al., 2002). OPA1 has a complicated regulation (MacVicar and Langer, 2016). It contains eight splice variants that encodes proteins of 924-1014 amino acids along with a transmembrane domain collectively known as large 'L-OPA1' (Del Dotto et al., 2017; Olichon et al., 2007). OPA1 also contains two cleavage sites recognized by IM proteases, YME1L and OMA1 (Anand et al., 2014; Sprenger and Langer, 2019). The processing of OPA1 at the cleavage site by these proteases removes its transmembrane domain and leads to accumulation of short 'S-OPA1' in the IMS (Anand et al., 2014; MacVicar and Langer, 2016). At the steady state levels, different cell types contain a mixture of L-OPA1 and S-OPA1. During stress conditions, OPA1 undergoes stress-induced processing (Duvezin-Caubet et al., 2006; Ishihara et al., 2006), which is mediated by OMA1 that promotes conversion of L-OPA1 to S-OPA1 and lead to mitochondrial fragmentation and cristae loss (Baker et al., 2014; Gilkerson et al., 2021). It has been proposed that L-OPA1 is sufficient to support mitochondrial fusion (Anand et al., 2014; Ishihara et al., 2006) and S-OPA1 can bind and remodel membranes (Zhang et al., 2020). The combined action of L-OPA1 and S-OPA1 regulate width of cristae lumen and CJ such that CJs are kept closed, limiting the release of cytochrome *c* (Frezza et al., 2006; Olichon et al., 2003). Studies in Baker's yeast, *Saccharomyces cerevisiae*, proposed that the formation of lamellar cristae is mediated by IM fusion by Mgm-1 (OPA1 homolog) (Harner et al., 2016; Kojima et al., 2019). There have been several studies describing the differential or combinational roles of both the forms of OPA1 and no uniform consensus has been achieved so far, which highlights the complexity of this regulation (MacVicar and Langer, 2016;

Quintana-Cabrera and Scorrano, 2023). OPA1 interacts with MIC60 (Barrera et al., 2016) and acts epistatic to MICOS in regulating CJs and cristae curvature (Glytsou et al., 2016). OPA1 also affects the stability of F_1F_O-ATP synthase as loss of OPA1 causes reduced dimerization of F_1F_O-ATP synthase (Quintana-Cabrera et al., 2021). Heterozygous mutations in *OPA1* cause dominant optic atrophy (Alexander et al., 2000; Delettre et al., 2000).

The F_1F_O-ATP Synthase Complex

The F_1F_O-ATP synthase complex that generates ATP using electrochemical proton gradient is involved in sculpturing the positive curvature of membrane at the cristae tips (Davies et al., 2012; He et al., 2018; Kuhlbrandt, 2015; Paumard et al., 2002; Strauss et al., 2008) (Figure 3.1). Monomeric F_1F_O-ATP synthase has a stalk-like structure with an F_1 head which contains the catalytic domain and faces matrix and F_O stem embedded in the membrane (Davies et al., 2011). The F_1F_O-ATP synthase is enriched at the tips of the cristae and its assembly into a higher order oligomers drives the bending of the membrane at the cristae tips (Blum et al., 2019). The dimerization/oligomerization involves connection between adjacent F_O domains of F_1F_O-ATP synthase at a definitive tilted angle to form wedges or V-shape structures which lead to membrane bending and formation of cristae tips (Arnold et al., 1998; Davies et al., 2011; Strauss et al., 2008; Wagner et al., 2009). Loss of the components required for the F_1F_O-ATP synthase dimerization causes altered cristae structure (Arselin et al., 2004; Paumard et al., 2002; Stephan et al., 2020; Strauss et al., 2008). The antagonistic relationship between yeast Mic60 and F_1F_O-ATP synthase regulates the balance between the number of CJs and cristae tips, respectively (Rabl et al., 2009). Contrary to Mic60, Mic10 and Mic27 promote the formation of F_1F_O-ATP synthase oligomers in yeast (Eydt et al., 2017; Rampelt et al., 2017; Rampelt et al., 2018).

Phospholipid Remodelling

Not only proteins but lipids also play a very important role in shaping the cristae architecture. Considering the sharp turns in the IM at the cristae tips and CJs, remodelling of lipids plays important roles in formation and maintenance of these membrane domains. CL is the signature lipid of mitochondria which contains double glycerophosphate backbone and four acyl side chains resulting in a cone-like structure (Horvath and Daum, 2013; Schlame, 2008). Phosphatidylethanolamine (PE) is another cone-shaped phospholipid enriched in mitochondrial IM. In fact, CL and PE account for around 50% of phospholipids in the mitochondria (Ikon and Ryan, 2017). Owing to their shape, PE and CL can be tailored into the negatively curved inner leaflet of the CM to reduce the tension and to confer stability to curved membranes (Ikon and Ryan, 2017; Renner and Weibel, 2011). Any disturbances in the amount of PE or CL causes abnormal cristae structure (Acehan et al., 2011; Tasseva et al., 2013). Mutations in tafazzin (*TAZ*) causes Barth syndrome which is associated with aberrant cristae that are highly branched or arranged as concentric rings (Acehan et al., 2007). *TAZ* encodes an enzyme required for CL remodelling. A direct interaction between the CL-synthesizing enzymes and several cristae modulators supports an intricate relationship between membrane lipid composition and cristae shape (Serricchio et al., 2018; Tatsuta and Langer, 2017). MIC27 binds CL *in vitro* (Weber et al., 2013) and deletion of both *MIC26* and *MIC27* leads to decreased CL levels accompanied by loss of CJs, accumulation of swirling cristae as well as reduced stability of RCs (Anand et al., 2020). SAMM50 together with MIC60 also binds cardiolipin, and loss of SAMM50 in mouse hepatocytes causes cristae remodelling and cardiolipin externalization leading to release of mtDNA to initiate the cGAS-STING pathway and liver inflammation (Chen et al., 2022). The location of MICOS and MIB complexes at the contact sites between the OM and IM facilitates the synthesis of PE in trans at OM via Psd1 enzyme, which is present in IM (Aaltonen et al., 2016), showing a link between IM architecture and phospholipid synthesis.

CRISTAE REMODELLING ADAPTS TO METABOLIC CHANGES

The relationship between cristae morphology and bioenergetic status of mitochondria was proposed as early as 1966 by Hackenbrock. These classical experiments showed that the isolated mitochondria in state III with more ADP have 'condensed' (matrix compacted) morphology that changes to 'orthodox' (matrix expanded) morphology in state IV with low ADP amounts (Hackenbrock, 1966). ET studies of mitochondria during this transition showed IM remodelling (Mannella, 2006b). 'Condensed' mitochondria contained large cristae compartments with multiple tubular connections to peripheral IM or each other. 'Orthodox' mitochondria contain cristae with short flat lamellar or tubular structures. It was proposed that this interconversion could only be achieved by fusion or fission of IM (Mannella, 2006b). The transition between orthodox and condensed forms of mitochondria was proposed to drive ATP production (Afzal et al., 2021). Due to limited diffusion of ADP via CJs, the local concentration was proposed to be modulated depending on the shape of cristae in orthodox or condensed mitochondria and hence rate of ATP production could be controlled this way (Mannella, 2006a; Mannella et al., 2013; Mannella et al., 2001). Though this hypothesis needs experimental proof, the link between cristae remodelling and mitochondrial function has been reinforced in several other studies (Baker et al., 2019; Cogliati et al., 2016). The assembly and stability of RCs and its efficiency was shown to be dependent on cristae shape (Cogliati et al., 2013). The width of the cristae lumen was reduced in response to metabolic demands during starvation in an OPA1-dependent manner (Patten et al., 2014) (Figure 3.2). These structural changes along with hypertubulation of mitochondria protects them from autophagy during starvation (Gomes et al., 2011). Rapid and reversible oligomerization of OPA1 promotes narrowing of cristae and ramps up

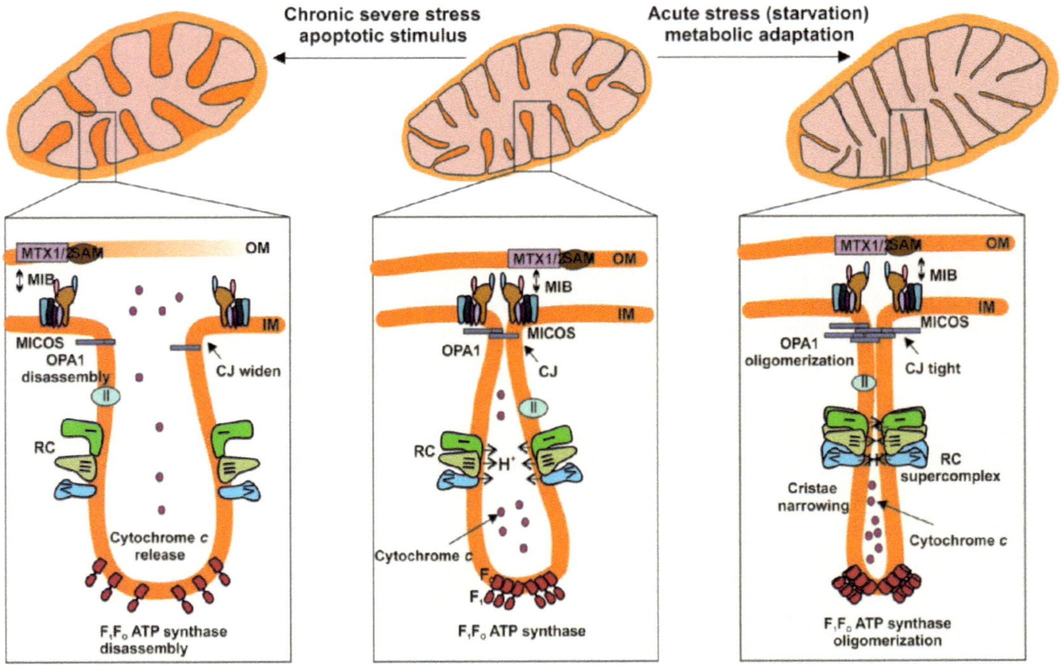

FIGURE 3.2 Cristae remodelling during metabolic changes. The scheme describes cristae remodelling during various metabolic conditions. During acute stress or metabolic adaptation, cristae become narrow, and CJs become tight, which is mediated by oligomerization of OPA1 and F_1F_O-ATP synthase complex. It is also accompanied by increase in respiratory chain (RC) supercomplex formation and respiratory capacity. On the other hand, severe chronic stress or apoptotic stimuli leads to widening of cristae lumen and width of CJs. It is mediated by disassembly of OPA1 complex and is associated with loss of respiratory capacity and deceased ATP production. Subsequently, cytochrome *c* is accessible and could be released through CJs.

respiratory capacity by enhancing supercomplex formation (Cogliati et al., 2013; Khacho et al., 2014; Patten et al., 2014). OPA1-mediated cristae remodelling occurs independent of its mitochondrial fusion role (Patten et al., 2014). The protective effect of OPA1 is possibly mediated by its interaction with F_1F_O-ATP synthase that favours F_1F_O-ATP synthase oligomerization during inhibition of the respiratory chain (Quintana-Cabrera et al., 2018). On the other hand, widening of cristae lumen and CJ width was seen during hypoxic stress (Dlasková et al., 2019; Khacho et al., 2014; Plecita-Hlavata et al., 2016) and accompanied by 20% reduction in MIC60 with concomitant clustering of MICOS complex and lower levels of F_1F_O-ATP synthase dimers (Plecita-Hlavata et al., 2016) (Figure 3.2). Cristae defects during hypoxia were reversed by two different approaches: by inducing acidosis during hypoxia (Khacho et al., 2014) or by adding cell permeable derivate of OXPHOS substrate 2-oxoglutarate, dimethyl-2-oxoglutrate to cells pre-adapted to hypoxia (Dlasková et al., 2019). Adaptation to hypoxia decreases ATP synthesis and inhibits pyruvate dehydrogenase which prevents entry of acetyl-CoA to the Krebs cycle (Plecitá-Hlavatá et al., 2015; Plecitá-Hlavatá et al., 2016). Increase in respiratory substrate load led to reversal of hypoxia defects with narrowing of cristae in already hypoxia-widened cristae within 10 min (Dlasková et al., 2019). Though widening of cristae lumen is associated with substrate deficiency, cristae lumen narrowing occurred during conditions with different energetic status: starvation and addition of excess substrate in pre-adapted hypoxic cells, making it difficult to draw a uniform theme between cristae remodelling and metabolic status of cells. The hypoxia-adapted *Drosophila* flies and Tibetan sheep adapted to high altitude hypoxia conditions showed overall dense cristae, suggesting that the cristae remodelling could be more complex and can depend on cell type and extent of metabolic changes (Perkins et al., 2012; Wang et al., 2019b). Indeed, a chronic hypoxia response can protect against mitochondrial toxicity and is proposed as a therapy for mitochondrial diseases (Jain et al., 2016). Accumulation of another metabolite, 2-hydroxyglutrate (D-2HG), which is structurally similar to 2-oxoglutarate, can cause 2-hydroxyglutaric aciduria (2HGA) (Du and Hu, 2021). Loss of enzyme D-2HG dehydrogenase (DHGD-1) causes accumulation of D-2HG which can inhibit 3-hydroxypropionate dehydrogenase (HPHD-1) and lead to accumulation of a toxic metabolite 3-hydroxypropionate (3-HP). *Caenorhabditis elegans* mutants of *dhgd-1* and *hphd-1* contained enlarged mitochondria with small vacuoles which were reminiscent of mitochondria that lacked MIC60 (Zhou et al., 2022). Indeed, 3-HP binds MIC60 and inhibits membrane-binding and membrane-shaping activity of MIC60 (Zhou et al., 2022). A large spectrum of abnormal cristae was seen in mitochondrial myopathy perhaps due to the perturbed metabolic and energy homeostasis in these conditions (Vincent et al., 2016). The general ability of IM to show a dynamic remodelling has been illustrated in many other physiological conditions including apoptosis and ageing, which are described below.

Cristae Remodels during Apoptosis for Release of Cytochrome *c*

Cytochrome *c* is a pro-apoptotic molecule which activates caspases in the cytosol to initiate the apoptotic cascade (Li et al., 1997; Zou et al., 1999). However, majority of the cytochrome *c* (85%) is trapped in the cristae lumen within the mitochondria and is inaccessible due to the diffusion barrier imposed by CJs, whereas another 15% of cytochrome *c* is present in the IMS (Bernardi and Azzone, 1981). How is the cytochrome *c* released into the cytosol during apoptosis? In the early 2000s, it was shown that cristae undergo remodelling during apoptosis which mobilize and release the cytochrome *c* trapped in cristae lumen (Scorrano et al., 2002). A serial time course of ET images after apoptosis induction showed remodelling of IM morphology (Scorrano et al., 2002; Sun et al., 2007). Cristae changed from normal to a highly interconnected organization such that they could not be separated as individual entities but rather formed few large compartments. Widening of CJs to a diameter of 57–71 nm opened up ICS into the IMS and made cytochrome *c* accessible to OM (Scorrano et al., 2002). Later studies showed that OPA1 oligomers are required for the tightening of CJs (Frezza et al., 2006; Olichon et al., 2002; Yamaguchi et al., 2008). Apoptosis induction leads to disruption of OPA1 oligomers which widens the CJs and facilitates release of cytochrome *c* (Cipolat and Scorrano, 2006;

Cogliati et al., 2016; Frezza et al., 2006). Induction of apoptosis activates stress-responsive protease OMA1 that facilitates cleavage of L-OPA1 and disassembly of OPA1 complex (Gilkerson et al., 2021; Jiang et al., 2014). The complete release of cytochrome *c* depends on two coordinated steps, opening of the CJs and formation of an OM pore (Yamaguchi and Perkins, 2009). On the other hand, overexpression of *OPA1* protects against apoptosis and is proposed as a therapeutic target in disease models (Dong et al., 2021; Varanita et al., 2015).

Cristae Shape Changes during Physiological Processes Like Ageing

Mitochondria play a crucial role in age-related changes as their IM structure is altered during ageing. Moreover, changes in mitochondrial cristae shape can modulate the ageing process in some organisms. In an ageing model organism of *Podospora anserina*, cristae change their structure from lamellar arrangement to big vesicles and widening of crista junctions (Daum et al., 2013). The curvature of IM becomes more concave without the cristae tip and concomitant loss of F_1F_O-ATP synthase dimers. *Drosophila* accelerated ageing model shows subsets of swirling cristae along with normal cristae. The swirling cristae membrane lacked F_1F_O-ATP synthase dimers and had impaired energy production capacity compared to normal cristae in the same mitochondria (Jiang et al., 2017). The age-related cristae remodelling differs among various organs, mouse heart being more resilient to cristae changes compared to liver and kidney (Brandt et al., 2017). These alterations in cristae are associated with defective F_1F_O-ATP synthase dimerization (Bou-Teen et al., 2022). Do these cristae alterations directly influence ageing process or are these (mere) the consequence of age-related mitochondrial structures? Apparently, aberrant cristae due to MICOS disruption could lead to extension of lifespan in *P. anserina* (Warnsmann et al., 2022). Both MICOS subcomplexes affect longevity by at least partially independent pathways. Loss of MIC10-subcomplex leads to lifespan extension induced by mitohormesis (Warnsmann et al., 2022). Longevity in MIC60-subcomplex mutants depend on CL remodelling (Marschall et al., 2022). Similarly, other physiological conditions like embryo implantation shows drastic cristae remodelling from tubular or vesicular cristae during late morula stage to lamellar and mature cristae by the end of neurulation (Shepard et al., 2000). Presynaptic mitochondria in boutons with high synaptic activity have more cristae density and high rate of metabolism compared to boutons with lower activity, suggesting IM remodelling of presynaptic mitochondria adjust to the metabolic demands (Cserép et al., 2018).

CRISTAE UNDERGO FAST FISSION AND FUSION CYCLES IN A REGULAR MANNER

The EM technique was crucial in providing insights into the mitochondrial ultrastructure details. Though many cristae remodelling events which are described in previous sections suggest dynamic behaviour of cristae membrane, the cristae dynamics were not studied in real time until recently. Recently, SR techniques overcame the diffraction barrier as IM structures could be imaged at very high resolution even in a live cell (Anand et al., 2021; Huang et al., 2018; Kondadi et al., 2020a; Kondadi et al., 2020b; Stephan et al., 2019). Staining of the MICOS proteins showed a rail-like pattern across the two sides of mitochondria that resembles their localization at CJs (Jans et al., 2013; Kondadi et al., 2020a; Stoldt et al., 2019). Contrary to the static view of CJs and cristae, live cell STED nanoscopy shows that cristae and CJs move dynamically within seconds (Kondadi et al., 2020a). Tracking CJs in live showed that CJs come together and move apart in a rapid and apparently regular fashion (Kondadi et al., 2020a). MIC13-SNAP that dually marked both CJs and cristae showed that the movement of CJs could bring adjoining cristae together such that two cristae can 'merge/fuse' or 'split' in a balanced and reversible manner. This pattern of cristae dynamics was confirmed using the bona fide cristae markers like COX8a and ATP5I, leading to the proposal of Cristae Fission and Fusion (CriFF) model (Kondadi et al., 2020a). According to the model, cristae

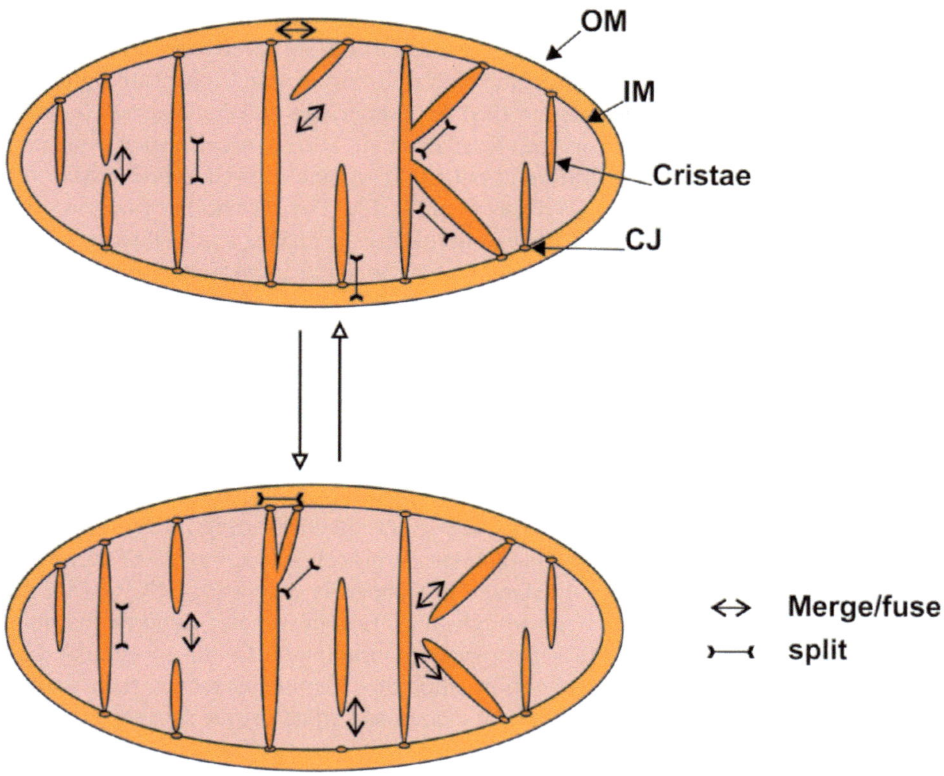

FIGURE 3.3 Cristae undergo fast dynamics accompanied by merging and splitting cycles. Diagram depicts that cristae undergo merging or splitting events. Inward arrow shows merging events and outward arrow shows splitting events. There is a balance between the merging and splitting events and usually a merging event is followed by a splitting event or vice versa at the same site or in the vicinity.

undergo continuous cycles of fusion and fission such that cristae fusion occurs in continuation to a cristae fission event either at the same site or close to the earlier event (Figure 3.3). Cristae appeared as letter 'X' or 'Y' depending on the fusion along the length of the cristae or with the IBM. Isolated cristae separated from IBM were also observed in time-lapse movies of cristae dynamics. CriFF was dependent on assembled MICOS complex as deletion of *MIC13* caused drastic reduction in cristae merging and splitting events (Kondadi et al., 2020a). Another study also confirmed a direct link between MICOS and cristae dynamics as loss of MIC10, MIC19, MIC60 and SAMM50 showed reduced cristae dynamics (Hu et al., 2020). It is possible that a fully assembled MICOS complex at individual CJ can provide a stable anchor to facilitate cristae fusion and fission cycles. Another possibility could be that CJ formation is dynamically regulated via MICOS in an equilibrium such that a fission event at the IBM will lead to an isolated cristae stack that can re-fuse to the IBM again or to another cristae. This is supported by the fact that loss of MICOS components are accompanied by accumulation of isolated cristae stacks within matrix (Kondadi et al., 2020a; Stephan et al., 2020), which might occur due to shift in equilibrium and increased fission of cristae from the IBM. Other factors that affect cristae dynamics include YME1L and ATAD3A (Hu et al., 2020). Cristae events of detachment from IBM are decreased in *YME1L* KO, whereas *ATAD3A* KO shows increased cristae dynamics (Hu et al., 2020).

Considering that STED SR microscopy still has a limit of resolution at 50 nm in living cells, cristae merging or fusion events cannot ascertain actual fusion or exchange of content between the cristae. Still, an experiment using ATP5I tagged to a photoactivable-GFP (ATP5I-PAGFP) and cell fusion assay strongly suggested content mixing between cristae. Usually photoactivating ATP5I-PAGFP at the tip of a mitochondrion leads to instant spread of fluorescence between neighbouring cristae due to diffusion. Still, few isolated cristae vesicles became fluorescent after a time delay only when they are re-fused to IBM or other cristae and could access the fluorescent GFP due to content mixing between cristae (Kondadi et al., 2020a). A cell fusion assay employing the two kinds of cells labelled with different colours of cristae showed hybrid cristae formation in accordance with an earlier study (Wilkens et al., 2013). The formation of hybrid cristae was drastically reduced in *MIC13* KO cells, indicating that cristae mixing and exchange is impaired in *MIC13* KOs. In few instances, a contact among the cristae led to an instantaneous redistribution of membrane potential dye (TMRM) corroborating an exchange of material between cristae during fusion events (Kondadi et al., 2020a).

To determine the bioenergetic parameters that regulate fast cristae dynamics, cristae dynamics was measured after inhibiting individual RCs or F_1F_O-ATP synthase or by uncoupling respiration using classical OXPHOS inhibitors for short durations of 10–30 min (Golombek et al., 2023). Surprisingly, no drastic change in cristae dynamics including cristae fusion or fission events was observed after acute mitochondrial stress, even though ATP levels were reduced in the same treatment, suggesting that cristae dynamics is not negatively affected during inhibition of respiratory activity. These inhibitors lead to formation of enlarged mitochondria, but no difference in cristae dynamics was specifically observed in enlarged mitochondria except for CCCP treatment which rather caused a moderate increase in cristae dynamics. One hypothesis could be that increase in cristae dynamics occurs due to increase in oxygen consumption due to uncoupling via CCCP. Additionally, cristae dynamics was analysed by inhibiting adenine nucleotide translocator (ANT) using bongkrekic acid (BKA). BKA treatment causes perturbed cristae morphology and cristae dynamics was reduced in mitochondria with perturbed cristae, indicating a connection between cristae morphology and dynamics (Golombek et al., 2023). All the molecules which show defects in cristae dynamics studied so far also show some defects in cristae morphology (Hu et al., 2020; Kondadi et al., 2020a) and the direct role or mechanism of how cristae dynamics is regulated has to be still analysed. It could also be that the cristae morphology defects manifest due to defective cristae dynamics or vice versa.

The field of cristae dynamics and cristae fission-fusion is in its initial stages, and further studies will help to understand its relevance for optimizing mitochondrial function. We saw that cristae change shape and size depending on their metabolic or bioenergetic needs. It can be envisaged that the fast cristae dynamics might fulfil this purpose and in fact required to bring about changes in cristae structure. In 2006, Carmen Mannella speculated that cristae remodelling between orthodox and condensed mitochondrial morphology occurs via fusion and fission of IM (Mannella, 2006a). The content mixing between the individual cristae could be a quality control mechanism by locally mixing already accumulated dysfunctional RCs within a mitochondrion. Cristae dynamics might help in optimizing OXPHOS efficiency by trapping protons inside the isolated cristae stacks to promote ATP production. On the other hand, the metabolite exchange such as ATP, ADP, amino acids or Fe-S clusters could be promoted upon fusion of cristae vesicles to IBM and hence increasing the area exposed to the IMS or cytosol.

CONCLUSIONS

The dynamic remodelling of cristae membranes is a very important process which mitochondria employ to modulate its function according to the cellular or biological demands. Current advances in the field of cristae dynamics pose many interesting questions and opens up new avenues to understand the mechanism and function of fast cristae dynamics. How do these fast cristae dynamics function during mitochondrial pathological conditions? What are the metabolic factors that determine the cristae dynamics? Many of these exciting yet challenging questions need to be addressed in the coming years.

REFERENCES

Aaltonen, M.J., J.R. Friedman, C. Osman, B. Salin, J.P. di Rago, J. Nunnari, T. Langer, and T. Tatsuta. 2016. MICOS and phospholipid transfer by Ups2-Mdm35 organize membrane lipid synthesis in mitochondria. *Journal of Cell Biology*. 213:525–534.

Acehan, D., F. Vaz, R.H. Houtkooper, J. James, V. Moore, C. Tokunaga, W. Kulik, J. Wansapura, M.J. Toth, A. Strauss, and Z. Khuchua. 2011. Cardiac and skeletal muscle defects in a mouse model of human Barth syndrome. *Journal of Biological Chemistry*. 286:899–908.

Acehan, D., Y. Xu, D.L. Stokes, and M. Schlame. 2007. Comparison of lymphoblast mitochondria from normal subjects and patients with Barth syndrome using electron microscopic tomography. *Laboratory Investigation*. 87:40–48.

Afzal, N., W.J. Lederer, M.S. Jafri, and C.A. Mannella. 2021. Effect of crista morphology on mitochondrial ATP output: A computational study. *Current Research in Physiology*. 4:163–176.

Alexander, C., M. Votruba, U.E.A. Pesch, D.L. Thiselton, S. Mayer, A. Moore, M. Rodriguez, U. Kellner, B. Leo-Kottler, G. Auburger, S.S. Bhattacharya, and B. Wissinger. 2000. OPA1, encoding a dynamin-related GTPase, is mutated in autosomal dominant optic atrophy linked to chromosome 3q28. *Nature Genetics*. 26:211–215.

An, J., J. Shi, Q. He, K. Lui, Y. Liu, Y. Huang, and M.S. Sheikh. 2012. CHCM1/CHCHD6, novel mitochondrial protein linked to regulation of mitofilin and mitochondrial cristae morphology. *The Journal of Biological Chemistry*. 287:7411–7426.

Anand, R., A.K. Kondadi, J. Meisterknecht, M. Golombek, O. Nortmann, J. Riedel, L. Peifer-Weiss, N. Brocke-Ahmadinejad, D. Schlutermann, B. Stork, T.O. Eichmann, I. Wittig, and A.S. Reichert. 2020. MIC26 and MIC27 cooperate to regulate cardiolipin levels and the landscape of OXPHOS complexes. *Life Science Alliance*. 3:e202000711.

Anand, R., A.S. Reichert, and A.K. Kondadi. 2021. Emerging roles of the MICOS complex in cristae dynamics and biogenesis. *Biology (Basel)*. 10:600.

Anand, R., T. Wai, M.J. Baker, N. Kladt, A.C. Schauss, E. Rugarli, and T. Langer. 2014. The i-AAA protease YME1L and OMA1 cleave OPA1 to balance mitochondrial fusion and fission. *Journal of Cell Biology*. 204:919–929.

Anand, R., V. Strecker, J. Urbach, I. Wittig, and A.S. Reichert. 2016. Mic13 is essential for formation of crista junctions in mammalian cells. *PLOS ONE*. 11:e0160258.

Arnold, I., K. Pfeiffer, W. Neupert, R.A. Stuart, and H. Schagger. 1998. Yeast mitochondrial F1F0-ATP synthase exists as a dimer: Identification of three dimer-specific subunits. *The EMBO Journal*. 17:7170–7178.

Arselin, G., J. Vaillier, B. Salin, J. Schaeffer, M.F. Giraud, A. Dautant, D. Brèthes, and J. Velours. 2004. The modulation in subunits e and g amounts of yeast ATP synthase modifies mitochondrial cristae morphology. *Journal of Biological Chemistry*. 279:40392–40399.

Baker, M.J., P.A. Lampe, D. Stojanovski, A. Korwitz, R. Anand, T. Tatsuta, and T. Langer. 2014. Stress-induced OMA1 activation and autocatalytic turnover regulate OPA1-dependent mitochondrial dynamics. *EMBO Journal*. 33:578–593.

Baker, N., J. Patel, and M. Khacho. 2019. Linking mitochondrial dynamics, cristae remodeling and supercomplex formation: How mitochondrial structure can regulate bioenergetics. *Mitochondrion*. 49:259–268.

Barbot, M., D.C. Jans, C. Schulz, N. Denkert, B. Kroppen, M. Hoppert, S. Jakobs, and M. Meinecke. 2015. Mic10 oligomerizes to bend mitochondrial inner membranes at cristae junctions. *Cell Metabolism*. 21:756–763.

Barrera, M., S. Koob, D. Dikov, F. Vogel, and A.S. Reichert. 2016. OPA1 functionally interacts with MIC60 but is dispensable for crista junction formation. *FEBS Letters*. 590:3309–3322.

Beninca, C., V. Zanette, M. Brischigliaro, M. Johnson, A. Reyes, D.A.D. Valle, A.J. Robinson, A. Degiorgi, A. Yeates, B.A. Telles, J. Prudent, E. Baruffini, M.L. S.F. Santos, R.L. R. de Souza, E. Fernandez-Vizarra, A.J. Whitworth, and M. Zeviani. 2021. Mutation in the MICOS subunit gene APOO (MIC26) associated with an X-linked recessive mitochondrial myopathy, lactic acidosis, cognitive impairment and autistic features. *Journal of Medical Genetics*. 58:155–167.

Bernardi, P., and G.F. Azzone. 1981. Cytochrome c as an electron shuttle between the outer and inner mitochondrial membranes. *Journal of Biological Chemistry*. 256:7187–7192.

Blum, T.B., A. Hahn, T. Meier, K.M. Davies, and W. Kuhlbrandt. 2019. Dimers of mitochondrial ATP synthase induce membrane curvature and self-assemble into rows. *Proceedings of the National Academy of Sciences USA*. 116:4250–4255.

Bock-Bierbaum, T., K. Funck, F. Wollweber, E. Lisicki, K. von der Malsburg, A. von der Malsburg, J. Laborenz, J.K. Noel, M. Hessenberger, S. Jungbluth, C. Bernert, S. Kunz, D. Riedel, H. Lilie, S. Jakobs, M. van der Laan, and O. Daumke. 2022. Structural insights into crista junction formation by the Mic60-Mic19 complex. *Science Advances*. 8:eabo4946.

Bohnert, M., R.M. Zerbes, K.M. Davies, A.W. Muhleip, H. Rampelt, S.E. Horvath, T. Boenke, A. Kram, I. Perschil, M. Veenhuis, W. Kuhlbrandt, I.J. van der Klei, N. Pfanner, and M. van der Laan. 2015. Central role of Mic10 in the mitochondrial contact site and cristae organizing system. *Cell Metabolism*. 21:747–755.

Bou-Teen, D., C. Fernandez-Sanz, E. Miro-Casas, Z. Nichtova, E. Bonzon-Kulichenko, K. Casós, J. Inserte, A. Rodriguez-Sinovas, B. Benito, S.S. Sheu, J. Vázquez, I. Ferreira-González, and M. Ruiz-Meana. 2022. Defective dimerization of FoF1-ATP synthase secondary to glycation favors mitochondrial energy deficiency in cardiomyocytes during aging. *Aging Cell*. 21:e13564.

Brandt, T., A. Mourier, L.S. Tain, L. Partridge, N.G. Larsson, and W. Kühlbrandt. 2017. Changes of mitochondrial ultrastructure and function during ageing in mice and *Drosophila*. *eLife*. 6.

Chen, L., J. Dong, S. Liao, S. Wang, Z. Wu, M. Zuo, B. Liu, C. Yan, Y. Chen, H. He, Q. Meng, and Z. Song. 2022. Loss of Sam50 in hepatocytes induces cardiolipin-dependent mitochondrial membrane remodeling to trigger mtDNA release and liver injury. *Hepatology*. 76:1389–1408.

Cipolat, S., and L. Scorrano. 2006. To fuse and to protect. A novel role for CED-9 in mitochondrial morphology reveals an ancient function. *Cell Death & Differentiation*. 13:1833–1834.

Cogliati, S., C. Frezza, M.E. Soriano, T. Varanita, R. Quintana-Cabrera, M. Corrado, S. Cipolat, V. Costa, A. Casarin, L.C. Gomes, E. Perales-Clemente, L. Salviati, P. Fernandez-Silva, J.A. Enriquez, and L. Scorrano. 2013. Mitochondrial cristae shape determines respiratory chain supercomplexes assembly and respiratory efficiency. *Cell*. 155:160–171.

Cogliati, S., J.A. Enriquez, and L. Scorrano. 2016. Mitochondrial cristae: Where beauty meets functionality. *Trends in Biochemical Sciences*. 41:261–273.

Cserép, C., B. Pósfai, A.D. Schwarcz, and Á. Dénes. 2018. Mitochondrial ultrastructure is coupled to synaptic performance at axonal release sites. *eNeuro*. 5. https://doi.org/10.1523/ENEURO.0390-17.2018

Daems, W.T., and E. Wisse. 1966. Shape and attachment of the cristae mitochondriales in mouse hepatic cell mitochondria. *Journal of Ultrastructure Research*. 16:123–140.

Daum, B., A. Walter, A. Horst, H.D. Osiewacz, and W. Kühlbrandt. 2013. Age-dependent dissociation of ATP synthase dimers and loss of inner-membrane cristae in mitochondria. *Proceedings of the National Academy of Sciences USA*. 110:15301–15306.

Davies, K.M., C. Anselmi, I. Wittig, J.D. Faraldo-Gomez, and W. Kuhlbrandt. 2012. Structure of the yeast F_1F_O-ATP synthase dimer and its role in shaping the mitochondrial cristae. *Proceedings of the National Academy of Sciences USA*. 109:13602–13607.

Davies, K.M., M. Strauss, B. Daum, J.H. Kief, H.D. Osiewacz, A. Rycovska, V. Zickermann, and W. Kuhlbrandt. 2011. Macromolecular organization of ATP synthase and complex I in whole mitochondria. *Proceedings of the National Academy of Sciences USA*. 108:14121–14126.

Del Dotto, V., P. Mishra, S. Vidoni, M. Fogazza, A. Maresca, L. Caporali, J.M. McCaffery, M. Cappelletti, E. Baruffini, G. Lenaers, D. Chan, M. Rugolo, V. Carelli, and C. Zanna. 2017. OPA1 isoforms in the hierarchical organization of mitochondrial functions. *Cell Reports*. 19:2557–2571.

Delettre, C., G. Lenaers, J.M. Griffoin, N. Gigarel, C. Lorenzo, P. Belenguer, L. Pelloquin, J. Grosgeorge, C. Turc-Carel, E. Perret, C. Astarie-Dequeker, L. Lasquellec, B. Arnaud, B. Ducommun, J. Kaplan, and C.P. Hamel. 2000. Nuclear gene OPA1, encoding a mitochondrial dynamin-related protein, is mutated in dominant optic atrophy. *Nature Genetics*. 26:207–210.

Dlasková, A., T. Špaček, H. Engstová, J. Špačková, A. Schröfel, B. Holendová, K. Smolková, L. Plecitá-Hlavatá, and P. Ježek. 2019. Mitochondrial cristae narrowing upon higher 2-oxoglutarate load. *Biochimica et Biophysica Acta (BBA) — Bioenergetics*. 1860:659–678.

Dong, T., X. Zhang, Y. Liu, S. Xu, H. Chang, F. Chen, L. Pan, S. Hu, M. Wang, and M. Lu. 2021. Opa1 prevents apoptosis and cisplatin-induced ototoxicity in murine cochleae. *Frontiers in Cell and Developmental Biology*. 9:744838.

Du, X., and H. Hu. 2021. The roles of 2-hydroxyglutarate. *Frontiers in Cell and Developmental Biology*. 9:651317.

Duvezin-Caubet, S., R. Jagasia, J. Wagener, S. Hofmann, A. Trifunovic, A. Hansson, A. Chomyn, M.F. Bauer, G. Attardi, N.G. Larsson, W. Neupert, and A.S. Reichert. 2006. Proteolytic processing of OPA1 links mitochondrial dysfunction to alterations in mitochondrial morphology. *Journal of Biological Chemistry*. 281:37972–37979.

Elouej, S., K. Harhouri, M. Le Mao, G. Baujat, S. Nampoothiri, H. Kayserili, N.A. Menabawy, L. Selim, A.L. Paneque, C. Kubisch, D. Lessel, et al. 2020. Loss of MTX2 causes mandibuloacral dysplasia and links mitochondrial dysfunction to altered nuclear morphology. *Nature Communications*. 11:4589.

Eydt, K., K.M. Davies, C. Behrendt, I. Wittig, and A.S. Reichert. 2017. Cristae architecture is determined by an interplay of the MICOS complex and the F_1F_O ATP synthase via Mic27 and Mic10. *Microbial Cell*. 4:259–272.

Frey, T.G., and C.A. Mannella. 2000. The internal structure of mitochondria. *Trends in Biochemical Sciences*. 25:319–324.

Frey, T.G., C.W. Renken, and G.A. Perkins. 2002. Insight into mitochondrial structure and function from electron tomography. *Biochimica et Biophysica Acta (BBA) Bioenergetics*. 1555:196–203.

Frezza, C. 2017. Mitochondrial metabolites: Undercover signalling molecules. *Interface Focus*. 7:20160100.

Frezza, C., S. Cipolat, O. Martins de Brito, M. Micaroni, G.V. Beznoussenko, T. Rudka, D. Bartoli, R.S. Polishuck, N.N. Danial, B. De Strooper, and L. Scorrano. 2006. OPA1 controls apoptotic cristae remodeling independently from mitochondrial fusion. *Cell*. 126:177–189.

Gilkerson, R., P. De La Torre, and S. St Vallier. 2021. Mitochondrial OMA1 and OPA1 as gatekeepers of organellar structure/function and cellular stress response. *Frontiers in Cell and Developmental Biology*. 9:626117.

Gilkerson, R.W., J.M. Selker, and R.A. Capaldi. 2003. The cristal membrane of mitochondria is the principal site of oxidative phosphorylation. *FEBS Letters*. 546:355–358.

Glancy, B., Y. Kim, P. Katti, and T.B. Willingham. 2020. The functional impact of mitochondrial structure across subcellular scales. *Frontiers in Physiology*. 11:541040.

Glytsou, C., E. Calvo, S. Cogliati, A. Mehrotra, I. Anastasia, G. Rigoni, A. Raimondi, N. Shintani, M. Loureiro, J. Vazquez, L. Pellegrini, J.A. Enriquez, L. Scorrano, and M.E. Soriano. 2016. Optic atrophy 1 is epistatic to the core MICOS component MIC60 in mitochondrial cristae shape control. *Cell Reports*. 17:3024–3034.

Golombek, M., T. Tsigaras, Y. Schaumkessel, S. Hänsch, S. Weidtkamp-Peters, R. Anand, A.S. Reichert, and A.K. Kondadi. 2023. Cristae dynamics is modulated in bioenergetically compromised mitochondria. *Life Science Alliance*. 13:7(2):e202302386. doi: 10.26508/lsa.202302386. PMID: 37957016; PMCID: PMC10643176.

Gomes, L.C., G. Di Benedetto, and L. Scorrano. 2011. During autophagy mitochondria elongate, are spared from degradation and sustain cell viability. *Nature Cell Biology*. 13:589–598.

Guarani, V., C. Jardel, D. Chretien, A. Lombes, P. Benit, C. Labasse, E. Lacene, A. Bourillon, A. Imbard, J.F. Benoist, I. Dorboz, M. Gilleron, E.S. Goetzman, P. Gaignard, A. Slama, M. Elmaleh-Berges, N.B. Romero, P. Rustin, H. Ogier de Baulny, J.A. Paulo, J.W. Harper, and M. Schiff. 2016. QIL1 mutation causes MICOS disassembly and early onset fatal mitochondrial encephalopathy with liver disease. *eLife*. 5:e17163.

Guarani, V., E.M. McNeill, J.A. Paulo, E.L. Huttlin, F. Frohlich, S.P. Gygi, D. Van Vactor, and J.W. Harper. 2015. QIL1 is a novel mitochondrial protein required for MICOS complex stability and cristae morphology. *eLife*. 4. https://doi.org/10.7554/eLife.06265

Hackenbrock, C.R. 1966. Ultrastructural bases for metabolically linked mechanical activity in mitochondria. I. Reversible ultrastructural changes with change in metabolic steady state in isolated liver mitochondria. *The Journal of Cell Biology*. 30:269–297.

Harmon, H.J., J.D. Hall, and F.L. Crane. 1974. Structure of mitochondrial cristae membranes. *Biochimica et Biophysica Acta*. 344:119–155.

Harner, M., C. Korner, D. Walther, D. Mokranjac, J. Kaesmacher, U. Welsch, J. Griffith, M. Mann, F. Reggiori, and W. Neupert. 2011. The mitochondrial contact site complex, a determinant of mitochondrial architecture. *The EMBO Journal*. 30:4356–4370.

Harner, M.E., A.K. Unger, W.J. Geerts, M. Mari, T. Izawa, M. Stenger, S. Geimer, F. Reggiori, B. Westermann, and W. Neupert. 2016. An evidence based hypothesis on the existence of two pathways of mitochondrial crista formation. *eLife*. 5:e18853.

He, J., H.C. Ford, J. Carroll, C. Douglas, E. Gonzales, S. Ding, I.M. Fearnley, and J.E. Walker. 2018. Assembly of the membrane domain of ATP synthase in human mitochondria. *Proceedings of the National Academy of Sciences USA*. 115:2988–2993.

Hessenberger, M., R.M. Zerbes, H. Rampelt, S. Kunz, A.H. Xavier, B. Purfurst, H. Lilie, N. Pfanner, M. van der Laan, and O. Daumke. 2017. Regulated membrane remodeling by Mic60 controls formation of mitochondrial crista junctions. *Nature Communications*. 8:15258.

Hoppins, S., S.R. Collins, A. Cassidy-Stone, E. Hummel, R.M. Devay, L.L. Lackner, B. Westermann, M. Schuldiner, J.S. Weissman, and J. Nunnari. 2011. A mitochondrial-focused genetic interaction map reveals a scaffold-like complex required for inner membrane organization in mitochondria. *The Journal of Cell Biology*. 195:323–340.

Horvath, S.E., and G. Daum. 2013. Lipids of mitochondria. *Progress in Lipid Research*. 52:590–614.

Hu, C., L. Shu, X. Huang, J. Yu, L. Li, L. Gong, M. Yang, Z. Wu, Z. Gao, Y. Zhao, L. Chen, and Z. Song. 2020. OPA1 and MICOS Regulate mitochondrial crista dynamics and formation. *Cell Death & Disease*. 11:940.

Huang, X., J. Fan, L. Li, H. Liu, R. Wu, Y. Wu, L. Wei, H. Mao, A. Lal, P. Xi, L. Tang, Y. Zhang, Y. Liu, S. Tan, and L. Chen. 2018. Fast, long-term, super-resolution imaging with Hessian structured illumination microscopy. *Nature Biotechnology*. 36:451–459.

Ikon, N., and R.O. Ryan. 2017. Cardiolipin and mitochondrial cristae organization. *Biochimica et Biophysica Acta (BBA) Biomembranes*. 1859:1156–1163.

Ioakeimidis, F., C. Ott, V. Kozjak-Pavlovic, F. Violitzi, V. Rinotas, E. Makrinou, E. Eliopoulos, C. Fasseas, G. Kollias, and E. Douni. 2014. A splicing mutation in the novel mitochondrial protein DNAJC11 causes motor neuron pathology associated with cristae disorganization, and lymphoid abnormalities in mice. *PLOS ONE*. 9:e104237.

Ishihara, N., Y. Fujita, T. Oka, and K. Mihara. 2006. Regulation of mitochondrial morphology through proteolytic cleavage of OPA1. *The EMBO Journal*. 25:2966–2977.

Jain, I.H., L. Zazzeron, R. Goli, K. Alexa, S. Schatzman-Bone, H. Dhillon, O. Goldberger, J. Peng, O. Shalem, N.E. Sanjana, F. Zhang, W. Goessling, W.M. Zapol, and V.K. Mootha. 2016. Hypoxia as a therapy for mitochondrial disease. *Science (New York, N.Y.)*. 352:54–61.

Jans, D.C., C.A. Wurm, D. Riedel, D. Wenzel, F. Stagge, M. Deckers, P. Rehling, and S. Jakobs. 2013. STED super-resolution microscopy reveals an array of MINOS clusters along human mitochondria. *Proceedings of the National Academy of Sciences USA*. 110:8936–8941.

Jiang, X., H. Jiang, Z. Shen, and X. Wang. 2014. Activation of mitochondrial protease OMA1 by Bax and Bak promotes cytochrome c release during apoptosis. *Proceedings of the National Academy of Sciences USA*. 111:14782–14787.

Jiang, Y.-f., S.-s. Lin, J.-m. Chen, H.-z. Tsai, T.-s. Hsieh, and C.-y. Fu. 2017. Electron tomographic analysis reveals ultrastructural features of mitochondrial cristae architecture which reflect energetic state and aging. *Scientific Reports*. 7:45474.

Kang, H.W., S. Ribich, B.W. Kim, S.J. Hagen, A.C. Bianco, and D.E. Cohen. 2009. Mice lacking Pctp/StarD2 exhibit increased adaptive thermogenesis and enlarged mitochondria in brown adipose tissue. *Journal of Lipid Research*. 50:2212–2221.

Khacho, M., M. Tarabay, D. Patten, P. Khacho, J.G. MacLaurin, J. Guadagno, R. Bergeron, S.P. Cregan, M.E. Harper, D.S. Park, and R.S. Slack. 2014. Acidosis overrides oxygen deprivation to maintain mitochondrial function and cell survival. *Nature Communications*. 5:3550.

Klecker, T., and B. Westermann. 2021. Pathways shaping the mitochondrial inner membrane. *Open Biology*. 11:210238.

Kojima, R., Y. Kakimoto, S. Furuta, K. Itoh, H. Sesaki, T. Endo, and Y. Tamura. 2019. Maintenance of cardiolipin and crista structure requires cooperative functions of mitochondrial dynamics and phospholipid transport. *Cell Reports*. 26:518–528 e516.

Kondadi, A.K., R. Anand, S. Hansch, J. Urbach, T. Zobel, D.M. Wolf, M. Segawa, M. Liesa, O.S. Shirihai, S. Weidtkamp-Peters, and A.S. Reichert. 2020a. Cristae undergo continuous cycles of membrane remodelling in a MICOS-dependent manner. *EMBO Reports*. 21:e49776.

Kondadi, A.K., R. Anand, and A.S. Reichert. 2020b. Cristae membrane dynamics - a paradigm change. *Trends in Cell Biology*. 30:923–936.

Kühlbrandt, W. 2015. Structure and function of mitochondrial membrane protein complexes. *BMC Biology*. 13:89.

Li, P., D. Nijhawan, I. Budihardjo, S.M. Srinivasula, M. Ahmad, E.S. Alnemri, and X. Wang. 1997. Cytochrome c and dATP-dependent formation of Apaf-1/caspase-9 complex initiates an apoptotic protease cascade. *Cell*. 91:479–489.

MacVicar, T., and T. Langer. 2016. OPA1 processing in cell death and disease - the long and short of it. *Journal of Cell Science*. 129:2297–2306.

Mannella, C.A. 2006a. The relevance of mitochondrial membrane topology to mitochondrial function. *Biochimica et Biophysica Acta*. 1762:140–147.

Mannella, C.A. 2006b. Structure and dynamics of the mitochondrial inner membrane cristae. *Biochimica et Biophysica Acta*. 1763:542–548.

Mannella, C.A., D.R. Pfeiffer, P.C. Bradshaw, Moraru, II, B. Slepchenko, L.M. Loew, C.E. Hsieh, K. Buttle, and M. Marko. 2001. Topology of the mitochondrial inner membrane: Dynamics and bioenergetic implications. *IUBMB Life*. 52:93–100.

Mannella, C.A., W.J. Lederer, and M.S. Jafri. 2013. The connection between inner membrane topology and mitochondrial function. *Journal of Molecular and Cellular Cardiology*. 62:51–57.

Marco-Hernández, A.V., M. Tomás-Vila, A. Montoya-Filardi, H. Barranco-González, J.J. Vilchez Padilla, I. Azorín, P. Smeyers Dura, S. Monfort-Membrado, I. Pitarch-Castellano, and F. 2022. Martínez-Castellano. Mitochondrial developmental encephalopathy with bilateral optic neuropathy related to homozygous variants in IMMT gene. *Clinical Genetics*. 101(2):233–241. doi: 10.1111/cge.14093. Epub 2021 Dec 5. PMID: 34842280.

Marschall, L.M., V. Warnsmann, A.C. Meeßen, T. Löser, and H.D. Osiewacz. 2022. Lifespan extension of *Podospora anserina* Mic60-subcomplex mutants depends on cardiolipin remodeling. *International Journal of Molecular Sciences*. 23:4741.

Meeusen, S., R. DeVay, J. Block, A. Cassidy-Stone, S. Wayson, J.M. McCaffery, and J. Nunnari. 2006. Mitochondrial inner-membrane fusion and crista maintenance requires the dynamin-related GTPase Mgm1. *Cell*. 127:383–395.

Monzel, A.S., J.A. Enríquez, and M. Picard. 2023. Multifaceted mitochondria: Moving mitochondrial science beyond function and dysfunction. *Nature Metabolism*. 5:546–562.

Olichon, A., G. Elachouri, L. Baricault, C. Delettre, P. Belenguer, and G. Lenaers. 2007. OPA1 alternate splicing uncouples an evolutionary conserved function in mitochondrial fusion from a vertebrate restricted function in apoptosis. *Cell Death & Differentiation*. 14:682–692.

Olichon, A., L. Baricault, N. Gas, E. Guillou, A. Valette, P. Belenguer, and G. Lenaers. 2003. Loss of OPA1 perturbates the mitochondrial inner membrane structure and integrity, leading to cytochrome c release and apoptosis. *The Journal of Biological Chemistry*. 278:7743–7746.

Olichon, A., L.J. Emorine, E. Descoins, L. Pelloquin, L. Brichese, N. Gas, E. Guillou, C. Delettre, A. Valette, C.P. Hamel, B. Ducommun, G. Lenaers, and P. Belenguer. 2002. The human dynamin-related protein OPA1 is anchored to the mitochondrial inner membrane facing the inter-membrane space. *FEBS Letters*. 523:171–176.

Ott, C., E. Dorsch, M. Fraunholz, S. Straub, and V. Kozjak-Pavlovic. 2015. Detailed analysis of the human mitochondrial contact site complex indicate a hierarchy of subunits. *PLOS ONE*. 10:e0120213.

Ott, C., K. Ross, S. Straub, B. Thiede, M. Gotz, C. Goosmann, M. Krischke, M.J. Mueller, G. Krohne, T. Rudel, and V. Kozjak-Pavlovic. 2012. Sam50 functions in mitochondrial intermembrane space bridging and biogenesis of respiratory complexes. *Molecular and Cellular Biology*. 32:1173–1188.

Patten, D.A., J. Wong, M. Khacho, V. Soubannier, R.J. Mailloux, K. Pilon-Larose, J.G. MacLaurin, D.S. Park, H.M. McBride, L. Trinkle-Mulcahy, M.E. Harper, M. Germain, and R.S. Slack. 2014. OPA1-dependent cristae modulation is essential for cellular adaptation to metabolic demand. *EMBO Journal*. 33:2676–2691.

Paumard, P., J. Vaillier, B. Coulary, J. Schaeffer, V. Soubannier, D.M. Mueller, D. Brethes, J.P. di Rago, and J. Velours. 2002. The ATP synthase is involved in generating mitochondrial cristae morphology. *The EMBO Journal*. 21:221–230.

Peifer-Weiß, L., M. Kurban, C. David, M. Lubeck, A. K. Kondadi, G. Nemer, A.S. Reichert, and R. Anand. 2023. A X-linked nonsense APOO/MIC26 variant causes a lethal mitochondrial disease with progeria-like phenotypes. *Clinical Genetics*. 104(6):659–668. doi: 10.1111/cge.14420. Epub 2023 Aug 30. PMID: 37649161.

Perkins, G., C. Renken, M.E. Martone, S.J. Young, M. Ellisman, and T. Frey. 1997. Electron tomography of neuronal mitochondria: Three-dimensional structure and organization of cristae and membrane contacts. *Journal of Structural Biology*. 119:260–272.

Perkins, G., Y.H. Hsiao, S. Yin, J. Tjong, M.T. Tran, J. Lau, J. Xue, S. Liu, M.H. Ellisman, and D. Zhou. 2012. Ultrastructural modifications in the mitochondria of hypoxia-adapted *Drosophila* melanogaster. *PLOS ONE*. 7:e45344.

Phillips, D., R. Covian, A.M. Aponte, B. Glancy, J.F. Taylor, D. Chess, and R.S. Balaban. 2012. Regulation of oxidative phosphorylation complex activity: Effects of tissue-specific metabolic stress within an allometric series and acute changes in workload. *American Journal of Physiology Regulatory Integrative and Comparative Physiology*. 302:R1034–1048.

Plecitá-Hlavatá, L., H. Engstova, L. Alan, T. Spacek, A. Dlaskova, K. Smolkova, J. Spackova, J. Tauber, V. Stradalova, J. Malinsky, M. Lessard, J. Bewersdorf, and P. Jezek. 2016. Hypoxic HepG2 cell adaptation decreases ATP synthase dimers and ATP production in inflated cristae by mitofilin down-regulation concomitant to MICOS clustering. *FASEB Journal: Official Publication of the Federation of American Societies for Experimental Biology*. 30:1941–1957.

Plecitá-Hlavatá, L., J. Ježek, and P. Ježek. 2015. Aglycemia keeps mitochondrial oxidative phosphorylation under hypoxic conditions in HepG2 cells. *Journal of Bioenergetics and Biomembranes.* 47:467–476.

Quintana-Cabrera, R., and L. Scorrano. 2023. Determinants and outcomes of mitochondrial dynamics. *Molecular Cell.* 83:857–876.

Quintana-Cabrera, R., C. Quirin, C. Glytsou, M. Corrado, A. Urbani, A. Pellattiero, E. Calvo, J. Vazquez, J.A. Enriquez, C. Gerle, M.E. Soriano, P. Bernardi, and L. Scorrano. 2018. The cristae modulator Optic atrophy 1 requires mitochondrial ATP synthase oligomers to safeguard mitochondrial function. *Nature Communications.* 9:3399.

Quintana-Cabrera, R., I. Manjarrés-Raza, C. Vicente-Gutiérrez, M. Corrado, J.P. Bolaños, and L. Scorrano. 2021. Opa1 relies on cristae preservation and ATP synthase to curtail reactive oxygen species accumulation in mitochondria. *Redox Biology.* 41:101944.

Rabl, R., V. Soubannier, R. Scholz, F. Vogel, N. Mendl, A. Vasiljev-Neumeyer, C. Korner, R. Jagasia, T. Keil, W. Baumeister, M. Cyrklaff, W. Neupert, and A.S. Reichert. 2009. Formation of cristae and crista junctions in mitochondria depends on antagonism between Fcj1 and Su e/g. *The Journal of Cell Biology.* 185:1047–1063.

Rampelt, H., F. Wollweber, C. Gerke, R. de Boer, I.J. van der Klei, M. Bohnert, N. Pfanner, and M. van der Laan. 2018. Assembly of the mitochondrial cristae organizer Mic10 is regulated by Mic26-Mic27 antagonism and cardiolipin. *Journal of Molecular Biology.* 430:1883–1890.

Rampelt, H., M. Bohnert, R.M. Zerbes, S.E. Horvath, B. Warscheid, N. Pfanner, and M. van der Laan. 2017. Mic10, a core subunit of the mitochondrial contact site and cristae organizing system, interacts with the dimeric F(1)F(o)-ATP synthase. *Journal of Molecular Biology.* 429:1162–1170.

Renner, L.D., and D.B. Weibel. 2011. Cardiolipin microdomains localize to negatively curved regions of Escherichia coli membranes. *Proceedings of the National Academy of Sciences USA.* 108:6264–6269.

Sakowska, P., D.C. Jans, K. Mohanraj, D. Riedel, S. Jakobs, and A. Chacinska. 2015. The oxidation status of Mic19 regulates MICOS assembly. *Molecular and Cellular Biology.* 35:4222–4237.

Salewskij, K., B. Rieger, F. Hager, T. Arroum, P. Duwe, J. Villalta, S. Colgiati, C.P. Richter, O.E. Psathaki, J.A. Enriquez, T. Dellmann, and K.B. Busch. 2020. The spatio-temporal organization of mitochondrial F(1) F(O) ATP synthase in cristae depends on its activity mode. *Biochimica Biophysica Acta Bioenergetics.* 1861:148091.

Schlame, M. 2008. Cardiolipin synthesis for the assembly of bacterial and mitochondrial membranes. *Journal of Lipid Research.* 49:1607–1620.

Scorrano, L., M. Ashiya, K. Buttle, S. Weiler, S.A. Oakes, C.A. Mannella, and S.J. Korsmeyer. 2002. A distinct pathway remodels mitochondrial cristae and mobilizes cytochrome c during apoptosis. *Developmental Cell.* 2:55–67.

Serricchio, M., A. Vissa, P.K. Kim, C.M. Yip, and G.A. McQuibban. 2018. Cardiolipin synthesizing enzymes form a complex that interacts with cardiolipin-dependent membrane organizing proteins. *Biochimica et Biophysica Acta.* 1863:447–457.

Shami, G.J., D. Cheng, P. Verhaegh, G. Koek, E. Wisse, and F. Braet. 2021. Three-dimensional ultrastructure of giant mitochondria in human non-alcoholic fatty liver disease. *Scientific Reports.* 11:3319.

Shepard, T.H., L.A. Muffley, and L.T. Smith. 2000. Mitochondrial ultrastructure in embryos after implantation. *Human Reproduction.* 15:218–228.

Spinelli, J.B., and M.C. Haigis. 2018. The multifaceted contributions of mitochondria to cellular metabolism. *Nature Cell Biology.* 20:745–754.

Sprenger, H.G., and T. Langer. 2019. The good and the bad of mitochondrial breakups. *Trends in Cell Biology.* 29:888–900.

Stenton, S.L., and H. Prokisch. 2020. Genetics of mitochondrial diseases: Identifying mutations to help diagnosis. *EBioMedicine.* 56:102784.

Stephan, T., A. Roesch, D. Riedel, and S. Jakobs. 2019. Live-cell STED nanoscopy of mitochondrial cristae. *Scientific Reports.* 9:12419.

Stephan, T., C. Brüser, M. Deckers, A.M. Steyer, F. Balzarotti, M. Barbot, T.S. Behr, G. Heim, W. Hübner, P. Ilgen, F. Lange, D. Pacheu-Grau, J.K. Pape, S. Stoldt, T. Huser, S.W. Hell, W. Möbius, P. Rehling, D. Riedel, and S. Jakobs. 2020. MICOS assembly controls mitochondrial inner membrane remodeling and crista junction redistribution to mediate cristae formation. *The EMBO Journal.* 39:e104105.

Stoldt, S., T. Stephan, D.C. Jans, C. Bruser, F. Lange, J. Keller-Findeisen, D. Riedel, S.W. Hell, and S. Jakobs. 2019. Mic60 exhibits a coordinated clustered distribution along and across yeast and mammalian mitochondria. *Proceedings of the National Academy of Sciences USA.* 116: 9853–9858.

Strauss, M., G. Hofhaus, R.R. Schroder, and W. Kuhlbrandt. 2008. Dimer ribbons of ATP synthase shape the inner mitochondrial membrane. *The EMBO Journal*. 27:1154–1160.

Sun, M.G., J. Williams, C. Munoz-Pinedo, G.A. Perkins, J.M. Brown, M.H. Ellisman, D.R. Green, and T.G. Frey. 2007. Correlated three-dimensional light and electron microscopy reveals transformation of mitochondria during apoptosis. *Nature Cell Biology*. 9:1057–1065.

Tarasenko, D., M. Barbot, D.C. Jans, B. Kroppen, B. Sadowski, G. Heim, W. Mobius, S. Jakobs, and M. Meinecke. 2017. The MICOS component Mic60 displays a conserved membrane-bending activity that is necessary for normal cristae morphology. *The Journal of Cell Biology*. 216:889–899.

Tasseva, G., H.D. Bai, M. Davidescu, A. Haromy, E. Michelakis, and J.E. Vance. 2013. Phosphatidylethanolamine deficiency in Mammalian mitochondria impairs oxidative phosphorylation and alters mitochondrial morphology. *The Journal of Biological Chemistry*. 288:4158–4173.

Tatsuta, T., and T. Langer. 2017. Intramitochondrial phospholipid trafficking. *Biochimica et Biophysica Acta*. 1862:81–89.

Tsai, P.I., C.H. Lin, C.H. Hsieh, A.M. Papakyrikos, M.J. Kim, V. Napolioni, C. Schoor, J. Couthouis, R.M. Wu, Z.K. Wszolek, D. Winter, M.D. Greicius, O.A. Ross, and X. Wang. 2018. PINK1 phosphorylates MIC60/Mitofilin to control structural plasticity of mitochondrial crista junctions. *Molecular Cell*. 69:744–756 e746.

Urbach, J., A.K. Kondadi, C. David, R. Naha, K. Deinert, A.S. Reichert, and R. Anand. 2021. Conserved GxxxG and WN motifs of MIC13 are essential for bridging two MICOS subcomplexes. *Biochimica et Biophysica Acta Biomembranes*. 1863:183683.

Vafai, S.B., and V.K. Mootha. 2012. Mitochondrial disorders as windows into an ancient organelle. *Nature*. 491:374–383.

Varanita, T., M.E. Soriano, V. Romanello, T. Zaglia, R. Quintana-Cabrera, M. Semenzato, R. Menabò, V. Costa, G. Civiletto, P. Pesce, C. Viscomi, M. Zeviani, F. Di Lisa, M. Mongillo, M. Sandri, and L. Scorrano. 2015. The OPA1-dependent mitochondrial cristae remodeling pathway controls atrophic, apoptotic, and ischemic tissue damage. *Cell Metabolism*. 21:834–844.

Vincent, A.E., Y.S. Ng, K. White, T. Davey, C. Mannella, G. Falkous, C. Feeney, A.M. Schaefer, R. McFarland, G.S. Gorman, R.W. Taylor, D.M. Turnbull, and M. Picard. 2016. The spectrum of mitochondrial ultrastructural defects in mitochondrial myopathy. *Scientific Reports*. 6:30610.

Vogel, F., C. Bornhovd, W. Neupert, and A.S. Reichert. 2006. Dynamic subcompartmentalization of the mitochondrial inner membrane. *Journal of Cell Biology*. 175:237–247.

von der Malsburg, K., J.M. Muller, M. Bohnert, S. Oeljeklaus, P. Kwiatkowska, T. Becker, A. Loniewska-Lwowska, S. Wiese, S. Rao, D. Milenkovic, D.P. Hutu, R.M. Zerbes, A. Schulze-Specking, H.E. Meyer, J.C. Martinou, S. Rospert, P. Rehling, C. Meisinger, M. Veenhuis, B. Warscheid, I.J. van der Klei, N. Pfanner, A. Chacinska, and M. van der Laan. 2011. Dual role of mitofilin in mitochondrial membrane organization and protein biogenesis. *Developmental Cell*. 21:694–707.

Wagner, K., P. Rehling, L.K. Sanjuán Szklarz, R.D. Taylor, N. Pfanner, and M. van der Laan. 2009. Mitochondrial F1Fo-ATP synthase: The small subunits e and g associate with monomeric complexes to trigger dimerization. *Journal of Molecular Biology*. 392:855–861.

Wang, C., M. Taki, Y. Sato, Y. Tamura, H. Yaginuma, Y. Okada, and S. Yamaguchi. 2019a. A photostable fluorescent marker for the superresolution live imaging of the dynamic structure of the mitochondrial cristae. *Proceedings of the National Academy of Sciences USA*. 116:15817–15822.

Wang, G., Y. He, and Y. Luo. 2019b. Expression of OPA1 and Mic60 genes and their association with mitochondrial cristae morphology in Tibetan sheep. *Cell and Tissue Research*. 376:273–279.

Warnsmann, V., L.M. Marschall, A.C. Meeßen, M. Wolters, L. Schürmanns, M. Basoglu, S. Eimer, and H.D. Osiewacz. 2022. Disruption of the MICOS complex leads to an aberrant cristae structure and an unexpected, pronounced lifespan extension in *Podospora anserina*. *Journal of Cellular Biochemistry*. 123:1306–1326.

Weber, T.A., S. Koob, H. Heide, I. Wittig, B. Head, A. van der Bliek, U. Brandt, M. Mittelbronn, and A.S. Reichert. 2013. APOOL is a cardiolipin-binding constituent of the Mitofilin/MINOS protein complex determining cristae morphology in mammalian mitochondria. *PLOS ONE*. 8:e63683.

Wilkens, V., W. Kohl, and K. Busch. 2013. Restricted diffusion of OXPHOS complexes in dynamic mitochondria delays their exchange between cristae and engenders a transitory mosaic distribution. *Journal of Cell Science*. 126:103–116.

Wurm, C.A., and S. Jakobs. 2006. Differential protein distributions define two sub-compartments of the mitochondrial inner membrane in yeast. *FEBS Letters*. 580:5628–5634.

Yamaguchi, R., and G. Perkins. 2009. Dynamics of mitochondrial structure during apoptosis and the enigma of Opa1. *Biochimica et Biophysica Acta*. 1787:963–972.

Yamaguchi, R., L. Lartigue, G. Perkins, R.T. Scott, A. Dixit, Y. Kushnareva, T. Kuwana, M.H. Ellisman, and D.D. Newmeyer. 2008. Opa1-mediated cristae opening is Bax/Bak and BH3 dependent, required for apoptosis, and independent of Bak oligomerization. *Molecular Cell*. 31:557–569.

Zhang, D., Y. Zhang, J. Ma, C. Zhu, T. Niu, W. Chen, X. Pang, Y. Zhai, and F. Sun. 2020. Cryo-EM structures of S-OPA1 reveal its interactions with membrane and changes upon nucleotide binding. *eLife*. 9. https://doi.org/10.7554/eLife.50294

Zhou, J., M. Duan, X. Wang, F. Zhang, H. Zhou, T. Ma, Q. Yin, J. Zhang, F. Tian, G. Wang, and C. Yang. 2022. A feedback loop engaging propionate catabolism intermediates controls mitochondrial morphology. *Nature Cell Biology*. 24:526–537.

Zou, H., Y. Li, X. Liu, and X. Wang. 1999. An APAF-1 cytochrome c multimeric complex is a functional apoptosome that activates procaspase-9. *Journal of Biological Chemistry*. 274:11549–11556.

4 Emerging Roles of Phospholipids in the Structure and Function of the Oxidative Phosphorylation Complexes

*Nanami Senoo and Steven M. Claypool**

INTRODUCTION

Lipid bilayers are critical platforms for cellular homeostasis. The main components of lipid bilayers are phospholipids, which contain various combinations of headgroup, and acyl chains that self-organize to ensure membrane integrity while promoting the structure and function of membrane-embedded proteins. Mitochondria contain two membranes: the inner and outer mitochondrial membranes. The inner mitochondrial membrane contains the oxidative phosphorylation (OXPHOS) machinery that is crucial for cellular energy homeostasis. The OXPHOS machinery consists of electron transport chain complexes and the F_1F_O-ATP synthase:

- Complex I (NADH: ubiquinone oxidoreductase)
- Complex II (succinate dehydrogenase)
- Complex III (cytochrome bc_1 oxidoreductase)
- Complex IV (cytochrome *c* oxidase)
- Complex V (ATP synthase)

All of which are embedded in the inner mitochondrial membrane.

Mitochondrial membranes are characterized by unique phospholipid compositions [1]. In particular, the presence of cardiolipin (CL), a dimeric phospholipid consisting of two headgroups and four acyl chains, is the signature mitochondrial phospholipid, accounting for 11%–14% of the total phospholipid therein [1]. Other major phospholipids in mitochondria include phosphatidylcholine (PC, 40%–44%) and phosphatidylethanolamine (PE, 27%–34%) [1]. Phosphatidylinositol (PI) is also abundant in yeast mitochondria (~15%) [1]. The phospholipid compositions of the inner and outer mitochondrial membranes differ significantly. In yeast, the inner membrane contains ~40% of PC, ~30% of PE, and ~20% of CL, whereas the outer membrane contains ~50% of PC, ~30% of PE, and <10% of CL [1, 2]. The distribution of phospholipids across the membrane bilayers is also asymmetric [1]. Particularly, mammalian inner mitochondrial membranes exhibit asymmetric phospholipid distributions that are tissue-specific [3, 4]. Investigations of phospholipid biosynthetic pathways have extended our knowledge of how the specific phospholipid environments of mitochondrial membranes are created, although these details are omitted in this chapter as this topic has been extensively discussed in several excellent recent reviews [5–8].

As critical components of mitochondrial membranes, the contribution of phospholipids to the functionality of OXPHOS complexes has been appreciated since the 1980s [9, 10]. Subsequently, biochemical characterizations in the context of genetic models with altered mitochondrial membrane

* Corresponding author: sclaypo1@jhmi.edu

DOI: 10.1201/9781003221180-6

phospholipid profiles have emphasized that mitochondrial phospholipids are critical factors for the assembly and activities of the OXPHOS machinery. Moreover, defects in mitochondrial phospholipid biosynthesis have been discovered to be pathogenic (MedGen UID: 107893 [11], MedGen UID: 395228 [12], and Refs. [13–17]). A typical example is Barth syndrome, which is characterized by cardiac and skeletal myopathy and neutropenia [11]. Barth syndrome is caused by mutations in the *TAFAZZIN* gene, which encodes an enzyme that controls CL acylation [16, 17]. Taken together, these studies indicate that phospholipids play a significant role in the integrity and functionality of the inner membrane OXPHOS machinery. However, our knowledge of the full range of mechanisms by which phospholipids promote and enable OXPHOS function is by and large rudimentary in scope.

Structural studies have shown that specific phospholipids interact directly with proteins in OXPHOS complexes. In particular, recent advances in cryo-electron microscopy (EM) have expanded this view and begun to provide mechanistic insights into the roles exerted by phospholipids with respect to the OXPHOS machinery. This chapter reviews lipid-protein interactions and lipid-mediated protein-protein interactions based on the expanding catalog of OXPHOS complex structures. In outlining these protein-phospholipid binding events, we discuss the potential functionalities of these phospholipids to frame and guide future investigations designed to unravel their exact mechanistic roles.

COMPLEX I-ASSOCIATED PHOSPHOLIPIDS

Complex I couples the transfer of electrons from NADH to ubiquinone to the translocation of protons across the inner mitochondrial membrane. Complex I is structurally divided into two domains: the peripheral arm and the membrane arm. The peripheral arm is located in the matrix and contains modules necessary for NADH binding and ubiquinone binding. The membrane arm contains two major units: the ND1 module and antiporter-like units consisting of ND2, ND4, and ND5 modules. The ND1 module is attached to the peripheral arm and links electron transfer to proton translocation. The antiporter-like units mediate proton translocation. The first crystal structure of Complex I (from *Yarrowia lipolytica*) was solved in 2015 [18]. In the interim, the atomic structures of mammalian Complex I determined by cryo-EM have been reported by several groups [19–24], some of which have emphasized the involvement of lipids in the structures [20–24]. Final resolutions have improved over time, and the recent ovine structure has finally uncovered the mechanism by which Complex I couples quinone binding and reduction to proton translocation [25]. Figure 4.1 shows the latest ovine Complex I structure along with the identified interacting phospholipids.

Catalytic Function

High-resolution structures of ovine Complex I in different states corresponding to its catalytic cycle have revealed that Complex I dynamically moves the peripheral arm by repeated tilting between open and closed states [25]. When NADH binds to the peripheral arm, electrons are ultimately transferred to a quinone at the interface between the peripheral and membrane arms. Accompanying this, the orientations of NDUFS2 and NDUFS7 (peripheral arm subunits) and ND1, ND3, and ND6 (membrane arm subunits) switch to create a cavity in which quinone enters and binds. Ubiquinone binding to the cavity, and its subsequent reduction to ubiquinol, initiates a cascade of conformational changes that allow ubiquinol movement and proton pumping as the peripheral arm transits into the closed state. When the reduced ubiquinol diffuses to the shallow entrance en route to the bulk inner membrane bilayer, the peripheral arm tilts back to the open. In both the open and closed states, PC and PE surround the site at the interface of the peripheral and membrane arms [25] (Figure 4.1, Table 4.1). Notably, one PC directly interacts with NDUFS2 and ND1, which are themselves associated with NDUFS8 and NDUFA3. Similar interactions are observed in other reported Complex I structures (porcine [20], mouse [22], and human [21]). When comparing the open and closed states, several differences in phospholipid interactions with the associated subunits

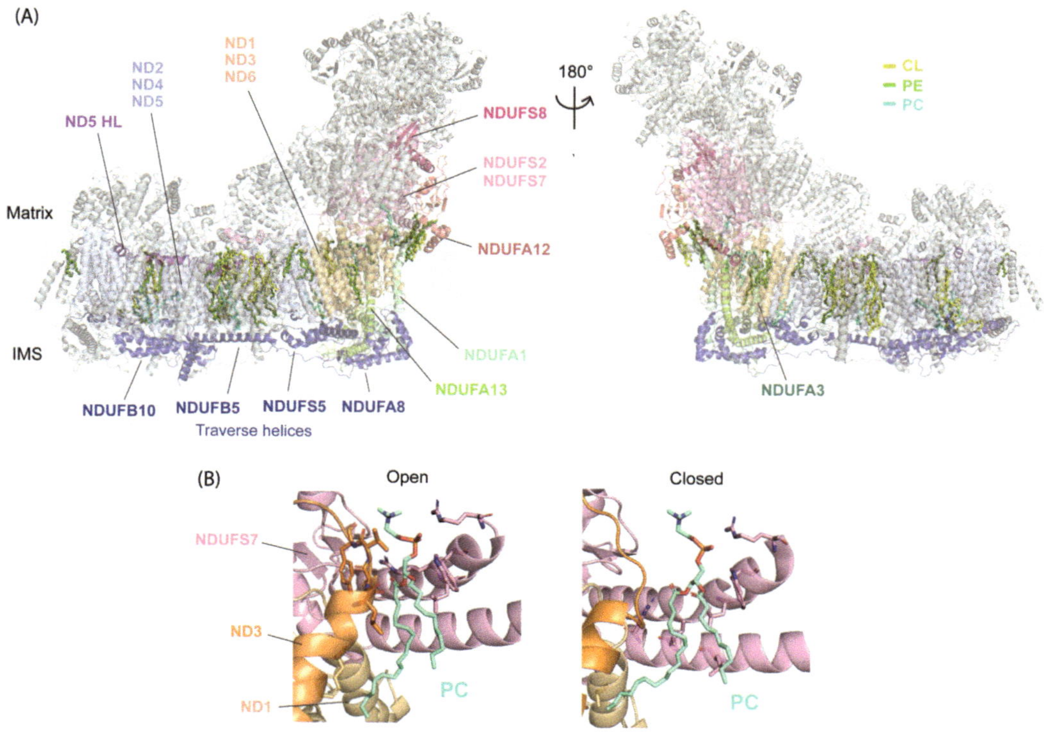

FIGURE 4.1 (A) Phospholipids associated with the open state Complex I of ovine (PDB: 6ZKE). Subunits related to catalytic function and assembly mechanism are highlighted as indicated. (B) Enlarged view of PC interaction with ND1, ND3, and NDUFS7 in open (PDB: 6ZKE) and closed (PDB: 6ZKC) states. Residues directly associated with PC are shown as sticks.

are observed (Table 4.1). For example, on the matrix side, PC interacts directly with NDUFS7 and ND3 via its headgroup-acyl chains in the open state (Figure 4.1B, left). Upon the transition to the closed state, this PC maintains direct interactions with NDUFS7 but is slightly distant from ND3 (Figure 4.1B, right). One end of an acyl chain of this PC is associated with ND1 in both states (Figure 4.1B). These open-closed cycle-associated patterns of phospholipid interactions appear to correspond to dynamic structural transitions. The putative roles of the phospholipids at the interface of the peripheral and matrix arms would likely provide flexibility around the ubiquinone cavity to enable the dynamic turnover cycle.

The membrane arm of Complex I consists of three core antiporter-like subunits (ND2, ND4, ND5) that perform proton pumping. The orientation of the three subunits is well coordinated, supported by the ND5 lateral helix (HL) on the matrix side and the traversing helices of four accessary subunits (NDUFB10, NDUFB5, NDUFS5, and NDUFA8) on the intermembrane space (IMS) side. A number of phospholipids are associated with both the matrix and IMS sides of the membrane arm of Complex I as if to complement the grooves between the three antiporter-like subunits (ovine [19, 25], porcine [20], mouse [22], and human [21]). Several matrix-side phospholipids interact with the ND5 HL. IMS side phospholipids have interactions with the traversing helices of NDUFA8 and NDUFB5. A possible role of these membrane arm phospholipids would be to provide flexibility for optimal movement of the transmembrane (TM) helices during proton pumping. Although it has been indicated that the membrane arm is relatively static during the catalytic cycle [25], several phospholipids alter their lipid-protein interaction patterns as Complex I transitions between the open and closed states [25]. Alternatively, these phospholipids may reinforce the orientation of the membrane arm by facilitating protein-protein interactions between the core subunits and surrounding accessory subunits.

TABLE 4.1
Phospholipids Located at the Interface of the Peripheral and Membrane Arms of Complex I

		Interacted Subunits			
		Open State (PDB: 6ZKE)		Closed State (PDB: 6ZKC)	
Phospholipid	**Side**	**Peripheral Arm**	**Membrane Arm**	**Peripheral Arm**	**Membrane Arm**
PE	matrix	NDUFS7 NDUFA12	ND1	NDUFS7 NDUFA12	ND1
PE*	matrix	NDUFA12	ND1	NDUFA12	ND1 NDUFA1
PE	matrix	NDUFA3	ND1 ND3	NDUFA3	ND1 ND3
PC	matrix	NDUFS2 NDUFS8 NDUFA3	ND1	NDUFS2 NDUFS8 NDUFA3	ND1
PC*	matrix	NDUFS7	ND1 ND3	NDUFS7	ND1
CL*	matrix	NDUFA7 NDUFA12	ND1 NDUFA1	NDUFA7 NDUFA12	NDUFA1
PE*	IMS		ND1 ND3 ND6 NDUFA13		ND1 ND3 ND6 NDUFA1 NDUFA13
PC*	IMS		ND3 ND6		

* Phospholipids altering their subunit interactions along with the open/closed transition.

These striking observations indicate that Complex I-associated phospholipids are intimately involved in its catalytic cycle, as illustrated in References [10, 26–30]. Previous studies have indicated that CL is more tightly bound to Complex I than PC or PE [31, 32]. A biochemical study showed that CLs are tightly and constantly bound to Complex I; in contrast, PC and PE binding events are activity-dependent and less tight [31]. Molecular dynamics (MD) simulations also showed that CL strongly interacts with the membrane arm, whereas PC and PE interact transiently [32]. A simple consideration of the functionality of these Complex I-associated phospholipids would be that PC and PE are able to behave in a flexible manner in response to conformational changes of the neighboring subunits, whereas CL provides a stable structural platform. As mentioned earlier, given the locations and associated subunits to which each lipid binds, the flexibility of PC and PE would be better suited to support the movement of the peripheral arm during the open-closed cycle, whereas CL would be expected to tightly integrate the membrane arm. However, the numerous phospholipid interactions of this gigantic protein complex make it difficult to fully understand their individual and collective functionalities. Based on the high-resolution structures of ovine Complex I in different states and the deep mechanistic insight they provided [25], the stage is set to interrogate the specific roles of each phospholipid and their integrated functionality in the coming years.

Assembly Mechanism

CL is found at the interface of the peripheral and membrane arms in ovine [19, 25] and mouse [22] structures. This CL tightly interacts with the N-terminal helix of NDUFA12 via its headgroup-acyl chains and with NDUFA1 via its acyl chains, and contacts NDUFA7 and ND1. This interaction likely pulls NDUFA12 inward to the ND1 side edge. NDUFA12 is part of the peripheral arm and docks at a late stage of Complex I assembly [33, 34]. Several studies suggest a potential role for NDUFA12 in Complex I assembly [35–37]. NDUFA1 is an accessory subunit of the membrane arm but is also assembled at a late stage [33, 34]. Potentially, CL in this position may have a role in completing Complex I assembly, although the precise function of CL interactions with NDUFA12 and NDUFA1 requires detailed investigation.

A recent cryo-EM study of yeast *Y. lipolytica* Complex I reported Tafazzin as a novel assembly factor that functions in its early assembly [24]. Tafazzin is a transacylase that catalyzes acylation of monolyso-CL (MLCL), using PC and/or PE as acyl chain donors [38, 39], to synthesize fully acylated CL. In an early Complex I assembly step, Tafazzin is associated with the matrix-side TM helices of ND2 and NDUFC2. MLCL and PC are tightly associated with these TM helices via their headgroup-acyl chains in a position proximal to Tafazzin. At a later assembly stage, Tafazzin is dissociated from Complex I, and MLCL and PC are missing; instead, multiple CLs are found. This suggests that Tafazzin-mediated CL synthesis is required for early-stage Complex I assembly and facilitates its subsequent maturation. This putative biogenic role is consistent with the repeated observation that Complex I abundance and subunit expression are reduced in mammalian cells and tissues that have defective Tafazzin or lack Tafazzin [40–43] or other upstream CL metabolism enzymes [30, 44]. However, there are still several things to consider, particularly in the context of interspecies conservation. For example, although the main subunits ND2 and NDUFC2 tied to this putative Tafazzin function are conserved between yeast *Y. lipolytica* and mammals, subunits associated with ND2 and NDUFC2 are not fully conserved, such as NUXM. The positioning of CLs in the fully assembled Complex I is also slightly different. The reduction of Complex I subunit expression has also been seen in PE-deficient models [45, 46]. Detailed investigations are needed to determine whether the role of Tafazzin in Complex I assembly is conserved in mammals, whether Tafazzin activity is required for this Complex I assembly-related function, and whether enzymes for other lipids can act in a similar manner.

COMPLEX II-ASSOCIATED PHOSPHOLIPIDS

Complex II mediates the oxidation of succinate to fumarate and the reduction of ubiquinone to ubiquinol; the former is shared with the TCA cycle. Complex II consists of four subunits. SDHA and SDHB are assembled in the matrix and connected to the membrane by virtue of their association with the membrane-embedded subunits, SDHC and SDHD. Three eukaryotic structures of Complex II have been reported to date, including X-ray crystallographic structures of porcine [47] and avian [48], and a cryo-EM structure of human [49] (Figure 4.2); all feature several resolved phospholipids, indicating the significance of lipids for Complex II integrity. Consistently, in conditions of reduced CL [41, 42, 44, 50, 51] or PE [27, 45, 46], the expression of Complex II subunits, as well as the abundance and activity of Complex II, are reduced.

Stability of the Membrane Domain

In most Complex II structures, a density for a lipid molecule is observed at the interface of the membrane subunits, SDHC and SDHD. This lipid is assigned as PE in porcine [47] and human [49], whereas CL occupies this location in *Escherichia coli* [52, 53]. Regardless of identity, this lipid is located on the IMS side, directly below the heme b, and associated with the TM helices of SDHC and SDHD. The difference in lipid type may be due to the degree of conservation of the associated helices of SDHC and SDHD [47, 49, 54]. The orientation of the lipid-associated helices in porcine and human structures appears to allow them to fit PE, whereas their orientation in *E. coli* structures is more spacious, allowing them to accommodate the larger dimeric lipid, CL. MD simulations of the human SDHC-SDHD membrane domain demonstrated that the binding of PE regulates the conformational stability of the Complex II membrane domain [49]. The presence

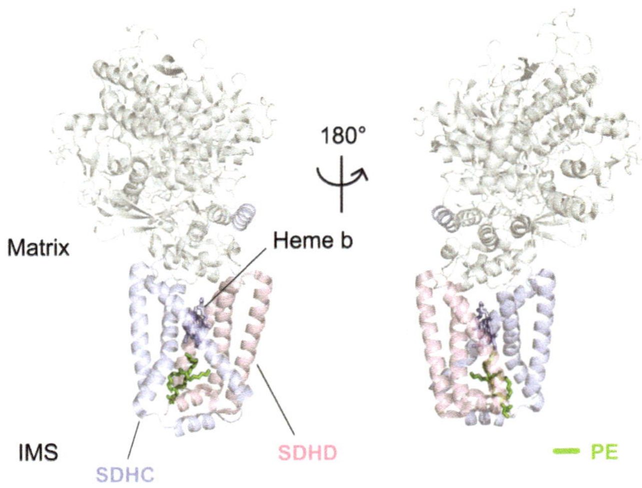

FIGURE 4.2 Phospholipids associated with Complex II of human (PDB: 8GS8). SDHC and SDHD are highlighted as indicated.

of inhibitors 3-nitropropionate (NPA) or 2-thenoyltrifluoroacetone (TTFA) caused the porcine Complex II structure to lose the membrane domain PE [47]. NPA and TTFA inhibit succinate oxidation and ubiquinone reduction, respectively, which suggests that this PE is also involved in the catalytic actions of Complex II.

Assembly of Matrix and Membrane Side Subunits

The porcine Complex II contains PE outside of the SDHD subunit on the matrix side membrane boundary [47]. This outside PE is also missing in the presence of the inhibitors NPA and TTFA [47], indicating that it too is involved in the catalytic actions of Complex II. Another possible function of this PE may be to help integrate the assembly of the matrix subunits with the membrane subunits. It has been shown that a biomimetic lipid composition is required for the stable assembly of the matrix subunits with the membrane subunits and Complex II activity [55]. Using lipid nanodiscs consisting of PC, PE, and CL, this study highlighted the importance of CL in Complex II structure and function, rather than PE. A recent study has demonstrated that macrophage CL promotes the disassembly of Complex II during inflammatory metabolic remodeling [56]. During this remodeling, SDHA dissociates from the other subunits which are subsequently sequestered in vesicles and released from mitochondria [56]. However, the underlying mechanism for this CL-dependent metabolic regulation is unclear as none of the reported Complex II structures contain CL at the matrix side [47–49, 53, 54]. The specific role of each lipid in the structure and function of Complex II still requires experimental validation ideally in the context of distinct cell models.

COMPLEX III-ASSOCIATED PHOSPHOLIPIDS

Complex III oxidizes the ubiquinol received from Complexes I and II while simultaneously reducing the electron carrier, cytochrome *c*. These catalytic activities are coupled with proton translocation through the Q-cycle. Complex III consists of ten subunits, three of which form core modules: cytochrome *b*, cytochrome c_1, and Rip1/UQCRFS1 (yeast/mammals; when subunit names differ, the yeast and mammalian names will both be presented and separated by /). Cytochrome *b* is a mitochondrial DNA-encoded subunit. The number and assembly of the subunits are relatively conserved across species. Phospholipid interactions with Complex III have been reported in several distinct contexts, including as a monomer [57], or homodimer alone [21, 23, 58] or in the context of supercomplexes (III_2-IV_1 in yeast [59–61], I_1-III_2-IV_1 in porcine [20]). Representative phospholipids associated with Complex III are shown in Figure 4.3. The patterns of lipid interactions suggest that phospholipids may be involved in the stepwise assembly of Complex III, including its catalytically essential dimerization [62, 63]. In the Complex III homodimer in yeast supercomplex III_2-IV_1, there are slight differences in phospholipid associations between the III-IV interaction side versus the inner membrane-exposed side [59] (Figure 4.3A).

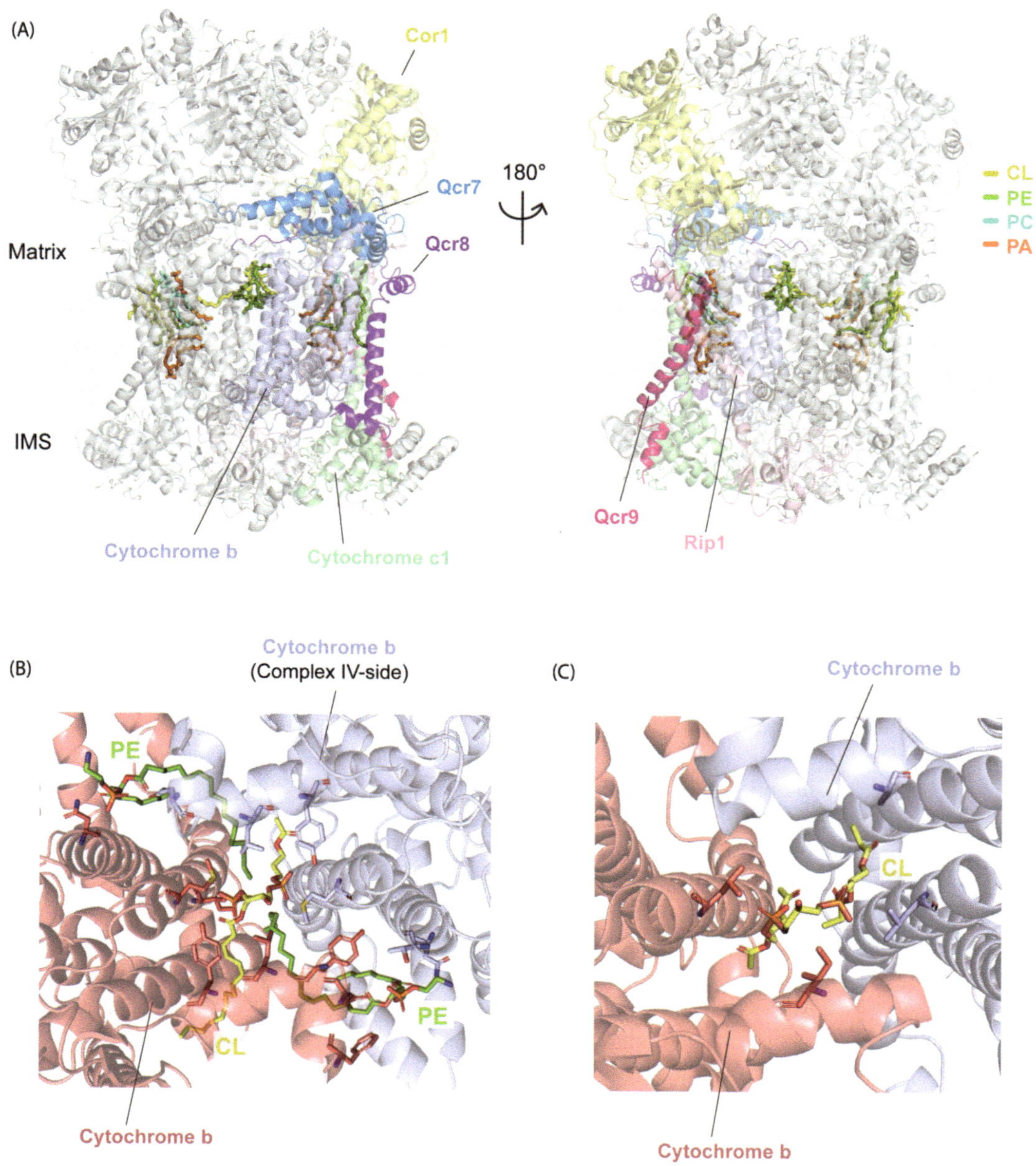

FIGURE 4.3 Phospholipid interactions with Complex III of yeast (A and B, PDB: 6YMX) and ovine (C, PDB: 6Q9E). (A) Side view of yeast Complex III. Subunits involved in the assembly mechanism in one monomer (Complex IV-side) are highlighted as indicated. Enlarged matrix view of yeast (B) and ovine (C) Complex III. Each monomer of dimeric cytochrome *b* is highlighted in blue or red. Residues directly associated with indicated phospholipids are shown as sticks.

Early Assembly

In an early step of Complex III assembly, two accessary subunits, Qcr7/UQCRB and Qcr8/UQCRQ, are attached to the core membrane-inserted subunit cytochrome *b*, which stabilizes the cytochrome *b* module. Qcr7/UQCRB is tangential to cytochrome *b* at the membrane boundary on the matrix side. Qcr8/UQCRQ has a TM helix that runs parallel to the peripheral TM region of cytochrome *b*. It bends into an L-shape in the proximity of Qcr7/UQCRB on the matrix side. Proximal to Qcr7/UQCRB and Qcr8/UQCRQ are several phospholipids that are mainly associated with the bent Qcr8/UQCRQ helix and cytochrome *b* TM helices (Figure 4.3A). The number of phospholipids varies slightly from structure to structure (yeast [59–61], porcine [20], ovine [23], human [21]). In the yeast supercomplex-associated Complex III dimer [59], the Complex IV-facing monomer contains PE at this location, which only interacts with cytochrome *b*. In the non-Complex IV-associated monomer, PE and CL occupy the equivalent position. Here, PE interacts with cytochrome *b* and Qcr8, and CL interacts with cytochrome *b*, Qcr7, Qcr8, and the neighboring cytochrome c_1 [59]. Of note, the functionality of these Qcr8/UQCRQ-aligned lipids has been tested [57, 64]. A yeast cytochrome c_1 mutant, in which three lysine residues associated with CL at this site were mutated to disrupt this lipid interaction, has reduced Complex III activity [57]. Similarly, when a glutamate residue of Qcr7 is mutated to dissociate PE from this site, Complex III activity is impaired while its assembly is unaffected [64]. These studies highlight the functional importance of individual Complex III-associated phospholipids. The other Complex III cryo-EM structures have similar phospholipid interactions aligned over Qcr8/UQCRQ but with slightly different patterns that involve PC-CL-CL (yeast [60]) and PE-CL-CL (porcine [20] and ovine [23]). Potentially, the membrane-embedded cytochrome *b* uses the surrounding phospholipids to recruit accessory subunits to complete the initial assembly of this module.

Middle Assembly

After the cytochrome *b* module is formed, Cor1/UQCRC1 and Cor2/UQCRC2 are attached to the matrix side of the cytochrome *b* module. Simultaneously, cytochrome c_1 and Qcr6/UQCRH stabilize the IMS side of the module. The attachment of these subunits promotes the formation of the premature Complex III homodimer [34, 62]. Consequently, on the matrix-side interface of the homodimer, cytochrome *b* in one monomer forms direct interactions with cytochrome *b* in the other monomer. There are two types of phospholipid interactions related to matrix-side cytochrome *b* dimerization which appear to reflect an interspecies difference. In yeast structures, CL is located in the center of the two cytochrome *b* molecules [59, 61], and PE symmetrically interacts with each side of cytochrome *b* via its acyl chains [59–61] (Figure 4.3B). The acyl chains of the two PEs also approach the sandwiched CL. It appears that the PE-CL-PE interaction pulls each monomer-associated cytochrome *b* inward to facilitate homodimer formation. Interestingly, the sandwiched CL is missing in Complex III isolated from *crd1*Δ (CL-depleted) yeast, whereas the two PEs remain associated [65]. In contrast, mammalian Complex III structures include only one CL sandwiched between the cytochrome *b* dimers (ovine [23] and human [21]) (Figure 4.3C). Each headgroup-upper acyl chain pair of this CL symmetrically interacts with the helices of the two cytochrome *b* subunits facing the dimer boundary surface.

In addition, several phospholipids are found at the matrix-side boundary where Cor1/UQCRC1 is attached to the cytochrome *b* module (Figure 4.3A). Although the number and type of lipids vary to some degree, two similar interaction patterns are observed in both yeast and mammalian Complex III structures. PA (yeast [59, 61]) or PE and CL (yeast [60] and porcine [20]) align along the TM helices of cytochrome *b* and are associated with cytochrome c_1 and Cor1/UQCRC1. This suggests that the docking of cytochrome c_1 and Cor1/UQCRC1, which occurs upon dimer formation, involves these lipid interactions. Another phospholipid nearby is a conserved PC (yeast [59–61] and porcine [20]), which is located at the boundary of Cor1/UQCRC1 in the matrix and cytochrome *b* within the membrane and associated with the Rip1 module subunits Rip1/UQCRFS1 and Qcr9/UQCR10. This PC binds to Cor1/UQCRC1, Rip1/UQCRFS1, and Qcr9/UQCR10 via its headgroup. Given the associated assembly step of Complex III, it is likely that this protein-lipid-protein interaction has a dual role: first, to help recruit Cor1/UQCRC1 to the membrane; second, to promote the docking of the Rip1 module at this position.

Later Assembly

Ultimately, Rip1/UQCRFS1 is inserted into the premature Complex III dimer and firmly assembled with the other core subunits, cytochrome *b* and cytochrome c_1, on the IMS side [34, 62]. This assembly step is likely mediated by a specific phospholipid (Figure 4.3A). Most reported structures contain PA (yeast [59, 61] or PE (yeast [60], porcine [20], and ovine [23]) at this site. Each monomer in the yeast supercomplex-associated Complex III dimer has a different pattern of interactions [59]. PA in the Complex IV-associated monomer interacts with cytochrome c_1 and Rip1, whereas in the non-Complex IV-associated monomer, PA interacts with cytochrome *b* and Rip1. In the other structures, PA or PE has a direct interaction with all three subunits [20, 23, 60]. In any case, the headgroup and acyl chains of PA or PE are tightly associated with the TM helix of Rip1/UQCRFS1, as if coiling around it. The headgroup of this lipid (PA or PE) is near cytochrome c_1. The acyl chains are also associated with cytochrome c_1 and/or cytochrome *b*. A tentative role of this phospholipid may be to stabilize the insertion of Rip1/UQCRFS1 into the premature Complex III dimer.

Upon the insertion of the Rip1 module into the premature Complex III homodimer, Rip1/UQCRFS1 and its accessory subunit Qcr10/UQCR11 form a direct interaction with cytochrome *b* present in the other dimer-associated Complex III monomer. In mammalian Complex III homodimers, CL (human [21]) or PE (ovine [23]) is found at the interface of these subunits. In the human Complex III homodimer [21], the cytochrome *b* interaction is mediated via an acyl chain of CL, while its interaction with the Rip1 module involves a head group interaction with UQCRFS1 and acyl chain interactions with UQCR11. PE at this position in the ovine Complex III homodimer interacts similarly but also engages cytochrome c_1 [23]. This CL- or PE-mediated cytochrome *b*-UQCRFS1 interaction may reinforce the stability and functionality of the Complex III homodimer. Dimerization is a critical step in the function of Complex III, which couples the oxidation of ubiquinol to the reduction of cytochrome *c*. At the dimer interface, Rip1/UQCRFS1 receives electrons from ubiquinone bound to cytochrome *b* within one monomer, and transfers the electrons to cytochrome c_1 in the other monomer [62, 63]. Together with the aforementioned phospholipids that associate with dimerized cytochrome *b*, these phospholipids at the dimer interface may have a critical role in stabilizing dimerization which is necessary for efficient electron transfer and the Complex III catalytic cycle.

The interpretations of the potential function of Complex III-associated phospholipids drawn from the above structural insights are consistent with experimental results. De-lipidation of purified Complex III from bovine heart or yeast diminishes Complex III activity, which is restored by reconstitution with CL [10, 66]. Similarly, Complex III activity is decreased in the absence of CL in yeast [67]. Furthermore, Complex III activity is dependent on PE, especially PE generated in the inner mitochondrial membrane [64].

COMPLEX IV-ASSOCIATED PHOSPHOLIPIDS

Complex IV receives electrons from cytochrome *c* to reduce oxygen to water, a process that is coupled to the translocation of protons into the IMS. Membrane integral core subunits of Complex IV are encoded by mitochondrial DNA, including Cox1/COX1, Cox2/COX2, and Cox3/COX3, and form assembly intermediates called the MT-CO1, MT-CO2, and MT-CO3 modules. Initial Complex IV assembly steps include the formation of MT-CO1 and MT-CO2 modules [34, 62]. After MT-CO1 and MT-CO2 modules are docked, the MT-CO3 module is added to them to form mature Complex IV. The number of subunits and pattern of assembly remain controversial. In particular, there is a notable difference in mammalian Complex IV structures between previous X-ray crystallographic [68–82] and recent cryo-EM studies [83]. Accordingly, the pattern of associated phospholipids with Complex IV differs depending on the reported structure [59, 60, 83–85]. This section focuses primarily on the phospholipids identified in recent cryo-EM structures (Figure 4.4) [83] and then compares the bound phospholipids with the reported X-ray structures [68–82, 85].

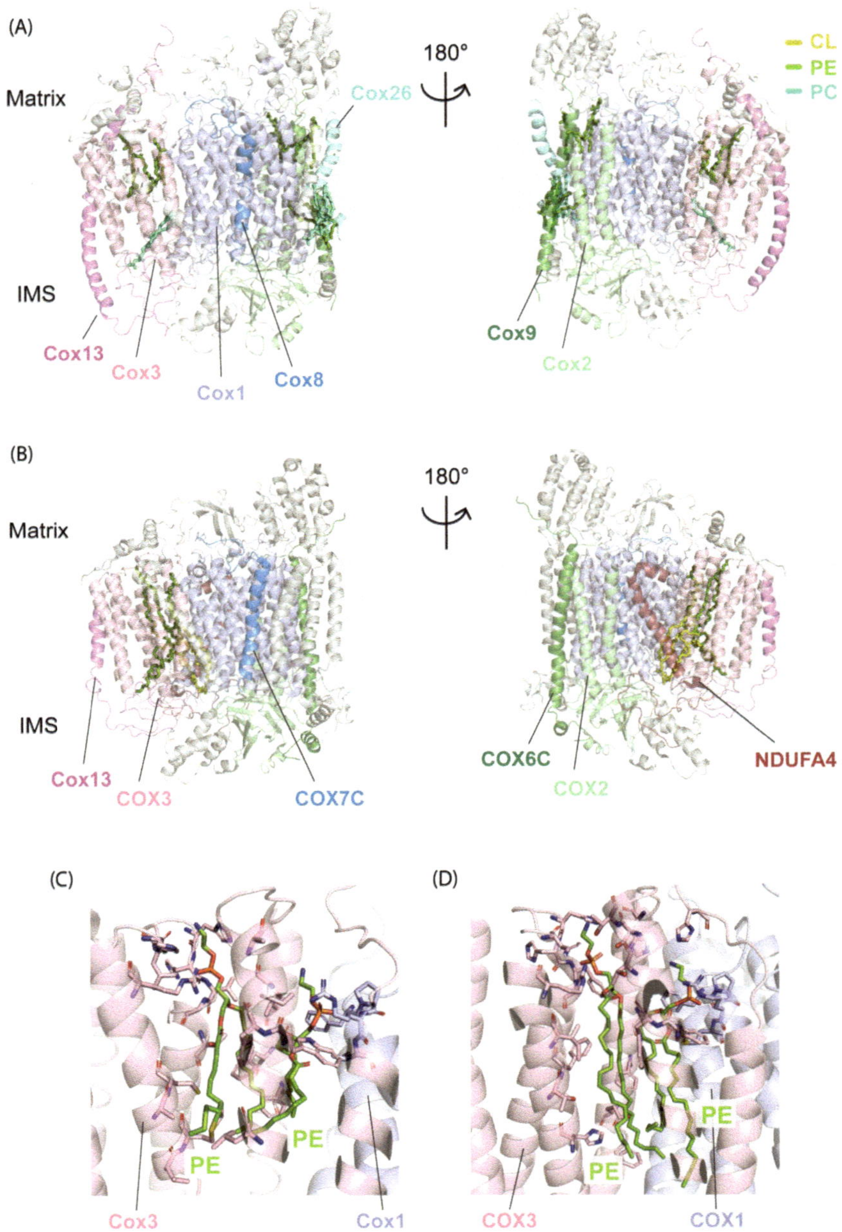

FIGURE 4.4 (A, B) Phospholipids associated with Complex IV of yeast (A, PDB: 6YMX) and human (B, PDB: 5Z62). Core subunits and NDUFA4 are highlighted as indicated. (C, D) Enlarged view of PE interactions with Cox1/COX1 and Cox3/COX3 of yeast (C) and human (D). Residues directly associated with PEs are shown as sticks.

MT-CO1 and MT-CO2 Modules

Phospholipids in Complex IV are frequently found at the interfaces of the TM core modules. The interface of the MT-CO1 and MT-CO2 modules consistently contains several phospholipids in the reported Complex IV monomer structures (yeast [59, 60] and bovine [85]). A phospholipid, PE or CL, interacts with Cox1/COX1 (MT-CO1 module core), Cox2/COX2 (MT-CO2 module core), and the MT-CO2 module accessory subunit Cox9/COX6C at the IMS side. This phospholipid is assigned as PE in yeast cryo-EM Complex IV monomers [59, 60] and as CL in the X-ray structure of the bovine Complex IV monomer [85]. Acyl chains of PE or CL interact with the TM helices of Cox1/COX1, Cox2/COX2, and Cox9/COX6C. Similarly, the matrix side in yeast Complex IV monomers includes Cox1-Cox2-Cox9 interactions which involve a PE with a tight interaction with Cox9 [59, 60].

In the vicinity of Cox1, Cox2, and Cox9, another subunit, Cox26, also associates with lipids. Cox26 is a subunit that was only recently assigned to yeast Complex IV [60, 86, 87]. It is located near the interface of Complexes III-IV. The Cox26 TM helix interacts with Cox1 and Cox2. Yeast IMS-side PEs that interact with Cox1, Cox2, and Cox9 also engage Cox26 [59, 60] These phospholipids appear to provide a link for the assembly of Cox26 to this IMS-side MTCO1-MTCO2 interface and solidify the association of Cox26 with the Complex IV core modules.

The interface of MT-CO1 and MT-CO2 involves an additional lipid interaction through Cox1/COX1-Cox8/COX7C on the matrix side (yeast [60, 84] and bovine [85]) (this lipid was not identified in the yeast structure shown in Figure 4.4A [59]). This lipid, assigned as PE or CL, directly interacts with Cox1/COX1 and Cox8/COX7C. The interaction with Cox8/COX7C is tight in that it involves both headgroup and acyl chain contacts. Cox8/COX7C is an accessory subunit that binds to the MT-CO2 module during an early assembly step [34, 62]. Successful docking of the assembled MT-CO1 and MT-CO2 modules may be facilitated by this bound lipid.

MT-CO3 Module

The matrix side of Cox3/COX3 TM helices is associated with two PEs in both yeast [59, 60, 84] and mammalian (human [83]) cryo-EM Complex IV monomer structures. One PE is tightly associated with the folded TM helices of Cox3/COX3 (Figure 4.4C, D). The second PE exists nearby at the interface of Cox1/COX1 and Cox3/COX3; it is associated with Cox1/COX1 via its headgroup and with Cox3/COX3 via its headgroup-acyl chains (Figure 4.4C, D). Given the Complex IV assembly order in which the MT-CO3 module is the last to be attached to the docked MT-CO1-MT-CO2 modules [34, 62], it is likely that these conserved, bound PEs help to anchor the MT-CO1 module core subunit Cox1/COX1 in this position and then facilitate the insertion of the MT-CO3 module core subunit Cox3/COX3 at this site.

In addition, the IMS-side interface of Cox1/COX1 and Cox3/COX3, to which an additional subunit Cox13/COX6A is bound, includes PC (yeast [59]) or PE (yeast [60] and human [21]). This PC or PE is tightly associated with Cox3/COX3 and Cox13/COX6A via headgroup-upper acyl chain contacts. Cox13/COX6A is the most peripheral subunit of Complex IV. While the length of Cox13/COX6A is different between yeast and mammals [60], the IMS-side Cox13/COX6A interaction with the core modules is potentially enhanced by the inclusion of a conserved phospholipid. In mammals, COX6A is also implicated as involved in the NDUFA4-related dimer-monomer transition (discussed below).

The catalytic cycle of Complex IV includes two reactions: electron transfer from cytochrome *c* to the heme cofactors and the reduction of oxygen to water coupled to the proton pumping. The core of the catalytic site is at the center of Cox1/COX1, which includes the important heme and copper cofactors. While there are no phospholipids that are directly associated with the center of Cox1/COX1, the aforementioned phospholipids preferentially surround the peripheral edge of Cox1/COX1 TM helices. The potential role of these lipids is to stabilize the catalytic core by integrating its associated subunits, which may explain the repeated observation that the absence/reduction of CL [50, 88–90] or PE [27, 64, 91] diminishes Complex IV activity in a wide range of species.

NDUFA4 and CL

Biochemical assays indicate that Complex IV, either as a monomer or supercomplex-associated, also includes NDUFA4 [83, 92]. Yet structurally, NDUFA4 has only been identified in one cryo-EM Complex IV monomer structure [83] and not in any reported cryo-EM supercomplex structures. The N- and C-termini of NDUFA4 reach the matrix side and IMS side, respectively. The middle region is a TM helix, which runs parallel to COX1 TM helices and is adjacent to COX2. On the IMS side, CL is inserted into a deep pocket formed by TM helices from COX1, COX3, and NDUFA4 (Figure 4.4B). This CL likely stabilizes NDUFA4 at this position. In contrast, most of the previous Complex IV X-ray structures are reported as dimers and all lack NDUFA4 and CL [68–82]. Interestingly, in these dimeric X-ray structures, the position occupied by CL and NDUFA4 is instead filled by a lipid assigned as phosphatidylglycerol (PG) and cholic acid as described in Ref. [83]. Cholic acid is a strong detergent widely used to prepare crystals, and perhaps it may have displaced CL in the dimeric X-ray structures. Considering that NDUFA4 and CL in this position are only observed in the monomer, the lack of the NDUFA4-CL interaction may cause NDUFA4 to dissociate from the Complex IV monomer, resulting in dimer formation. This possibility is supported by molecular dynamics (MD) simulation results that indicate that Complex IV does not self-assemble in the presence of CL but rather prefers to form supercomplexes with Complex III [93]. Collectively, it is suggested that the previously reported Complex IV dimers are a potential artifact stemming from the use of cholic acid for crystallization. Notably, the recent X-ray crystallographic study reporting the bovine Complex IV monomer used a newly established detergent, 3-oxatridecyl-α-D-mannoside (3OM) [85]. This X-ray resolved Complex IV monomer includes three associated CLs. Two of the CLs occupy the same positions as described above and engage Cox1/COX1-Cox8/COX7C on the IMS side and Cox1/COX1-Cox9/COX6C on the matrix side. The third CL is located at the predicted interface between Complex I and Complex IV. However, none of the CLs are close to the NDUFA4 assembly position. Clearly, details with respect to the association of NDUFA4 with Complex IV and the involvement of CL are needed. However, these recent studies highlight that there is still much to learn about this classic respiratory complex of Biochemistry textbooks.

SUPERCOMPLEX-ASSOCIATED PHOSPHOLIPIDS

Individual respiratory complexes can form higher order assemblies, termed supercomplexes. Yeast supercomplexes are relatively simple and only include III_2-IV_2 and III_2-IV_1 [59–61, 65]. In contrast, mammalian supercomplexes show diversity [94]; cryo-EM structures of I_1-III_2-IV_1 [95–99], I_1-III_2 [23], III_2-$IV_{1\text{-}2}$ [33, 100, 101], and I_2-III_2-IV_2 [21] have been reported. The functional roles of supercomplexes have been thoroughly discussed but a consensus has yet to be reached [62, 102]. Currently proposed functional benefits of supercomplexes include modulating reactive oxygen species (ROS) production [103–108], substrate channeling [100, 109, 110], facilitating electron transfer [33, 59, 111, 112], promoting membrane bending [113, 114], and maintaining inner membrane proteostasis [115]. Reported cryo-EM structures have identified phospholipids occupying the complex-complex interfaces within supercomplexes, which complements evidence generated over the past two decades that phospholipids are required for supercomplex assembly across species [27, 40, 45, 50, 51, 67, 90, 116–131].

Yeast III_2-IV_2 and III_2-IV_1

In the yeast III_2-IV_2 and III_2-IV_1 supercomplexes [59, 60, 65], subunit interactions at the boundary interface are similar. On the matrix side, Complex III subunits Qcr8 and Rip1 face the Complex IV subunit Cox5A. Here, CL is surrounded by Qcr8 (III), Rip1 (III), and Cox5A (IV) and forms direct lipid-protein interactions with each of these subunits [60, 65]. Nearby, another CL bound to Cox5A (IV) shares interactions with Rip1 (III) and Cox1 (IV) in one structure [65]. The IMS-side interface involves Qcr9 (III) and Cox5A (IV); in this position, PC is consistently observed, although there are slight structure-dependent differences in the distance of the PC from Qcr9 and Cox5A and the involvement of other phospholipids [59, 60, 65].

It is thought that the formation of III_2-IV_2 and III_2-IV_1 shortens the distance between Complex III and Complex IV and promotes efficient electron transfer by limiting cytochrome *c* diffusion [59]. The potential roles of the phospholipids located between Complexes III and IV interfaces, primarily CL and PC, are to mediate protein-protein interactions of interacting subunits and to strengthen the assembly of the two complexes, which would enhance electron transfer. A recent study has further highlighted the significance of CL in supercomplex formation and stability by structurally comparing supercomplexes isolated from wild type versus CL-lacking (*crd1*Δ) yeast [65]. In the presence of CL (wild type), the supercomplex is III_2-IV_2, whereas in the absence of CL (*crd1*Δ), it is III_2-IV_1. In the *crd1*Δ strain, CL is absent, and its precursor lipid PG accumulates [132]. The structure of the *crd1*Δ III_2-IV_1 supercomplex contains PG at identical positions occupied by CL in the wild type III_2-IV_2 supercomplex, which includes the interfaces of Qcr8-Rip1-Cox5A and Rip1-Cox5A-Cox1. However, the associated PG-protein and protein-protein interactions of the *crd1*Δ III_2-IV_1 supercomplexes are weakened, suggesting that PG can partially substitute for CL with respect to supercomplex assembly but the resulting supercomplex is less stable [65].

The stability of yeast supercomplexes is strongly perturbed by CL deficiency [67, 127, 129, 131]; in contrast, PE has been reported to have little effect [91]. These studies collectively indicate the significance of CL over PE in yeast supercomplex assembly, which is consistent with the observation that the complex-complex interfaces in the cryo-EM supercomplex structures preferentially involve CL versus PE. In addition, site-directed mutagenesis of cytochrome c_1 to disrupt CL binding (mentioned in the Complex III section) not only destabilizes Complex III [57] but also perturbs supercomplex assembly [66]. The proportion of supercomplexes in this yeast mutant shifts from III_2-IV_2 to III_2-IV_1 with dissociated III_2 and IV_1 detected; this shift is relatively small compared to CL-lacking yeast [66]. The targeted CL is not directly involved in protein-protein interactions or protein-lipid-protein interactions between Complex III and Complex IV, although the binding of this CL to Complex III is tight [20, 23, 59, 60]. Of interest, CL-binding at this position is conserved. Since the assembly of Complex III can occur in parallel with the formation of supercomplexes, as proposed in [33], the requirement of CL in Complex III assembly likely has another role to ensure the stability of supercomplexes.

Mammal I_1-III_2-IV_1 and I_1-III_2-IV_2

Mammalian Complex I is extensively assembled within supercomplexes of distinct compositions – I_1-III_2-IV_1 and I_1-III_2-IV_2 (often termed respirasomes) and the putative I_2-III_2-IV_2 [21, 23, 95–99]. The reported Complex I-including supercomplex structures (I_1-III_2-IV_1) have defined the interface between Complex I and Complex III as mediated by interactions involving NDUFA11 (I) and UQCRQ (III). In two I_1-III_2-IV_1 structures (porcine [20] and human [21]), PC is positioned at this interface on the matrix side. The acyl chains of PC interact with NDUFA11 and the neighboring NDUFB4 on the Complex I side. The headgroup of PC engages UQCRQ on the Complex III side. In another structure of bovine I_1-III_2-IV_1 [133], CL is located between NDUFA11 (I) and UQCRQ (III) at the matrix side. The acyl chains of CL have charge interactions with the TM helices of each subunit. The IMS side of the NDUFA11-UQCRQ position is also occupied by several lipids, including three CLs and one PE.

It is very likely that these phospholipids located at the complex-complex interface stabilize the interaction of Complexes I and III. Unlike yeast, mammalian supercomplexes appear to require a variety of lipids, including PC, PE, and CL [20, 21, 133]. This is generally consistent with previous studies in which supercomplex assembly is destabilized when CL is depleted [40, 50, 116–118,

120–126, 130], the CL acylation pattern is altered [51, 119], or PE is absent [27, 45, 128]. Observed differences in phospholipid interactions from structure to structure may suggest that the roles played by these phospholipids depend on the functions reflected by each structure, which is still hotly debated. These might include the protection against excessive ROS production [103–108] and substrate channeling [100, 109, 110] (but questioned recently [134, 135]). Further investigations, such as whether conformational differences reflect distinct hypothesized functional roles, are needed to fill these knowledge gaps. The reported mammalian Complex I-including supercomplex structures do not include any lipids at the interfaces between Complexes I and IV or Complexes III and IV. It will be of great interest to ascertain whether these complex-complex interactions indeed do not require associated lipids or whether current structural strategies have technical limitations that make it difficult to resolve functionally important but perhaps labile and/or conformation-sensitive, phospholipids.

Mammal III_2-IV_1

Although the overall proportion is lower than Complex I-including supercomplexes (I_1-III_2-IV_1 and I_1-III_2-IV_2), supercomplex III_2-IV_1 is present in mammals [33, 100, 101, 136]. The organization of mammalian III_2-IV_1 is different than the described yeast III_2-IV_1 supercomplex [65]. Characteristically, the mammalian III_2-IV_1 contains a supercomplex assembly factor, SCAF1, also known as COX7A2L, which does not exist in Complex I-including supercomplexes [33, 136].

The III_2-IV_1 supercomplex has two main conformational states: locked (immature) and unlocked (mature) [33]. Transitioning from the locked to unlocked states occurs during the assembly of the III_2-IV_1 supercomplex. In the locked state, Complex IV tightly contacts Complex III, which is mediated by boundary subunit protein-protein interactions, together with the assembly factor SCAF1. In the unlocked state, Complex IV is peripherally attached to Complex III, and SCAF1 changes orientation and moves off the boundary surface. In the locked state, SCAF1 may drive the attachment of Complex IV to the premature Complex III, which lacks the Rip1 module [33, 34, 62]. The insertion of the Rip1 module into the premature complex completes the formation of the supercomplex and allows its transition into the unlocked state [33]. This suggests that supercomplex formation may ensure the stability of individual complexes. In addition, similar to the yeast supercomplexes, the formation of III_2-IV_1 in mammals may enhance electron transfer mediated by cytochrome *c* between the two complexes. The measurement of Complex III and IV activity demonstrated that both activities are higher in the III_2-IV_1 supercomplex compared to a mixture of individual Complex III and Complex IV [33]. The locked state is inactive because Complex III is compacted, which causes cytochrome *c* to clash with UQCRH. The unlocked state retains a high degree of flexibility to allow functional cytochrome *c* movement [33].

Both the locked and unlocked states contain unique phospholipids at the boundary interface [33]. In the locked state, PC is located adjacent to the IMS side protein-protein interaction between UQCRQ (III) and COX7C (IV) near the SCAF1 C-terminus. The matrix side boundary contains three phospholipids: PC is positioned between UQCRQ (III) and SCAF1, PE is tightly associated with UQCRQ (III), and another PE is located between UQCRB (III) and COX7C (IV). In the unlocked state, PC is positioned near the IMS-side UQCR10 (III)-COX6A2 (IV) interactions. On the matrix side, CL sits between UQCRQ (III), UQCRFS1 (III), and COX6A2 (IV). These phospholipids probably orchestrate the formation of III_2 IV_1 and its transition from locked to unlocked states. Indeed, self-assembly MD simulations have shown that supercomplex formation uses CL molecules as molecular glue to sew Complex III and Complex IV together [93]. The idea that III_2-IV_1 supercomplex formation may stabilize each complex and that this is mediated by phospholipids is consistent with determinations that the absence of CL or PE decreases the steady-state expression of Complex III and Complex IV subunits [41, 44–46, 130]. However, this idea might be tissue specific [137] and requires further investigation.

COMPLEX V-ASSOCIATED PHOSPHOLIPIDS

In eukaryotic mitochondria, the F-type ATP synthase/Complex V generates ATP by using the proton gradient generated by complexes I, III, and IV to drive conformational transformations that phosphorylate ADP. Complex V can also work in reverse and use the energy provided by ATP hydrolysis to pump protons into the IMS. Complex V forms dimers and higher order oligomers, which through their membrane-bending capacity help organize the cristae structure [138–140]. Complex V has two main components, F_1 and F_O domains [34, 62]. The F_1 domain is in the matrix, comprising the catalytic head consisting of three α subunits and three β subunits. The F_O domain is membrane-embedded and includes the c-ring and associated membrane subunits. The F_1 and F_O domains are connected through the central stalk and the peripheral stalk. The central stalk (consisting of γ, δ, and ε subunits) holds the αβ catalytic head to stand on the membrane. The peripheral stalk consists of subunits b, OSCP, F6, and d. Subunit b is the core of the peripheral stalk; its N-terminal helices are folded within the membrane and its C-terminus reaches the tip of the αβ catalytic head via OSCP. Recent cryo-EM studies have obtained greater resolutions of the F_O membrane domain that include lipid-protein interactions and provide mechanistic insight into the coupling of proton translocation and ADP phosphorylation [141, 142]. Based on these, this section primarily discusses the functionality of the F_O domain and the potential roles of the associated phospholipids (Figure 4.5).

F_O Domain

The F_O domain contains a wedge-shaped structure, which is formed by the membrane-embedded helices of subunit b and contiguous membrane subunits e, f, g, and A6L. The wedge is a functionally critical unit of the Complex V dimer. As the Complex V monomers rotate in the inner membrane, the wedge provides a pivot that allows the formed dimer to flexibly adapt to the rotation of each

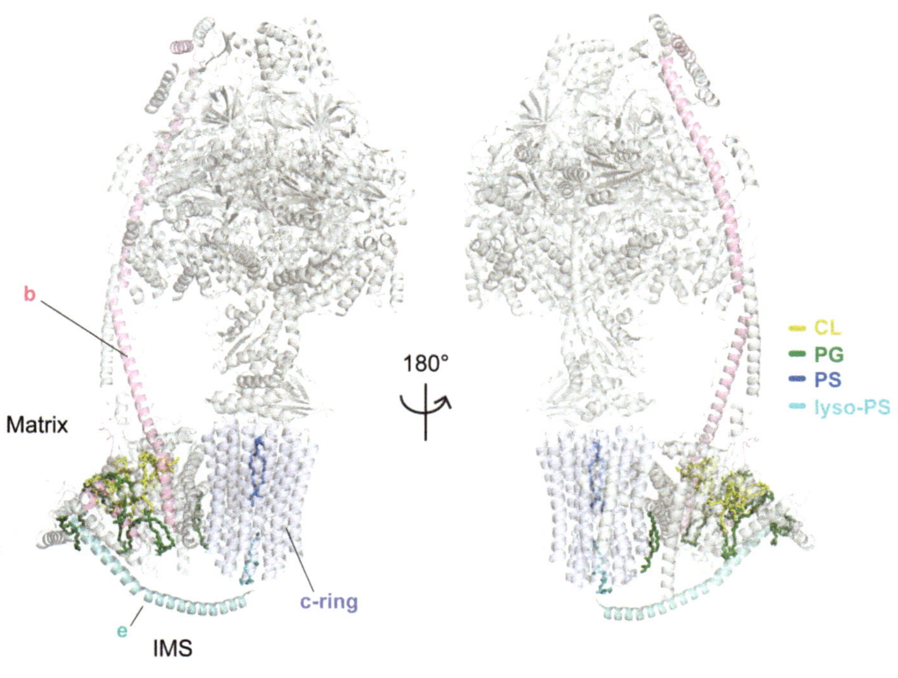

FIGURE 4.5 Phospholipids associated with Complex V of ovine (PDB: 6TT7). Subunit b, c-ring, and subunit e are highlighted as indicated.

monomer [141, 143]. Internal cavities in the wedge are filled by five lipids, including three CLs and two PGs. These lipid interactions at the wedge are consistent in two recently reported mammalian cryo-EM structures (bovine [141] and ovine [142]). These multiple phospholipid interactions surrounding the subunit b-wedge may enhance the stability of the wedge, which likely ensures the functionality of Complex V dimers.

C-Ring

The c-ring is responsible for proton translocation across the membrane. During ATP synthesis, the central stalk transmits the rotation of the c-ring driven by the proton motive force to the F1 domain. In every reported cryo-EM structure of mammalian Complex V, lipid-like densities are observed within the c-ring pore on both matrix and IMS sides. Except for one structure [142], these lipid-like densities are highly noisy, which did not allow precise mapping [141, 144–146]. One study modeled the internal c-ring lipids as phosphatidylserine (PS) at the matrix side and lyso-PS at the IMS side [142]. The matrix side PS fits tightly within the c-ring pore, indicating that it rotates with the ring. In contrast, the IMS side lyso-PS has enough space around it to be independent of c-ring rotation. The IMS side lyso-PS is directly attached to subunit e, which is itself anchored to the c-ring. The lipid-protein interaction occurs via the lipid headgroup and a highly conserved C-terminal lysine residue of subunit e. It was suggested that the IMS lyso-PS acts as a lipid plug within the c-ring. This lipid plug is thought to be involved in the opening of the permeability transition pore (PTP), which triggers cell death [147, 148]. In the steady state, the lipid plug is stable. However, in states that cause the PTP to open, the increased matrix load of Ca^{2+} leads to the movement of subunit e, which in turn unplugs the c-ring. This causes the destabilization of the c-ring and the peripheral stalk assembly, resulting in the opening of the pore. This proposed mechanism of PTP opening may suggest that this specific subunit e-lyso-PS interaction plays a role in mediating cell fate decisions, although its molecular details need to be tested experimentally.

Dimerization

Complex V generally exists as a dimer in mitochondria. The angle of dimerization varies across species, but dimer formation is conserved [138, 140]. In yeast and mammalian mitochondria, the angle of dimerization is approximately 90° [140]. Complex V dimers align in rows and the dimer-rows are placed in tightly curved membrane regions [138]. Dimer-row formation promotes high membrane curvatures found in tubular cristae as well as at the rim of lamellar cristae, which is required to organize cristae [138–140]. The dimerization interface of Complex V of *Euglena gracilis* contains nine bound phospholipids, five of which are CLs, although the angle of dimerization is narrow (45°) compared to yeast and mammals [149]. The presence of many CLs at the dimerization interface suggests that CL is required for dimerization. These structural insights likely underscore the mechanism(s) by which the absence of CL destabilizes Complex V dimer assembly [90, 116, 117, 150] and reduces Complex V abundance [40] and subunit expression [41, 151].

CONCLUSION

This chapter has discussed the functionality of phospholipids in OXPHOS complexes. As described, however, many details remain to be resolved, many of which likely reflect the fact that functional aspects and accompanying structural mechanisms of individual respiratory complexes and supercomplexes are not fully understood, especially in distinct physiological contexts. Moreover, the involvement of phospholipids in the mitochondrial bioenergetic machinery and mitochondrial biology is not limited to their functions in the five discussed OXPHOS complexes. Critical roles of phospholipids have been demonstrated for the mitochondrial protein import machinery [152], mitochondrial contact site and cristae organization system [153], mitochondrial carriers [46, 129,

154–156], and proteases that enforce mitochondrial quality control [157–159]. Future work to widely characterize the precise roles of phospholipids in these macromolecular structures will expand our appreciation of the multitude of hats worn by our membrane's major constituents, phospholipids.

ACKNOWLEDGEMENTS

This work was supported by NIH grants (R01HL108882 and R01HL165729 to S.M.C.), a post-doctoral fellowship from the Uehara Memorial Foundation (to N.S.), and a post-doctoral fellowship from the American Heart Association and the Barth Syndrome Foundation (Award ID: 828058 to N.S.).

REFERENCES

1. W. Basu Ball, J. K. Neff, V. M. Gohil, The role of nonbilayer phospholipids in mitochondrial structure and function. *FEBS Lett.* **592**, 1273–1290 (2018).
2. S. E. Horvath, G. Daum, Lipids of mitochondria. *Prog. Lipid Res.* **52**, 590–614 (2013).
3. J. S. Harb, J. Comte, D. C. Gautheron, Asymmetrical orientation of phospholipids and their interactions with marker enzymes in pig heart mitochondrial inner membrane. *Arch. Biochem. Biophys.* **208**, 305–318 (1981).
4. R. C. Crain, G. V. Marinetti, Phospholipid topology of the inner mitochondrial membrane of rat liver. *Biochemistry.* **18**, 2407–2414 (1979).
5. M. G. Acoba, N. Senoo, S. M. Claypool, Phospholipid ebb and flow makes mitochondria go. *J. Cell Biol.* **219**, e202003131 (2020).
6. Y. Tamura, S. Kawano, T. Endo, Lipid homeostasis in mitochondria. *Biol. Chem.* **401**, 821–833 (2020).
7. K. Funai, S. A. Summers, J. Rutter, Reign in the membrane: How common lipids govern mitochondrial function. *Curr. Opin. Cell Biol.* **63**, 162–173 (2020).
8. J. Dong, F. Ye, J. Lin, H. He, Z. Song, The metabolism and function of phospholipids in Mitochondria. *Mitochondrial Commun.* **1**, 2–12 (2023).
9. M. Fry, D. E. Green, Cardiolipin requirement by cytochrome oxidase and the catalytic role of phospholipid. *Biochem. Biophys. Res. Commun.* **93**, 1238–1246 (1980).
10. M. Fry, D. E. Green, Cardiolipin requirement for electron transfer in complex I and III of the mitochondrial respiratory chain. *J. Biol. Chem.* **256**, 1874–1880 (1981).
11. P. G. Barth, R. J. A. Wanders, P. Vreken, E. A. M. Janssen, J. Lam, F. Baas, X-linked cardioskeletal myopathy and neutropenia (Barth syndrome): An update. *Am. J. Med. Genet. A.* **126A**, 349–354 (2004).
12. R. C. Sengers, J. M. F. Trijbels, J. L. Willems, O. Daniels, A. M. Stadhouders, Congenital cataract and mitochondrial myopathy of skeletal and heart muscle associated with lactic acidosis after exercise. *J. Pediatr.* **86**, 873–880 (2006).
13. R. G. Lee, S. Balasubramaniam, M. Stentenbach, T. Kralj, T. McCubbin, B. Padman, J. Smith, L. G. Riley, A. Priyadarshi, L. Peng, M. R. Nuske, R. Webster, K. Peacock, P. Roberts, Z. Stark, G. Lemire, Y. A. Ito, Care4Rare Canada Consortium, K. M. Boycott, M. T. Geraghty, J. B. Klinken, S. Ferdinandusse, Y. Zhou, R. Walsh, E. Marcellin, D. R. Thorburn, T. Rosciolli, J. Fletcher, O. Rackham, F. M. Vaz, G. E. Reid, A. Filipovska, Deleterious variants in *CRLS1* lead to cardiolipin deficiency and cause an autosomal recessive multi-system mitochondrial disease. *Hum. Mol. Genet.* **31**(21), 3597–3612. (2022). doi: 10.1093/hmg/ddac040.
14. T. Zhao, C. M. Goedhart, P. N. Sam, R. Sabouny, S. Lingrell, A. J. Cornish, R. E. Lamont, F. P. Bernier, D. Sinasac, J. S. Parboosingh, Care4Rare Canada Consortium, J. E. Vance, S. M. Claypool, A. M. Innes, T. E. Shutt, *PISD* is a mitochondrial disease gene causing skeletal dysplasia, cataracts, and white matter changes. *Life Sci. Alliance.* **2**, e201900353 (2019).
15. K. M. Girisha, L. Von Elsner, K. Neethukrishna, M. Muranjan, A. Shukla, G. S. Bhavani, G. Nishimura, K. Kutsche, G. Mortier, The homozygous variant c.797G>A/p.(Cys266Tyr) in *PISD* is associated with a Spondyloepimetaphyseal dysplasia with large epiphyses and disturbed mitochondrial function. *Hum. Mutat.* **40**, 299–309 (2019).
16. A. F. Anzmann, S. M. Claypool, H. Vernon, Mitochondrial dysfunction and Barth syndrome. in *Handbook of Mitochondrial Dysfunction*, S. I. Ahmad, Ed. (CRC Press, ed. 1, 2019; https://www.taylorfrancis.com/books/9780429811777/chapters/10.1201/9780429443336-6), pp. 61–78.
17. J. Pang, Y. Bao, K. Mitchell-Silbaugh, J. Veevers, X. Fang, Barth syndrome cardiomyopathy: An update. *Genes.* **13**, 656 (2022).

18. V. Zickermann, C. Wirth, H. Nasiri, K. Siegmund, H. Schwalbe, C. Hunte, U. Brandt, Mechanistic insight from the crystal structure of mitochondrial complex I. *Science*. **347**, 44–49 (2015).
19. K. Fiedorczuk, J. A. Letts, G. Degliesposti, K. Kaszuba, M. Skehel, L. A. Sazanov, Atomic structure of the entire mammalian mitochondrial complex I. *Nature*. **538**, 406–410 (2016).
20. M. Wu, J. Gu, R. Guo, Y. Huang, M. Yang, Structure of Mammalian Respiratory Supercomplex I 1 III 2 IV 1. *Cell*. **167**, 1598–1609.e10 (2016).
21. R. Guo, S. Zong, M. Wu, J. Gu, M. Yang, Architecture of Human Mitochondrial Respiratory Megacomplex I2III2IV2. *Cell*. **170**, 1247–1257.e12 (2017).
22. A.-N. A. Agip, J. N. Blaza, H. R. Bridges, C. Viscomi, S. Rawson, S. P. Muench, J. Hirst, Cryo-EM structures of complex I from mouse heart mitochondria in two biochemically defined states. *Nat. Struct. Mol. Biol.* **25**, 548–556 (2018).
23. J. A. Letts, K. Fiedorczuk, G. Degliesposti, M. Skehel, L. A. Sazanov, Structures of respiratory supercomplex I+III2 reveal functional and conformational crosstalk. *Mol. Cell*. **75**, 1131–1146.e6 (2019).
24. J. Schiller, E. Laube, I. Wittig, W. Kühlbrandt, J. Vonck, V. Zickermann, Insights into complex I assembly: Function of NDUFAF1 and a link with cardiolipin remodeling. *Sci. Adv.* **8**, eadd3855 (2022).
25. D. Kampjut, L. A. Sazanov, The coupling mechanism of mammalian respiratory complex I. *Science*. **370**, eabc4209 (2020).
26. S. Dröse, K. Zwicker, U. Brandt, Full recovery of the NADH:ubiquinone activity of complex I (NADH:ubiquinone oxidoreductase) from *Yarrowia lipolytica* by the addition of phospholipids. *Biochim. Biophys. Acta BBA—Bioenerg.* **1556**, 65–72 (2002).
27. G. Tasseva, H. D. Bai, M. Davidescu, A. Haromy, E. Michelakis, J. E. Vance, Phosphatidylethanolamine deficiency in mammalian mitochondria impairs oxidative phosphorylation and alters mitochondrial morphology. *J. Biol. Chem.* **288**, 4158–4173 (2013).
28. G. Oemer, J. Koch, Y. Wohlfarter, K. Lackner, R. E. M. Gebert, S. Geley, J. Zschocke, M. A. Keller, *J. Inherit. Metab. Dis.*, in press, doi:10.1002/jimd.12433.
29. M. Vos, A. Geens, C. Böhm, L. Deaulmerie, J. Swerts, M. Rossi, K. Craessaerts, E. P. Leites, P. Seibler, A. Rakovic, T. Lohnau, B. De Strooper, S.-M. Fendt, V. A. Morais, C. Klein, P. Verstreken, Cardiolipin promotes electron transport between ubiquinone and complex I to rescue PINK1 deficiency. *J. Cell Biol.* **216**, 695–708 (2017).
30. J. Zhang, Z. Guan, A. N. Murphy, S. E. Wiley, G. A. Perkins, C. A. Worby, J. L. Engel, P. Heacock, O. K. Nguyen, J. H. Wang, C. R. H. Raetz, W. Dowhan, J. E. Dixon, Mitochondrial phosphatase PTPMT1 is essential for cardiolipin biosynthesis. *Cell Metab.* **13**, 690–700 (2011).
31. M. S. Sharpley, R. J. Shannon, F. Draghi, J. Hirst, Interactions between phospholipids and NADH:ubiquinone oxidoreductase (Complex I) from bovine mitochondria. *Biochemistry*. **45**, 241–248 (2006).
32. A. Jussupow, A. Di Luca, V. R. I. Kaila, How cardiolipin modulates the dynamics of respiratory complex I. *Sci. Adv.* **5**, eaav1850 (2019).
33. I. Vercellino, L. A. Sazanov, Structure and assembly of the mammalian mitochondrial supercomplex CIII2CIV. *Nature*. **598**, 364–367 (2021).
34. A. Signes, E. Fernandez-Vizarra, Assembly of mammalian oxidative phosphorylation complexes I–V and supercomplexes. *Essays Biochem.* **62**, 255–270 (2018).
35. M. Rak, P. Rustin, Supernumerary subunits NDUFA3, NDUFA5 and NDUFA12 are required for the formation of the extramembrane arm of human mitochondrial complex I. *FEBS Lett.* **588**, 1832–1838 (2014).
36. M. J. W. Adjobo-Hermans, R. De Haas, P. H. G. M. Willems, A. Wojtala, S. E. Van Emst-de Vries, J. A. Wagenaars, M. Van Den Brand, R. J. Rodenburg, J. A. M. Smeitink, L. G. Nijtmans, L. A. Sazanov, M. R. Wieckowski, W. J. H. Koopman, NDUFS4 deletion triggers loss of NDUFA12 in Ndufs4 mice and Leigh syndrome patients: A stabilizing role for NDUFAF2. *Biochim. Biophys. Acta BBA—Bioenerg.* **1861**, 148213 (2020).
37. A. Torraco, A. Nasca, D. Verrigni, A. Pennisi, M. S. Zaki, G. Olivieri, Z. Assouline, D. Martinelli, R. Maroofian, T. Rizza, M. Di Nottia, F. Invernizzi, E. Lamantea, D. Longo, H. Houlden, H. Prokisch, A. Rötig, C. Dionisi-Vici, E. Bertini, D. Ghezzi, R. Carrozzo, D. Diodato, Novel *NDUFA12* variants are associated with isolated complex I defect and variable clinical manifestation. *Hum. Mutat.* **42**, 699–710 (2021).
38. Y. Xu, A. Malhotra, M. Ren, M. Schlame, The enzymatic function of tafazzin. *J. Biol. Chem.* **281**, 39217–39224 (2006).
39. M. Schlame, *Thematic Review Series: Glycerolipids*. Cardiolipin synthesis for the assembly of bacterial and mitochondrial membranes. *J. Lipid Res.* **49**, 1607–1620 (2008).
40. F. Gonzalvez, M. D'Aurelio, M. Boutant, A. Moustapha, J.-P. Puech, T. Landes, L. Arnauné-Pelloquin, G. Vial, N. Taleux, C. Slomianny, R. J. Wanders, R. H. Houtkooper, P. Bellenguer, I. M.

Møller, E. Gottlieb, F. M. Vaz, G. Manfredi, P. X. Petit, Barth syndrome: Cellular compensation of mitochondrial dysfunction and apoptosis inhibition due to changes in cardiolipin remodeling linked to tafazzin (TAZ) gene mutation. *Biochim. Biophys. Acta BBA—Mol. Basis Dis.* **1832**, 1194–1206 (2013).
41. S. Wang, E. Yazawa, E. Keating, N. Mazumdar, A. Hauschild, Q. Ma, H. Wu, Y. Xu, X. Shi, D. Strathdee, R. E. Gerszten, M. Schlame, W. T. Pu, Genetic modifiers modulate phenotypic expression of tafazzin deficiency in a mouse model of Barth syndrome. *Hum. Mol. Genet.* **32**(12):2055–2067. (2023). doi: 10.1093/hmg/ddad041.
42. Y. Xu, H. Erdjument-Bromage, C. K. L. Phoon, T. A. Neubert, M. Ren, M. Schlame, Cardiolipin remodeling enables protein crowding in the inner mitochondrial membrane. *EMBO J.* **40**, e108428 (2021).
43. A. F. Anzmann, O. L. Sniezek, A. Pado, V. Busa, F. M. Vaz, S. D. Kreimer, L. R. DeVine, R. N. Cole, A. Le, B. J. Kirsch, S. M. Claypool, H. J. Vernon, Diverse mitochondrial abnormalities in a new cellular model of TAFFAZZIN deficiency are remediated by cardiolipin-interacting small molecules. *J. Biol. Chem.* **297**, 101005 (2021).
44. Z. Chen, S. Zhu, H. Wang, L. Wang, J. Zhang, Y. Gu, C. Tan, M. Dhanani, E. Wever, X. Wang, B. Xie, S. Wang, L. Huang, A. H. C. van Kampen, J. Liu, Z. Han, H. H. Patel, F. M. Vaz, X. Fang, J. Chen, K. Ouyang, PTPMT1 is required for embryonic cardiac cardiolipin biosynthesis to regulate mitochondrial morphogenesis and heart development. *Circulation.* **144**, 403–406 (2021).
45. A. Selathurai, G. M. Kowalski, S. A. Mason, D. L. Callahan, V. C. Foletta, P. A. Della Gatta, A. Lindsay, S. Hamley, G. Kaur, A. R. Curtis, M. L. Burch, T. Ang, S. L. McGee, C. R. Bruce, Phosphatidylserine decarboxylase is critical for the maintenance of skeletal muscle mitochondrial integrity and muscle mass. *Mol. Metab.* **27**, 33–46 (2019).
46. J. M. Johnson, A. D. Peterlin, E. Balderas, E. G. Sustarsic, J. A. Maschek, M. J. Lang, A. Jara-Ramos, V. Panic, J. T. Morgan, C. J. Villanueva, A. Sanchez, J. Rutter, I. J. Lodhi, J. E. Cox, K. H. Fisher-Wellman, D. Chaudhuri, Z. Gerhart-Hines, K. Funai, Mitochondrial phosphatidylethanolamine modulates UCP1 to promote brown adipose thermogenesis. *Sci. Adv.* **9**, eade7864 (2023).
47. F. Sun, X. Huo, Y. Zhai, A. Wang, J. Xu, D. Su, M. Bartlam, Z. Rao, Crystal structure of mitochondrial respiratory membrane protein complex II. *Cell.* **121**, 1043–1057 (2005).
48. L. Huang, G. Sun, D. Cobessi, A. C. Wang, J. T. Shen, E. Y. Tung, V. E. Anderson, E. A. Berry, 3-Nitropropionic acid is a suicide inhibitor of mitochondrial respiration that, upon oxidation by complex II, forms a covalent adduct with a catalytic base arginine in the active site of the enzyme. *J. Biol. Chem.* **281**, 5965–5972 (2006).
49. Z. Du, X. Zhou, Y. Lai, J. Xu, Y. Zhang, S. Zhou, Z. Feng, L. Yu, Y. Tang, W. Wang, L. Yu, C. Tian, T. Ran, H. Chen, L. W. Guddat, F. Liu, Y. Gao, Z. Rao, H. Gong, Structure of the human respiratory complex II. *Proc. Natl. Acad. Sci.* **120**, e2216713120 (2023).
50. T. Kasahara, M. Kubota-Sakashita, Y. Nagatsuka, Y. Hirabayashi, T. Hanasaka, K. Tohyama, T. Kato, Cardiolipin is essential for early embryonic viability and mitochondrial integrity of neurons in mammals. *FASEB J.* **34**, 1465–1480 (2020).
51. J. Dudek, I. Cheng, A. Chowdhury, K. Wozny, M. Balleininger, R. Reinhold, S. Grunau, S. Callegari, K. Toischer, R. J. Wanders, G. Hasenfuß, B. Brügger, K. Guan, P. Rehling, Cardiac-specific succinate dehydrogenase deficiency in Barth syndrome. *EMBO Mol. Med.* **8**, 139–154 (2016).
52. T. M. Iverson, C. Luna-Chavez, L. R. Croal, G. Cecchini, D. C. Rees, Crystallographic studies of the *Escherichia coli* quinol-fumarate reductase with inhibitors bound to the quinol-binding site. *J. Biol. Chem.* **277**, 16124–16130 (2002).
53. V. Yankovskaya, R. Horsefield, S. Törnroth, C. Luna-Chavez, H. Miyoshi, C. Léger, B. Byrne, G. Cecchini, S. Iwata, Architecture of succinate dehydrogenase and reactive oxygen species generation. *Science.* **299**, 700–704 (2003).
54. T. M. Iverson, Catalytic mechanisms of complex II enzymes: A structural perspective. *Biochim. Biophys. Acta BBA—Bioenerg.* **1827**, 648–657 (2013).
55. C. T. Schwall, V. L. Greenwood, N. N. Alder, The stability and activity of respiratory Complex II is cardiolipin-dependent. *Biochim. Biophys. Acta BBA—Bioenerg.* **1817**, 1588–1596 (2012).
56. M. B. Reynolds, H. S. Hong, B. C. Michmerhuizen, A.-L. E. Lawrence, L. Zhang, J. S. Knight, C. A. Lyssiotis, B. H. Abuaita, M. X. O'Riordan, Cardiolipin coordinates inflammatory metabolic reprogramming through regulation of Complex II disassembly and degradation. *Sci. Adv.* **9**, eade8701 (2023).
57. C. Lange, J. H. Nett, B. L. Trumpower, C. Hunte, Specific roles of protein-phospholipid interactions in the yeast cytochrome bc1 complex structure. *EMBO J.* **20**, 6591–6600 (2001).
58. S. R. N. Solmaz, C. Hunte, Structure of complex III with bound cytochrome c in reduced state and definition of a minimal core interface for electron transfer. *J. Biol. Chem.* **283**, 17542–17549 (2008).

59. J. Berndtsson, A. Aufschnaiter, S. Rathore, L. Marin-Buera, H. Dawitz, J. Diessl, V. Kohler, A. Barrientos, S. Büttner, F. Fontanesi, M. Ott, Respiratory supercomplexes enhance electron transport by decreasing cytochrome *c* diffusion distance. *EMBO Rep.* **21** (2020), doi:10.15252/embr.202051015.
60. A. M. Hartley, N. Lukoyanova, Y. Zhang, A. Cabrera-Orefice, S. Arnold, B. Meunier, N. Pinotsis, A. Maréchal, Structure of yeast cytochrome c oxidase in a supercomplex with cytochrome bc1. *Nat. Struct. Mol. Biol.* **26**, 78–83 (2019).
61. S. Rathore, J. Berndtsson, L. Marin-Buera, J. Conrad, M. Carroni, P. Brzezinski, M. Ott, Cryo-EM structure of the yeast respiratory supercomplex. *Nat. Struct. Mol. Biol.* **26**, 50–57 (2019).
62. I. Vercellino, L. A. Sazanov, The assembly, regulation and function of the mitochondrial respiratory chain. *Nat. Rev. Mol. Cell Biol.* (2021), doi:10.1038/s41580-021-00415-0.
63. D. Xia, L. Esser, W.-K. Tang, F. Zhou, Y. Zhou, L. Yu, C.-A. Yu, Structural analysis of cytochrome bc1 complexes: Implications to the mechanism of function. *Biochim. Biophys. Acta BBA—Bioenerg.* **1827**, 1278–1294 (2013).
64. E. Calzada, E. Avery, P. N. Sam, A. Modak, C. Wang, J. M. McCaffery, X. Han, N. N. Alder, S. M. Claypool, Phosphatidylethanolamine made in the inner mitochondrial membrane is essential for yeast cytochrome bc1 complex function. *Nat. Commun.* **10**, 1432 (2019).
65. C. F. Hryc, V. K. P. S. Mallampalli, E. I. Bovshik, S. Azinas, G. Fan, I. I. Serysheva, G. C. Sparagna, M. L. Baker, E. Mileykovskaya, W. Dowhan, Structural insights into cardiolipin replacement by phosphatidylglycerol in a cardiolipin-lacking yeast respiratory supercomplex. *Nat. Commun.* **14**, 2783 (2023).
66. T. Wenz, R. Hielscher, P. Hellwig, H. Schägger, S. Richers, C. Hunte, Role of phospholipids in respiratory cytochrome bc1 complex catalysis and supercomplex formation. *Biochim. Biophys. Acta BBA—Bioenerg.* **1787**, 609–616 (2009).
67. M. G. Baile, M. Sathappa, Y.-W. Lu, E. Pryce, K. Whited, J. M. McCaffery, X. Han, N. N. Alder, S. M. Claypool, Unremodeled and remodeled cardiolipin are functionally indistinguishable in yeast. *J. Biol. Chem.* **289**, 1768–1778 (2014).
68. T. Tsukihara, H. Aoyama, E. Yamashita, T. Tomizaki, H. Yamaguchi, K. Shinzawa-Itoh, R. Nakashima, R. Yaono, S. Yoshikawa, The whole structure of the 13-subunit oxidized cytochrome c oxidase at 2.8 Å. *Science.* **272**, 1136–1144 (1996).
69. S. Yoshikawa, K. Shinzawa-Itoh, R. Nakashima, R. Yaono, E. Yamashita, N. Inoue, M. Yao, M. J. Fei, C. P. Libeu, T. Mizushima, H. Yamaguchi, T. Tomizaki, T. Tsukihara, Redox-coupled crystal structural changes in bovine heart cytochrome c oxidase. *Science.* **280**, 1723–1729 (1998).
70. T. Tsukihara, K. Shimokata, Y. Katayama, H. Shimada, K. Muramoto, H. Aoyama, M. Mochizuki, K. Shinzawa-Itoh, E. Yamashita, M. Yao, Y. Ishimura, S. Yoshikawa, The low-spin heme of cytochrome *c* oxidase as the driving element of the proton-pumping process. *Proc. Natl. Acad. Sci.* **100**, 15304–15309 (2003).
71. K. Shinzawa-Itoh, H. Aoyama, K. Muramoto, H. Terada, T. Kurauchi, Y. Tadehara, A. Yamasaki, T. Sugimura, S. Kurono, K. Tsujimoto, T. Mizushima, E. Yamashita, T. Tsukihara, S. Yoshikawa, Structures and physiological roles of 13 integral lipids of bovine heart cytochrome c oxidase. *EMBO J.* **26**, 1713–1725 (2007).
72. K. Muramoto, K. Hirata, K. Shinzawa-Itoh, S. Yoko-o, E. Yamashita, H. Aoyama, T. Tsukihara, S. Yoshikawa, A histidine residue acting as a controlling site for dioxygen reduction and proton pumping by cytochrome *c* oxidase. *Proc. Natl. Acad. Sci.* **104**, 7881–7886 (2007).
73. V. R. I. Kaila, E. Oksanen, A. Goldman, D. A. Bloch, M. I. Verkhovsky, D. Sundholm, M. Wikström, A combined quantum chemical and crystallographic study on the oxidized binuclear center of cytochrome c oxidase. *Biochim. Biophys. Acta BBA — Bioenerg.* **1807**, 769–778 (2011).
74. H. Aoyama, K. Muramoto, K. Shinzawa-Itoh, K. Hirata, E. Yamashita, T. Tsukihara, T. Ogura, S. Yoshikawa, A peroxide bridge between Fe and Cu ions in the O_2 reduction site of fully oxidized cytochrome *c* oxidase could suppress the proton pump. *Proc. Natl. Acad. Sci.* **106**, 2165–2169 (2009).
75. K. Ohta, K. Muramoto, K. Shinzawa-Itoh, E. Yamashita, S. Yoshikawa, T. Tsukihara, X-ray structure of the NO-bound Cu_B in bovine cytochrome *c* oxidase. *Acta Crystallograph. Sect. F Struct. Biol. Cryst. Commun.* **66**, 251–253 (2010).
76. K. Muramoto, K. Ohta, K. Shinzawa-Itoh, K. Kanda, M. Taniguchi, H. Nabekura, E. Yamashita, T. Tsukihara, S. Yoshikawa, Bovine cytochrome *c* oxidase structures enable O_2 reduction with minimization of reactive oxygens and provide a proton-pumping gate. *Proc. Natl. Acad. Sci.* **107**, 7740–7745 (2010).
77. M. Suga, N. Yano, K. Muramoto, K. Shinzawa-Itoh, T. Maeda, E. Yamashita, T. Tsukihara, S. Yoshikawa, Distinguishing between Cl – and $O_2{}^{2-}$ as the bridging element between Fe^{3+} and Cu^{2+} in resting-oxidized cytochrome *c* oxidase. *Acta Crystallogr. D Biol. Crystallogr.* **67**, 742–744 (2011).

78. K. Hirata, K. Shinzawa-Itoh, N. Yano, S. Takemura, K. Kato, M. Hatanaka, K. Muramoto, T. Kawahara, T. Tsukihara, E. Yamashita, K. Tono, G. Ueno, T. Hikima, H. Murakami, Y. Inubushi, M. Yabashi, T. Ishikawa, M. Yamamoto, T. Ogura, H. Sugimoto, J.-R. Shen, S. Yoshikawa, H. Ago, Determination of damage-free crystal structure of an X-ray–sensitive protein using an XFEL. *Nat. Methods.* **11**, 734–736 (2014).
79. N. Yano, K. Muramoto, M. Mochizuki, K. Shinzawa-Itoh, E. Yamashita, S. Yoshikawa, T. Tsukihara, X-ray structure of cyanide-bound bovine heart cytochrome *c* oxidase in the fully oxidized state at 2.0 Å resolution. *Acta Crystallogr. Sect. F Struct. Biol. Commun.* **71**, 726–730 (2015).
80. S. Shimada, K. Shinzawa-Itoh, J. Baba, S. Aoe, A. Shimada, E. Yamashita, J. Kang, M. Tateno, S. Yoshikawa, T. Tsukihara, Complex structure of cytochrome *c* –cytochrome *c* oxidase reveals a novel protein–protein interaction mode. *EMBO J.* **36**, 291–300 (2017).
81. I. Ishigami, N. A. Zatsepin, M. Hikita, C. E. Conrad, G. Nelson, J. D. Coe, S. Basu, T. D. Grant, M. H. Seaberg, R. G. Sierra, M. S. Hunter, P. Fromme, R. Fromme, S.-R. Yeh, D. L. Rousseau, Crystal structure of CO-bound cytochrome *c* oxidase determined by serial femtosecond X-ray crystallography at room temperature. *Proc. Natl. Acad. Sci.* **114**, 8011–8016 (2017).
82. F. Luo, K. Shinzawa-Itoh, K. Hagimoto, A. Shimada, S. Shimada, E. Yamashita, S. Yoshikawa, T. Tsukihara, Structure of bovine cytochrome *c* oxidase crystallized at a neutral pH using a fluorinated detergent. *Acta Crystallogr. Sect. F Struct. Biol. Commun.* **73**, 416–422 (2017).
83. S. Zong, M. Wu, J. Gu, T. Liu, R. Guo, M. Yang, Structure of the intact 14-subunit human cytochrome c oxidase. *Cell Res.* **28**, 1026–1034 (2018).
84. G. Ing, A. M. Hartley, N. Pinotsis, A. Maréchal, Cryo-EM structure of a monomeric yeast *S. cerevisiae* complex IV isolated with maltosides: Implications in supercomplex formation. *Biochim. Biophys. Acta BBA—Bioenerg.* **1863**, 148591 (2022).
85. K. Shinzawa-Itoh, T. Sugimura, T. Misaki, Y. Tadehara, S. Yamamoto, M. Hanada, N. Yano, T. Nakagawa, S. Uene, T. Yamada, H. Aoyama, E. Yamashita, T. Tsukihara, S. Yoshikawa, K. Muramoto, Monomeric structure of an active form of bovine cytochrome *c* oxidase. *Proc. Natl. Acad. Sci.* **116**, 19945–19951 (2019).
86. M. Levchenko, J.-M. Wuttke, K. Römpler, B. Schmidt, K. Neifer, L. Juris, M. Wissel, P. Rehling, M. Deckers, Cox26 is a novel stoichiometric subunit of the yeast cytochrome c oxidase. *Biochim. Biophys. Acta BBA—Mol. Cell Res.* **1863**, 1624–1632 (2016).
87. V. Strecker, Z. Kadeer, J. Heidler, C.-M. Cruciat, H. Angerer, H. Giese, K. Pfeiffer, R. A. Stuart, I. Wittig, Supercomplex-associated Cox26 protein binds to cytochrome c oxidase. *Biochim. Biophys. Acta BBA—Mol. Cell Res.* **1863**, 1643–1652 (2016).
88. G. C. Sparagna, A. J. Chicco, R. C. Murphy, M. R. Bristow, C. A. Johnson, M. L. Rees, M. L. Maxey, S. A. McCune, R. L. Moore, Loss of cardiac tetralinoleoyl cardiolipin in human and experimental heart failure. *J. Lipid Res.* **48**, 1559–1570 (2007).
89. F. Jiang, M. T. Ryan, M. Schlame, M. Zhao, Z. Gu, M. Klingenberg, N. Pfanner, M. L. Greenberg, Absence of cardiolipin in the crd1 null mutant results in decreased mitochondrial membrane potential and reduced mitochondrial function. *J. Biol. Chem.* **275**, 22387–22394 (2000).
90. D. Schädeli, M. Serricchio, H. B. Hamidane, A. Loffreda, A. Hemphill, T. Beneke, E. Gluenz, J. Graumann, P. Bütikofer, Cardiolipin depletion-induced changes in the *Trypanosoma brucei* proteome. *FASEB J.* **33**, 13161–13175 (2019).
91. L. Böttinger, S. E. Horvath, T. Kleinschroth, C. Hunte, G. Daum, N. Pfanner, T. Becker, Phosphatidylethanolamine and cardiolipin differentially affect the stability of mitochondrial respiratory chain supercomplexes. *J. Mol. Biol.* **423**, 677–686 (2012).
92. E. Balsa, R. Marco, E. Perales-Clemente, R. Szklarczyk, E. Calvo, M. O. Landázuri, J. A. Enríquez, NDUFA4 is a subunit of complex IV of the mammalian electron transport chain. *Cell Metab.* **16**, 378–386 (2012).
93. C. Arnarez, S. J. Marrink, X. Periole, Molecular mechanism of cardiolipin-mediated assembly of respiratory chain supercomplexes. *Chem. Sci.* **7**, 4435–4443 (2016).
94. R. Acín-Pérez, P. Fernández-Silva, M. L. Peleato, A. Pérez-Martos, J. A. Enriquez, Respiratory active mitochondrial supercomplexes. *Mol. Cell.* **32**, 529–539 (2008).
95. T. Lobo-Jarne, C. Ugalde, Respiratory chain supercomplexes: Structures, function and biogenesis. *Semin. Cell Dev. Biol.* **76**, 179–190 (2018).
96. K. M. Davies, T. B. Blum, W. Kühlbrandt, Conserved in situ arrangement of complex I and III_2 in mitochondrial respiratory chain supercomplexes of mammals, yeast, and plants. *Proc. Natl. Acad. Sci.* **115**, 3024–3029 (2018).
97. J. A. Letts, K. Fiedorczuk, L. A. Sazanov, The architecture of respiratory supercomplexes. *Nature.* **537**, 644–648 (2016).

98. J. S. Sousa, D. J. Mills, J. Vonck, W. Kühlbrandt, Functional asymmetry and electron flow in the bovine respirasome. *eLife*. **5**, e21290 (2016).
99. J. Gu, M. Wu, R. Guo, K. Yan, J. Lei, N. Gao, M. Yang, The architecture of the mammalian respirasome. *Nature*. **537**, 639–643 (2016).
100. E. Calvo, S. Cogliati, P. Hernansanz-Agustín, M. Loureiro-López, A. Guarás, R. A. Casuso, F. García-Marqués, R. Acín-Pérez, Y. Martí-Mateos, Jc. Silla-Castro, M. Carro-Alvarellos, J. R. Huertas, J. Vázquez, J. A. Enríquez, Functional role of respiratory supercomplexes in mice: SCAF1 relevance and segmentation of the Q_{pool}. *Sci. Adv.* **6**, eaba7509 (2020).
101. T. Lobo-Jarne, E. Nývltová, R. Pérez-Pérez, A. Timón-Gómez, T. Molinié, A. Choi, A. Mourier, F. Fontanesi, C. Ugalde, A. Barrientos, Human COX7A2L regulates complex III biogenesis and promotes supercomplex organization remodeling without affecting mitochondrial bioenergetics. *Cell Rep.* **25**, 1786–1799.e4 (2018).
102. P. Brzezinski, A. Moe, P. Ädelroth, Structure and mechanism of respiratory III–IV supercomplexes in bioenergetic membranes. *Chem. Rev.* **121**, 9644–9673 (2021).
103. E. Maranzana, G. Barbero, A. I. Falasca, G. Lenaz, M. L. Genova, Mitochondrial respiratory supercomplex association limits production of reactive oxygen species from complex I. *Antioxid. Redox Signal.* **19**, 1469–1480 (2013).
104. I. Lopez-Fabuel, J. Le Douce, A. Logan, A. M. James, G. Bonvento, M. P. Murphy, A. Almeida, J. P. Bolaños, Complex I assembly into supercomplexes determines differential mitochondrial ROS production in neurons and astrocytes. *Proc. Natl. Acad. Sci.* **113**, 13063–13068 (2016).
105. C. L. Quinlan, I. V. Perevoshchikova, M. Hey-Mogensen, A. L. Orr, M. D. Brand, Sites of reactive oxygen species generation by mitochondria oxidizing different substrates. *Redox Biol.* **1**, 304–312 (2013).
106. T. Hou, R. Zhang, C. Jian, W. Ding, Y. Wang, S. Ling, Q. Ma, X. Hu, H. Cheng, X. Wang, NDUFAB1 confers cardio-protection by enhancing mitochondrial bioenergetics through coordination of respiratory complex and supercomplex assembly. *Cell Res.* **29**, 754–766 (2019).
107. G. Wang, B. Popovic, J. Tao, A. Jiang, Overexpression of COX7RP promotes tumor growth and metastasis by inducing ROS production in hepatocellular carcinoma cells.
108. K. Ikeda, K. Horie-Inoue, T. Suzuki, R. Hobo, N. Nakasato, S. Takeda, S. Inoue, Mitochondrial supercomplex assembly promotes breast and endometrial tumorigenesis by metabolic alterations and enhanced hypoxia tolerance. *Nat. Commun.* **10**, 4108 (2019).
109. C. Bianchi, M. L. Genova, G. Parenti Castelli, G. Lenaz, The mitochondrial respiratory chain is partially organized in a supercomplex assembly. *J. Biol. Chem.* **279**, 36562–36569 (2004).
110. E. Lapuente-Brun, R. Moreno-Loshuertos, R. Acín-Pérez, A. Latorre-Pellicer, C. Colás, E. Balsa, E. Perales-Clemente, P. M. Quirós, E. Calvo, M. A. Rodríguez-Hernández, P. Navas, R. Cruz, Á. Carracedo, C. López-Otín, A. Pérez-Martos, P. Fernández-Silva, E. Fernández-Vizarra, J. A. Enríquez, Supercomplex assembly determines electron flux in the mitochondrial electron transport chain. *Science*. **340**, 1567–1570 (2013).
111. A. Stuchebrukhov, J. Schäfer, J. Berg, P. Brzezinski, Kinetic advantage of forming respiratory supercomplexes. *Biochim. Biophys. Acta BBA—Bioenerg.* **1861**, 148193 (2020).
112. A. Moe, J. Di Trani, J. L. Rubinstein, P. Brzezinski, Cryo-EM structure and kinetics reveal electron transfer by 2D diffusion of cytochrome *c* in the yeast III-IV respiratory supercomplex. *Proc. Natl. Acad. Sci.* **118**, e2021157118 (2021).
113. A. Mühleip, R. K. Flygaard, R. Baradaran, O. Haapanen, T. Gruhl, V. Tobiasson, A. Maréchal, V. Sharma, A. Amunts, Structural basis of mitochondrial membrane bending by the I–II–III2–IV2 supercomplex. *Nature*. (2023), doi:10.1038/s41586-023-05817-y.
114. F. Han, Y. Hu, M. Wu, Z. He, H. Tian, L. Zhou, Structures of Tetrahymena thermophila respiratory megacomplexes on the tubular mitochondrial cristae. *Nat. Commun.* **14**, 2542 (2023).
115. D. Milenkovic, J. N. Blaza, N.-G. Larsson, J. Hirst, The enigma of the respiratory chain supercomplex. *Cell Metab.* **25**, 765–776 (2017).
116. R. Anand, A. K. Kondadi, J. Meisterknecht, M. Golombek, O. Nortmann, J. Riedel, L. Peifer-Weiß, N. Brocke-Ahmadinejad, D. Schlütermann, B. Stork, T. O. Eichmann, I. Wittig, A. S. Reichert, MIC26 and MIC27 cooperate to regulate cardiolipin levels and the landscape of OXPHOS complexes. *Life Sci. Alliance*. **3**, e202000711 (2020).
117. S. Zhu, Z. Chen, M. Zhu, Y. Shen, L. J. Leon, L. Chi, S. Spinozzi, C. Tan, Y. Gu, A. Nguyen, Y. Zhou, W. Feng, F. M. Vaz, X. Wang, A. B. Gustafsson, S. M. Evans, O. Kunfu, X. Fang, Cardiolipin remodeling defects impair mitochondrial architecture and function in a murine model of Barth syndrome cardiomyopathy. *Circ. Heart Fail.* **14** (2021), doi:10.1161/CIRCHEARTFAILURE.121.008289.
118. L. Peyta, K. Jarnouen, M. Pinault, C. Guimaraes, J.-P. Pais de Barros, S. Chevalier, J.-F. Dumas, F. Maillot, G. M. Hatch, P. Loyer, S. Servais, Reduced cardiolipin content decreases respiratory chain

capacities and increases ATP synthesis yield in the human HepaRG cells. *Biochim. Biophys. Acta BBA—Bioenerg.* **1857**, 443–453 (2016).
119. B. Rieger, A. Krajčová, P. Duwe, K. B. Busch, ALCAT1 overexpression affects supercomplex formation and increases ROS in respiring mitochondria. *Oxid. Med. Cell. Longev.* **2019**, 1–12 (2019).
120. C. H. Le, L. G. Benage, K. S. Specht, L. C. Li Puma, C. M. Mulligan, A. L. Heuberger, J. E. Prenni, S. M. Claypool, K. C. Chatfield, G. C. Sparagna, A. J. Chicco, Tafazzin deficiency impairs CoA-dependent oxidative metabolism in cardiac mitochondria. *J. Biol. Chem.* **295**, 12485–12497 (2020).
121. L. K. Cole, E. M. Mejia, G. C. Sparagna, M. Vandel, B. Xiang, X. Han, N. Dedousis, B. A. Kaufman, V. W. Dolinsky, G. M. Hatch, Cardiolipin deficiency elevates susceptibility to a lipotoxic hypertrophic cardiomyopathy. *J. Mol. Cell. Cardiol.* **144**, 24–34 (2020).
122. M. Elkes, M. Andonovski, D. Vidal, M. Farago, R. Modafferi, S. M. Claypool, P. J. LeBlanc, The influence of supplemental dietary linoleic acid on skeletal muscle contractile function in a rodent model of Barth syndrome. *Front. Physiol.* **12**, 731961 (2021).
123. J. Kim, K. Lee, H. Fujioka, B. Tandler, C. L. Hoppel, Cardiac mitochondrial structure and function in tafazzin-knockdown mice. *Mitochondrion.* **43**, 53–62 (2018).
124. I. A. Chatzispyrou, S. Guerrero-Castillo, N. M. Held, J. P. N. Ruiter, S. W. Denis, L. IJlst, R. J. Wanders, M. van Weeghel, S. Ferdinandusse, F. M. Vaz, U. Brandt, R. H. Houtkooper, Barth syndrome cells display widespread remodeling of mitochondrial complexes without affecting metabolic flux distribution. *Biochim. Biophys. Acta BBA—Mol. Basis Dis.* **1864**, 3650–3658 (2018).
125. J. Dudek, I.-F. Cheng, M. Balleininger, F. M. Vaz, K. Streckfuss-Bömeke, D. Hübscher, M. Vukotic, R. J. A. Wanders, P. Rehling, K. Guan, Cardiolipin deficiency affects respiratory chain function and organization in an induced pluripotent stem cell model of Barth syndrome. *Stem Cell Res.* **11**, 806–819 (2013).
126. C. Potting, T. Tatsuta, T. König, M. Haag, T. Wai, M. J. Aaltonen, T. Langer, TRIAP1/PRELI complexes prevent apoptosis by mediating intramitochondrial transport of phosphatidic acid. *Cell Metab.* **18**, 287–295 (2013).
127. K. Pfeiffer, V. Gohil, R. A. Stuart, C. Hunte, U. Brandt, M. L. Greenberg, H. Schägger, Cardiolipin stabilizes respiratory chain supercomplexes. *J. Biol. Chem.* **278**, 52873–52880 (2003).
128. T. D. Heden, J. M. Johnson, P. J. Ferrara, H. Eshima, A. R. P. Verkerke, E. J. Wentzler, P. Siripoksup, T. M. Narowski, C. B. Coleman, C.-T. Lin, T. E. Ryan, P. T. Reidy, L. E. de Castro Brás, C. M. Karner, C. F. Burant, J. A. Maschek, J. E. Cox, D. G. Mashek, G. Kardon, S. Boudina, T. N. Zeczycki, J. Rutter, S. R. Shaikh, J. E. Vance, M. J. Drummond, P. D. Neufer, K. Funai, Mitochondrial PE potentiates respiratory enzymes to amplify skeletal muscle aerobic capacity. *Sci. Adv.* **5** (2019), doi:10.1126/sciadv.aax8352.
129. S. M. Claypool, Y. Oktay, P. Boontheung, J. A. Loo, C. M. Koehler, Cardiolipin defines the interactome of the major ADP/ATP carrier protein of the mitochondrial inner membrane. *J. Cell Biol.* **182**, 937–950 (2008).
130. Y. Xu, C. K. L. Phoon, B. Berno, K. D'Souza, E. Hoedt, G. Zhang, T. A. Neubert, R. M. Epand, M. Ren, M. Schlame, Loss of protein association causes cardiolipin degradation in Barth syndrome. *Nat. Chem. Biol.* **12**, 641–647 (2016).
131. M. Zhang, E. Mileykovskaya, W. Dowhan, Gluing the respiratory chain together. Cardiolipin is required for supercomplex formation in the inner mitochondrial membrane. *J. Biol. Chem.* **277**, 43553–43556 (2002).
132. S.-C. Chang, P. N. Heacock, E. Mileykovskaya, D. R. Voelker, W. Dowhan, Isolation and characterization of the gene (CLS1) encoding cardiolipin synthase in *Saccharomyces cerevisiae. J. Biol. Chem.* **273**, 14933–14941 (1998).
133. T. J. Jeon, S.-G. Lee, S. H. Yoo, M. Kim, D. Song, J. Ryu, H. Park, D.-S. Kim, J. Hyun, H. M. Kim, S. E. Ryu, A dynamic substrate pool revealed by cryo-EM of a lipid-preserved respiratory supercomplex. *Antioxid. Redox Signal.* **36**, 1101–1118 (2022).
134. J. G. Fedor, J. Hirst, Mitochondrial supercomplexes do not enhance catalysis by quinone channeling. *Cell Metab.* **28**, 525–531.e4 (2018).
135. D. Milenkovic, J. Misic, J. F. Hevler, T. Molinié, I. Chung, I. Atanassov, X. Li, R. Filograna, A. Mesaros, A. Mourier, A. J. R. Heck, J. Hirst, N. G. Larsson. Preserved respiratory chain capacity and physiology in mice with profoundly reduced levels of mitochondrial respirasomes. Cell Metab. **35**(10):1799–1813.e7 (2023). PMID: 37633273.
136. S. Cogliati, E. Calvo, M. Loureiro, A. M. Guaras, R. Nieto-Arellano, C. Garcia-Poyatos, I. Ezkurdia, N. Mercader, J. Vázquez, J. A. Enriquez, Mechanism of super-assembly of respiratory complexes III and IV. *Nature.* **539**, 579–582 (2016).
137. L. K. Cole, J. H. Kim, A. A. Amoscato, Y. Y. Tyurina, H. Bayır, B. Karimi, T. J. Siddiqui, V. E. Kagan, G. M. Hatch, T. M. Kauppinen, Aberrant cardiolipin metabolism is associated with cognitive deficiency and hippocampal alteration in tafazzin knockdown mice. *Biochim. Biophys. Acta BBA—Mol. Basis Dis.* **1864**, 3353–3367 (2018).

138. M. Strauss, G. Hofhaus, R. R. Schröder, W. Kühlbrandt, Dimer ribbons of ATP synthase shape the inner mitochondrial membrane. *EMBO J.* **27**, 1154–1160 (2008).
139. K. M. Davies, C. Anselmi, I. Wittig, J. D. Faraldo-Gomez, W. Kuhlbrandt, Structure of the yeast F1Fo-ATP synthase dimer and its role in shaping the mitochondrial cristae. *Proc. Natl. Acad. Sci.* **109**, 13602–13607 (2012).
140. J. A. Nirody, I. Budin, P. Rangamani, ATP synthase: Evolution, energetics, and membrane interactions. *J. Gen. Physiol.* **152**, e201912475 (2020).
141. T. E. Spikes, M. G. Montgomery, J. E. Walker, Structure of the dimeric ATP synthase from bovine mitochondria. *Proc. Natl. Acad. Sci.* **117**, 23519–23526 (2020).
142. G. Pinke, L. Zhou, L. A. Sazanov, Cryo-EM structure of the entire mammalian F-type ATP synthase. *Nat. Struct. Mol. Biol.* **27**, 1077–1085 (2020).
143. T. E. Spikes, M. G. Montgomery, J. E. Walker, Interface mobility between monomers in dimeric bovine ATP synthase participates in the ultrastructure of inner mitochondrial membranes. *Proc. Natl. Acad. Sci.* **118**, e2021012118 (2021).
144. H. Guo, S. A. Bueler, J. L. Rubinstein, Atomic model for the dimeric F_O region of mitochondrial ATP synthase. *Science.* **358**, 936–940 (2017).
145. J. Gu, L. Zhang, S. Zong, R. Guo, T. Liu, J. Yi, P. Wang, W. Zhuo, M. Yang, Cryo-EM structure of the mammalian ATP synthase tetramer bound with inhibitory protein IF1. *Science.* **364**, 1068–1075 (2019).
146. A. V. Vlasov, K. V. Kovalev, S.-H. Marx, E. S. Round, I. Yu. Gushchin, V. A. Polovinkin, N. M. Tsoy, I. S. Okhrimenko, V. I. Borshchevskiy, G. D. Büldt, Yu. L. Ryzhykau, A. V. Rogachev, V. V. Chupin, A. I. Kuklin, N. A. Dencher, V. I. Gordeliy, Unusual features of the c-ring of F_1F_O ATP synthases. *Sci. Rep.* **9**, 18547 (2019).
147. X. Liu, C. N. Kim, J. Yang, R. Jemmerson, X. Wang, Induction of apoptotic program in cell-free extracts: Requirement for dATP and cytochrome c. *Cell.* **86**, 147–157 (1996).
148. M. Carraro, V. Checchetto, I. Szabó, P. Bernardi, F- ATP synthase and the permeability transition pore: Fewer doubts, more certainties. *FEBS Lett.* **593**, 1542–1553 (2019).
149. A. Mühleip, S. E. McComas, A. Amunts, Structure of a mitochondrial ATP synthase with bound native cardiolipin. *eLife.* **8**, e51179 (2019).
150. D. Acehan, Y. Xu, D. L. Stokes, M. Schlame, Comparison of lymphoblast mitochondria from normal subjects and patients with Barth syndrome using electron microscopic tomography. *Lab. Invest.* **87**, 40–48 (2007).
151. Y. Xu, M. Anjaneyulu, A. Donelian, W. Yu, M. L. Greenberg, M. Ren, E. Owusu-Ansah, M. Schlame, Assembly of the complexes of oxidative phosphorylation triggers the remodeling of cardiolipin. *Proc. Natl. Acad. Sci.* **116**, 11235–11240 (2019).
152. J. J. Hoffmann, T. Becker, Crosstalk between mitochondrial protein import and lipids. *Int. J. Mol. Sci.* **23**, 5274 (2022).
153. H. Rampelt, R. M. Zerbes, M. van der Laan, N. Pfanner, Role of the mitochondrial contact site and cristae organizing system in membrane architecture and dynamics. *Biochim. Biophys. Acta BBA—Mol. Cell Res.* **1864**, 737–746 (2017).
154. N. Senoo, S. Kandasamy, O. B. Ogunbona, M. G. Baile, Y. Lu, S. M. Claypool, Cardiolipin, conformation, and respiratory complex-dependent oligomerization of the major mitochondrial ADP/ATP carrier in yeast. *Sci. Adv.* **6**, eabb0780 (2020).
155. O. Jovanović, K. Chekashkina, S. Škulj, K. Žuna, M. Vazdar, P. V. Bashkirov, E. E. Pohl, Membrane lipid reshaping underlies oxidative stress sensing by the mitochondrial proteins UCP1 and ANT1. *Antioxidants.* **11**, 2314 (2022).
156. B. R. Alevriadou, A. Patel, M. Noble, S. Ghosh, V. M. Gohil, P. B. Stathopulos, M. Madesh, Molecular nature and physiological role of the mitochondrial calcium uniporter channel. *Am. J. Physiol.-Cell Physiol.* **320**, C465–C482 (2021).
157. H. Weidberg, A. Amon, MitoCPR—A surveillance pathway that protects mitochondria in response to protein import stress. *Science.* **360**, eaan4146 (2018).
158. D. Poveda-Huertes, A. A. Taskin, I. Dhaouadi, L. Myketin, A. Marada, L. Habernig, S. Büttner, F.-N. Vögtle, Increased mitochondrial protein import and cardiolipin remodelling upon early mtUPR. *PLOS Genet.* **17**, e1009664 (2021).
159. L. Merklinger, J. Bauer, P. A. Pedersen, R. B. Damgaard, J. P. Morth, Phospholipids alter activity and stability of mitochondrial membrane-bound ubiquitin ligase MARCH5. *Life Sci. Alliance.* **5**, e202101309 (2022).

Part III

Mitochondrial Control of Cellular Homeostasis

From Redox Signalling to Interorganellar Contact Sites

5 Mitochondrial Hydrogen Peroxide Dynamics

Jan Riemer and Helmut Sies**

INTRODUCTION

Mitochondrial H_2O_2 attracted research interest since the early 1970s (see reviews [1–3]). Much knowledge on the biological chemistry of mitochondrial H_2O_2 has accumulated but important aspects are still to be uncovered. With the advent of new technology using genetically encoded fluorescent H_2O_2 sensors [4–7], studying this exciting topic became amenable at a refined level. Here, we focus on some of these developments which help shed light on the dynamic role of mitochondrial H_2O_2 for cell biology in health and disease, to be viewed in the wider context of mitochondrial signal transduction [8].

CONTRIBUTION OF MITOCHONDRIAL H_2O_2 TO OVERALL CELLULAR H_2O_2

Much of our understanding of the role of mitochondria comes from a wealth of studies on (more or less) intact *isolated* mitochondria. Obviously, isolated mitochondria are deprived of their natural environment, which comprises the cytosol and contacts to other subcellular organelles. However, in recent years the study of mitochondria within the context of the *intact cell* has become feasible, not the least through the use of non-invasive probes such as were mentioned above. In addition to the mitochondrial source, other cell organelles produce H_2O_2, and about 40 enzymatic sources of cellular H_2O_2 have been identified [9], with the major enzymes being the NADPH oxidases (NOX) (Figure 5.1). The question: "Are mitochondria the main contributor of reactive oxygen species in cells?" was examined in a number of different cell types in culture, and the answer was: "The contribution of mitochondria varies between cell types but mitochondria are never the main source of cellular ROS. This indicates that although mitochondria are one of the significant sources of cellular ROS, they are not necessarily the main contributor under normal conditions. Intriguingly, similar findings were also observed in cells under a variety of stressors, life-history strategies and pathological stages, in which the rates of cellular ROS production were significantly enhanced" [10]. This result may be counterintuitive to many researchers investigating mitochondrial biology, but it appears plausible in view of the tightly controlled enzymatic production by NADPH oxidases, which were identified as the major producers of cellular H_2O_2 in Ref. [10]. Regardless of whether mitochondria are "the main source of reactive oxygen species (ROS) in cells" [11] or not, the impact of mitochondrial H_2O_2 on redox homeostasis is considerable (see below).

* Corresponding authors: jan.riemer@uni-koeln.de; sies@hhu.de

DOI: 10.1201/9781003221180-8

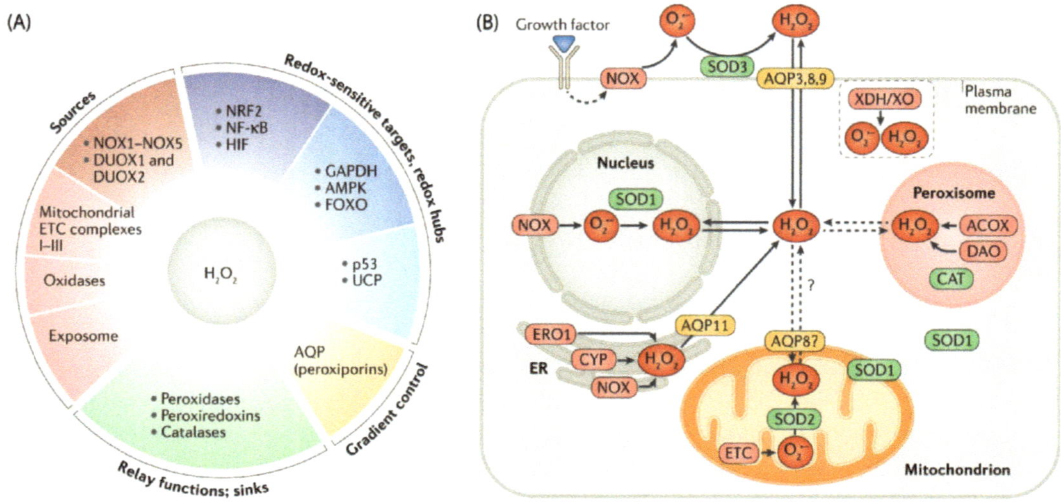

FIGURE 5.1 Key modulators and targets of H_2O_2. (A) H_2O_2 sources (red), redox-sensitive targets and hubs (blue), aquaporins (AQP/peroxiporins) (yellow) and sinks (green). (B) The major subcellular sites of generation of the superoxide anion radical and H_2O_2 are shown. (Originally published in Sies and Jones [9].)

MITOCHONDRIAL H_2O_2: UNAVOIDABLE WASTE PRODUCT OF THE ELECTRON TRANSPORT CHAIN OR USEFUL SIGNALLING AGENT?

The major source of mitochondrial H_2O_2 is the electron transport chain (ETC), which by one-electron reduction of oxygen produces the superoxide radical anion at three sites, Complexes I, II, and III, followed by dismutation to yield H_2O_2 [12, 13]. The concentrations of H_2O_2 in subcompartments of the mitochondria and their dynamics have been modelled for HeLa cells [14]. Among several mechanisms, regulation of mitochondrial H_2O_2 availability by protein S-glutathionylation plays a role [15]. In short, we posit that, under physiological conditions, mitochondrial H_2O_2 is a useful and regulated metabolite, and not a waste product of flux through the ETC, or a "leakage" of the ETC; see also Refs. [16, 17].

HANDLING OF MITOCHONDRIAL H_2O_2 AND SHAPING OF H_2O_2 GRADIENTS

Mitochondria generate varying amounts of oxidants, depending on the flux of electrons through different metabolic pathways, particularly the respiratory chain but also other redox cofactor-containing enzymes (e.g. flavin). Moreover, the subcellular spatiotemporal dynamics of mitochondrial oxidant generation and release and the resulting physiological outcomes depend on the shape and positioning of the mitochondrial network. In the intact cell, there is constant alert for redox homeostasis, called oxidative eustress [18]. The resulting heterogeneity of mitochondrial H_2O_2 release at different subcellular sites and also between different cells will be discussed in this section.

The proximal oxidant generated in mitochondria are superoxide anions. Superoxide anion generation depends strongly on the metabolic state of the cell, the flux of electrons through the ETC and the local availability of oxygen. Further determinants of fluctuating superoxide anion generation are the electron donors feeding the respiratory chain and the directionality of electron flow. For example, upon shifting tissue culture cells from glucose ("fermentation") to galactose ("increased use of the respiratory chain"), more superoxide anions are generated at the respiratory chain [19]. Upon exposure to low levels of oxygen in hypoxia, less superoxide anions are generated at the respiratory chain but succinate accumulates in the matrix [19, 20]. Upon reoxygenation (e.g. during

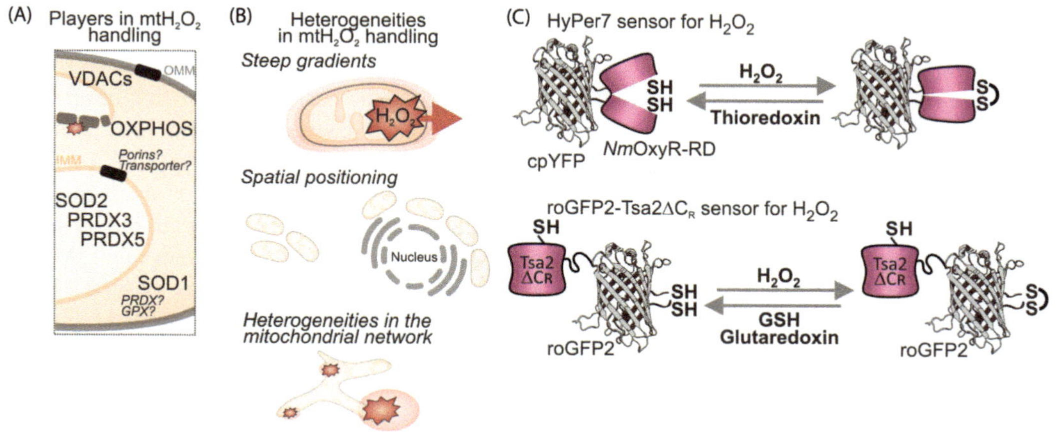

FIGURE 5.2 Sensors for H_2O_2 and their use in the assessment of mitochondrial H_2O_2 dynamics. (A) Major players in mitochondrial H_2O_2 handling. (B) The heterogeneity in mitochondrial H_2O_2 release. (C) Sensors of the HyPer and roGFP2 families.

ischemic reperfusion), succinate oxidation at Complex II does not only feed electrons via the Q pool into Complex III but also drives reverse electron transfer through Complex I and thereby massive superoxide anion generation at the flavin site of Complex I [20].

Superoxide anions are charged and cannot readily diffuse across membranes, being locally confined by membrane barriers. They are rapidly dismutated to H_2O_2 by matrix and intermembrane space (IMS)-localized SOD2 and SOD1, respectively (Figure 5.2A). The activity of SODs determines the local levels of superoxide anions and H_2O_2 and, consequently, the flux across membranes. For SOD1, which is localized both in the cytosol and the IMS, it has been shown that increased release of mitochondrial H_2O_2 leads to an accelerated import and thus increased levels of IMS SOD1 [21]. SOD2 levels can vary as well. SOD2 expression is higher in breast cancer compared to non-cancerous parental tissue [22]. Moreover, SOD activity is regulated by post-translational modifications. For example, SOD1 can become phosphorylated in an mTOR-mediated manner that attenuates its function [23], or SOD2 can become acetylated during tumour development that also attenuates its activity [24]. Neuronal deletion of SOD2 in mice leads to neuropathology which mimics phenotypes associated with multiple sclerosis [25].

Matrix H_2O_2 is reduced by peroxidases, including GSH peroxidases and peroxiredoxins 3 and 5 [26] (Figure 5.2A). The peroxiredoxins can undergo multiple post-translational modifications, which provides for specificity of redox signalling [27]. Matrix H_2O_2 can diffuse across the inner mitochondrial membrane(IMM) and the outer mitochondrial membrane (OMM) and thus has the potential to initiate redox signalling. H_2O_2 diffusion is modulated by the composition of the membranes to be crossed, the existence of dedicated H_2O_2 channels (porins/VDACs in the OMM), and potentially peroxiporins in the IMM like it has been described for the plasma and ER membranes. However, it remains disputed whether they also localize to mitochondria [28, 29]), the mitochondrial ultrastructure (i.e. the dynamics and form of cristae and cristae junctions), but even more so by the activity of local antioxidative machineries (peroxiredoxins, thioredoxin and glutathione systems as well as NADPH supply) (Figure 5.2A). For example, increasing the activity of antioxidative systems in the IMS by expressing an IMS-targeted PRDX5 strongly limits H_2O_2 release [30]. Likewise, cytosolic antioxidative machineries restrict the pathlength that H_2O_2 emanating from mitochondria can diffuse [19]. These steep gradients around mitochondria challenge classical views of mitochondrial H_2O_2 signalling and indicate that information transfer (i.e. oxidation of targets by H_2O_2) likely takes place close to the mitochondrial surface. This also opens the potential to signal at specific sites (foci) of the mitochondrial network or at contact sites to other organelles.

HETEROGENEITY OF H_2O_2 RELEASE

The intimate link of H_2O_2 generating and handling machineries with metabolisms as well as the resulting short diffusion distances of H_2O_2 form the mechanistic basis for interesting observations with respect to cellular spatiotemporal heterogeneity of H_2O_2 levels (Figure 5.2B).

How can spatiotemporal heterogeneity of H_2O_2 levels be assessed? A palette of genetically encoded ratiometric fluorescent sensors now allows the dynamic monitoring of small molecules within intact cells. One advantage of these sensors is the possibility to target them to highly specific sites within the cell, for example, to the ER-mitochondria contact sites or to the peripheral IMS (and not the cristae). For H_2O_2, two different families of genetically encoded sensors exist, the HyPer- and the roGFP2-based families of [31, 32] (Figure 5.2C). Both have in common that H_2O_2 mediates sensor oxidation, while different local reductive systems then reset the sensors. Importantly, this means that these sensors not only report on local H_2O_2 levels but integrate this information with the activity of local reducing system. They thereby represent information as proteins within this location would experience the interplay between oxidizing and reducing systems. roGFP2-based sensors, for example, roGFP2-Tsa2ΔC_R [7], have been used to assess H_2O_2 dynamics mainly in yeast. These studies unravelled distinct differences in H_2O_2 handling in mitochondria and a role of the mitochondria peroxiredoxin system in cell death mediation [33, 34]. The sensor also allowed monitoring of H_2O_2 dynamics under different metabolic conditions [7]. In mammalian cells, a different type of so-called HyPer sensors, with HyPer7 being the latest family member, appear to be superior in sensitivity [5]. These sensors allowed the monitoring of mitochondrial H_2O_2 release and the dynamics of intracellular H_2O_2 fluxes [5, 19].

Which insights did these sensors provide on mitochondrial H_2O_2 release and handling dynamics? Using the sensors, steep gradients of H_2O_2 around mitochondria were observed [19]. Since this rapid scavenging of H_2O_2 in the cytosol depends on cytosolic antioxidative machineries, a mechanism of regulation of H_2O_2 signalling might be the exclusion of the involved antioxidative enzymes from specific sites close to the mitochondrial surface. This could be, for example, at ER-mitochondria contact sites which are involved in different signalling modalities, including calcium signalling. Interestingly, crosstalk between H_2O_2 and calcium signalling was observed by specifically monitoring H_2O_2 levels and dynamics at ER-mitochondria contact sites. Calcium influx into mitochondria thereby led to a constriction of cristae that led to an expulsion of local H_2O_2 and the formation of H_2O_2 nanodomains. [35]. Furthermore, transfer of H_2O_2 from mitochondria to the ER was shown to occur in these ER-mitochondria contact sites through an ER membrane peroxiporin, aquaporin-11 [36]. This intimate interplay between mitochondria, ER and often also peroxisomes has been dubbed as "redox triangle" [37]. H_2O_2 signalling could also be influenced by the general positioning of mitochondria within the cell. During hypoxia, mitochondria cluster around the nucleus, and it has been shown that this generates an oxidative niche that is critical for hypoxia signalling [38]. Mitochondrial repositioning within the cell is guided by Miro1 which is also responsible for mitochondrial transport in larger cells such as neurons. It has been shown using HyPer7 that loss of Miro1 severely restricts cellular H_2O_2 signalling to the perinuclear space which directly affects intracellular responses [39, 40].

Not only in space but also in time, heterogeneity in H_2O_2 dynamics was observed using H_2O_2 sensors. Fluctuations in the metabolic state of the cell influence the H_2O_2 level [41]. During the so-called metabolic cycle in yeast cells (that might have equivalents in mammalian cells), the level of H_2O_2 strongly fluctuates in time and impact on cell cycle and division [42]. Moreover, there appears to exist intimate crosstalk between NADPH oxidases and mitochondria in H_2O_2 signalling [43], providing a further link between metabolism and redox signalling. Mitochondrial H_2O_2 dynamics is also closely linked to the local membrane potential in the mitochondrial network. Higher membrane potential was associated with increased superoxide generation. Since the membrane potential was shown to fluctuate within the mitochondrial network and individual mitochondria even within the same cell [44–47], local nanodomains of H_2O_2 release and signalling can occur.

Such membrane potential heterogeneity appears to be different between different cell types and mainly depend on local differences in respiratory chain activity [45]. Interestingly, not only the dynamics and activity of distinct sites in the mitochondrial network appear to cause differences in H_2O_2 release, the H_2O_2 level also affects mitochondrial redox network stability [48], giving rise to regulatory feedback loops. Such feedback loops can also be caused by changes in mitochondrial morphology and release capacity that are H_2O_2-induced. For example, oxidation of cysteines in OPA1 or VDACs have been shown to occur and might alter the dynamics of mitochondrial H_2O_2 handling and release [49, 50].

CONCLUSION AND PERSPECTIVE

The past years have significantly expanded our knowledge about the identity of reactive species, the intracellular mechanisms of H_2O_2 handling and redox signalling, and the targets and physiological outcomes of ROS signalling and damage. This was mainly due to the development of new tools, including genetically encoded fluorescent sensors that allowed monitoring fluxes of specific redox metabolites with extremely high spatiotemporal resolution, and tools to locally and acutely provoke the generation of reactive oxygen species without the need to use inhibitors. In the future, the expanding portfolio of sensors with different sensitivities and colours will allow multiplexing of sensors to better explore the crosstalk of redox metabolites between different compartments. Moreover, additional tools to engineer metabolism might be helpful to specifically and locally modulate, for example, $NADPH/NADP^+$ and $NADH/NAD^+$ ratios or to purposefully generate other reactive species.

REFERENCES

1. Chance B, Sies H, Boveris A (1979) Hydroperoxide metabolism in mammalian organs. *Physiol Rev* 59:527–605
2. Murphy MP (2009) How mitochondria produce reactive oxygen species. *Biochem J* 417:1–13
3. Flohe L (2020) Looking back at the early stages of redox biology. *Antioxidants (Basel)* 9:1254
4. Belousov VV, Fradkov AF, Lukyanov KA, Staroverov DB, Shakhbazov KS, Terskikh AV, Lukyanov S (2006) Genetically encoded fluorescent indicator for intracellular hydrogen peroxide. *Nat Methods* 3:281–286
5. Pak VV, Ezeriņa D, Lyublinskaya OG, Pedre B, Tyurin-Kuzmin PA, Mishina NM, Thauvin M, Young D, Wahni K, Martínez Gache SA, Demidovich AD, Ermakova YG, Maslova YD, Shokhina AG, Eroglu E, Bilan DS, Bogeski I, Michel T, Vriz S, Messens J, Belousov VV (2020) Ultrasensitive genetically encoded indicator for hydrogen peroxide identifies roles for the oxidant in cell migration and mitochondrial function. *Cell Metab* 31:642–653
6. Gutscher M, Sobotta MC, Wabnitz GH, Ballikaya S, Meyer AJ, Samstag Y, Dick TP (2009) Proximity-based protein thiol oxidation by H_2O_2-scavenging peroxidases. *J Biol Chem* 284:31532–31540
7. Morgan B, Van Laer K, Owusu TN, Ezeriņa D, Pastor-Flores D, Amponsah PS, Tursch A, Dick TP (2016) Real-time monitoring of basal H_2O_2 levels with peroxiredoxin-based probes. *Nat Chem Biol* 12:437–443
8. Picard M, Shirihai OS (2022) Mitochondrial signal transduction. *Cell Metab* 34:1620–1653
9. Sies H, Jones DP (2020) Reactive oxygen species (ROS) as pleiotropic physiological signalling agents. *Nat Rev Mol Cell Biol* 21:363–383
10. Zhang Y, Wong HS (2021) Are mitochondria the main contributor of reactive oxygen species in cells? *J Exp Biol* 224:221606
11. Zhang BY, Pan CY, Feng C, Yan CQ, Yu YJ, Chen ZL, Guo CJ, Wang XX (2022) Role of mitochondrial reactive oxygen species in homeostasis regulation. *Redox Rep* 27:45–52
12. Brand MD (2020) Riding the tiger – physiological and pathological effects of superoxide and hydrogen peroxide generated in the mitochondrial matrix. *Crit Rev Biochem Mol Biol* 55:592–661
13. Mailloux RJ (2020) An update on mitochondrial reactive oxygen species production. *Antioxidants (Basel)* 9:472
14. Stein KT, Moon SJ, Nguyen AN, Sikes HD (2020) Kinetic modeling of H_2O_2 dynamics in the mitochondria of HeLa cells. *PLoS Comput Biol* 16:e1008202

15. Mailloux RJ, Grayson C, Koufos O (2023) Regulation of mitochondrial hydrogen peroxide availability by protein S-glutathionylation. *Cells* 12:107
16. Giorgio M, Trinei M, Migliaccio E, Pelicci PG (2007) Hydrogen peroxide: a metabolic by-product or a common mediator of ageing signals? *Nat Rev Mol Cell Biol* 8:722–728
17. Sies H, Belousov VV, Chandel NS, Davies MJ, Jones DP, Mann GE, Murphy MP, Yamamoto M, Winterbourn C (2022) Defining roles of specific reactive oxygen species (ROS) in cell biology and physiology. *Nat Rev Mol Cell Biol* 23:499–515
18. Sies H (2021) Oxidative eustress: On constant alert for redox homeostasis. *Redox Biol* 41:101867
19. Hoehne MN, Jacobs L, Lapacz KJ, Calabrese G, Murschall LM, Marker T, Kaul H, Trifunovic A, Morgan B, Fricker M, Belousov VV, Riemer J (2022) Spatial and temporal control of mitochondrial H_2O_2 release in intact human cells. *EMBO J* 41:e109169
20. Chouchani ET, Pell VR, James AM, Work LM, Saeb-Parsy K, Frezza C, Krieg T, Murphy MP (2016) A unifying mechanism for mitochondrial superoxide production during ischemia-reperfusion injury. *Cell Metab.* 23:254–263
21. Suzuki Y, Ali M, Fischer M, Riemer J (2013) Human copper chaperone for superoxide dismutase 1 mediates its own oxidation-dependent import into mitochondria. *Nat Commun* 4:2430
22. Hart PC, Mao M, de Abreu ALP, Ansenberger-Fricano K, Ekoue DN, Ganini D, Kajdacsy-Balla A, Diamond AM, Minshall RD, Consolaro MEL, Santos JH, Bonini MG (2015) MnSOD upregulation sustains the Warburg effect via mitochondrial ROS and AMPK-dependent signalling in cancer. *Nat Commun* 6:6053
23. Tsang CK, Chen M, Cheng X, Qi YM, Chen Y, Das I, Li XX, Vallat B, Fu LW, Qian CN, Wang HY, White E, Burley SK, Zheng XFS (2018) SOD1 Phosphorylation by mTORC1 couples nutrient sensing and redox regulation. *Mol Cell* 70:502–515
24. He CX, Danes JM, Hart PC, Zhu YM, Huang YP, de Abreu AL, O'Brien J, Mathison AJ, Tang BW, Frasor JM, Wakefield LM, Ganini D, Stauder E, Zielonka J, Gantner BN, Urrutia RA, Gius D, Bonini MG (2019) SOD2 acetylation on lysine 68 promotes stem cell reprogramming in breast cancer. *P Natl Acad Sci USA* 116:23534–23541
25. Bhaskaran S, Kumar G, Thadathil N, Piekarz KM, Mohammed S, Lopez SD, Qaisar R, Walton D, Brown JL, Murphy A, Smith N, Saunders D, Beckstead MJ, Plafker S, Lewis TL, Jr., Towner R, Deepa SS, Richardson A, Axtell RC, Van Remmen H (2022) Neuronal deletion of MnSOD in mice leads to demyelination, inflammation and progressive paralysis that mimics phenotypes associated with progressive multiple sclerosis. *Redox Biol* 59:102550
26. Cox AG, Winterbourn CC, Hampton MB (2009) Mitochondrial peroxiredoxin involvement in antioxidant defence and redox signalling. *Biochem J* 425:313–325
27. Bolduc J, Koruza K, Luo T, Malo Pueyo J, Vo TN, Ezeriņa D, Messens J (2021) Peroxiredoxins wear many hats: Factors that fashion their peroxide sensing personalities. *Redox Biol* 42:101959
28. Ishibashi K, Tanaka Y, Morishita Y (2021) The role of mammalian superaquaporins inside the cell: An update. *Biochim Biophys Acta Biomembr* 1863:183617
29. Bienert GP, Møller AL, Kristiansen KA, Schulz A, Møller IM, Schjoerring JK, Jahn TP (2007) Specific aquaporins facilitate the diffusion of hydrogen peroxide across membranes. *J Biol Chem* 282:1183–1192
30. Sabharwal SS, Waypa GB, Marks JD, Schumacker PT (2013) Peroxiredoxin-5 targeted to the mitochondrial intermembrane space attenuates hypoxia-induced reactive oxygen species signalling. *Biochem J* 456:337–346
31. Bilan DS, Belousov VV (2018) In vivo imaging of hydrogen peroxide with HyPer probes. *Antioxid Redox Signal* 29:569–584
32. Roma LP, Deponte M, Riemer J, Morgan B (2018) Mechanisms and applications of redox-sensitive green fluorescent protein-based hydrogen peroxide probes. *Antioxid Redox Signal* 29:552–568
33. Pastor-Flores D, Talwar D, Pedre B, Dick TP (2020) Real-time monitoring of peroxiredoxin oligomerization dynamics in living cells. *Proc Natl Acad Sci USA* 117:16313–16323
34. Calabrese G, Peker E, Amponsah PS, Hoehne MN, Riemer T, Mai M, Bienert GP, Deponte M, Morgan B, Riemer J (2019) Hyperoxidation of mitochondrial peroxiredoxin limits H(2) O(2)-induced cell death in yeast. *EMBO J* 38:e101552
35. Booth DM, Enyedi B, Geiszt M, Varnai P, Hajnoczky G (2016) Redox nanodomains are induced by and control calcium signaling at the ER-mitochondrial interface. *Mol Cell* 63:240–248
36. Sorrentino I, Galli M, Medrano-Fernandez I, Sitia R (2022) Transfer of H_2O_2 from Mitochondria to the endoplasmic reticulum via Aquaporin-11. *Redox Biol.* 55: 102410
37. Yoboue ED, Sitia R, Simmen T (2018) Redox crosstalk at endoplasmic reticulum (ER) membrane contact sites (MCS) uses toxic waste to deliver messages. *Cell Death Dis* 9:331

38. Al-Mehdi AB, Pastukh VM, Swiger BM, Reed DJ, Patel MR, Bardwell GC, Pastukh VV, Alexeyev MF, Gillespie MN (2012) Perinuclear mitochondrial clustering creates an oxidant-rich nuclear domain required for hypoxia-induced transcription. *Sci Signal* 5:ra47
39. Alshaabi H, Shannon N, Gravelle R, Milczarek S, Messier T, Cunniff B (2021) Miro1-mediated mitochondrial positioning supports subcellular redox status. *Redox Biol* 38:101818
40. Saxton WM, Hollenbeck PJ (2012) The axonal transport of mitochondria. *J Cell Sci* 125:2095–2104
41. Amponsah PS, Yahya G, Zimmermann J, Mai M, Mergel S, Muhlhaus T, Storchova Z, Morgan B (2021) Peroxiredoxins couple metabolism and cell division in an ultradian cycle. *Nat Chem Biol* 17:477–484
42. Amponsah PS, Yahya G, Zimmermann J, Mai M, Mergel S, Muhlhaus T, Storchova Z, Morgan B (2021) Peroxiredoxins couple metabolism and cell division in an ultradian cycle. *Nat Chem Biol* 17:477–484
43. Fukai T, Ushio-Fukai M (2020) Cross-talk between NADPH oxidase and mitochondria: Role in ROS signaling and angiogenesis. *Cells* 9:1849
44. Koren S, De la Rosa L, Horn J, Wojtovich A (2022) Dual optogenetic generation and detection of reactive oxygen species throughout mitochondrial microdomains. *Free Radic Biol Med* 180:93–94
45. Rovini A, Heslop K, Hunt EG, Morris ME, Fang DN, Gooz M, Gerencser AA, Maldonado EN (2021) Quantitative analysis of mitochondrial membrane potential heterogeneity in unsynchronized and synchronized cancer cells. *Faseb J* 35:e21148
46. Wolf DM, Segawa M, Kondadi AK, Anand R, Bailey ST, Reichert AS, van der Bliek AM, Shackelford DB, Liesa M, Shirihai OS (2019) Individual cristae within the same mitochondrion display different membrane potentials and are functionally independent. *EMBO J* 38:e101056
47. Kondadi AK, Anand R, Hänsch S, Urbach J, Zobel T, Wolf DM, Segawa M, Liesa M, Shirihai OS, Weidtkamp-Peters S, Reichert AS (2020) Cristae undergo continuous cycles of membrane remodelling in a MICOS-dependent manner. *EMBO Rep* 21:e49776
48. Millare B, O'Rourke B, Trayanova N (2020) Hydrogen peroxide diffusion and scavenging shapes mitochondrial network instability and failure by sensitizing ROS-induced ROS release. *Sci Rep* 10: 15758–71308
49. Semenzato M, Kohr MJ, Quirin C, Menabò R, Alanova P, Alan L, Pellattiero A, Murphy E, Di Lisa F, Scorrano L (2023) Oxidization of optic atrophy 1 cysteines occurs during heart ischemia-reperfusion and amplifies cell death by oxidative stress. *Redox Biol* 63:102755
50. Reina S, Pittalà MGG, Guarino F, Messina A, De Pinto V, Foti S, Saletti R (2020) Cysteine oxidations in mitochondrial membrane proteins: The case of VDAC isoforms in mammals. *Front Cell Dev Biol* 8:397

6 Mitochondria-Lysosome Crosstalk, in Disease and in Health

Carla Lopes and Nuno Raimundo

INTRODUCTION: PARADIGMS OF MITOCHONDRIA-LYSOSOME COMMUNICATION

It may seem rather evident that all organelles inside one cell have to behave cooperatively in order to fulfil the cell's role. However, it is less evident how that cooperation is reached, and what are the underlying mechanisms. Mitochondria and lysosomes carry out the bulk of cellular catabolism while impacting signalling hubs such as AMPK and mTORC1, which then regulate the balance between anabolism and catabolism.

The metabolic roles of mitochondria and lysosomes are long imprinted as cellular power plant and recycling center, respectively (Ballabio, 2016; Pagliarini & Rutter, 2013). More recently, it became clear that beyond their roles in metabolism, these organelles also function as key cellular signalling platforms (Lawrence & Zoncu, 2019; Raimundo, 2014). Importantly, mitochondrial malfunction is an established hallmark of ageing (López-Otín et al., 2013) and lysosomes are progressively impaired during ageing (Carmona-Gutierrez et al., 2016). Defects in these two organelles have also been implicated in several age-related pathologies, particularly neurodegenerative diseases (Plotegher & Duchen, 2017). While mitochondria and lysosomes were mostly studied as individual entities, recent work has revealed that mitochondria and lysosomes are interdependent (Deus et al., 2020). Acute mitochondrial stress triggers lysosomal biogenesis (Fernández-Mosquera et al., 2017), while chronic mitochondrial malfunction blocks both lysosomal function and biogenesis, resulting in lysosomal saturation akin to lysosomal storage diseases (LSDs) (Fernandez-Mosquera et al., 2019). In many lysosomal storage diseases, mitochondrial biogenesis and function are repressed, and perturbation of the signalling pathways leading to that can restore basal mitochondrial biogenesis and function (Yambire, Fernandez-Mosquera et al., 2019). Furthermore, impaired lysosomal acidification (Yambire, Rostosky et al., 2019), a common feature in ageing cells (Carmona-Gutierrez et al., 2016), perturbs intracellular iron homeostasis, resulting in excessive mitophagy and repression of mitochondrial biogenesis (Yambire, Rostosky et al., 2019). Thus, while in physiological conditions, mitochondria and lysosomes seem to have a cooperative relationship, with each adapting to acute episodes of stress in the other. In chronic stress, the crosstalk is inhibitory: primary defects in either mitochondria or lysosomes result in impairment of the other organelle (Plotegher & Duchen, 2017).

There are several platforms for mitochondria and lysosomes to communicate with each other (see Figure 6.1). The most obvious one is mitophagy, in which mitochondrial contents are packaged inside autophagosomes and delivered to lysosomes for degradation. Mitochondria also release vesicles (mitochondria-derived vesicles, discussed in more detail further below), some of which are targeted to lysosomes. Both of these pathways result in the delivery of mitochondrial cargo to lysosomes. Furthermore, mitochondria and lysosomes can form membrane contact sites with each other, which keep the organelles in close proximity. Finally, mitochondria and lysosomes can impact each other remotely via signalling pathways, as mentioned above. How the delivery of mitochondrial cargo to lysosomes, or the close association between mitochondria and lysosomes, affects

 DOI: 10.1201/9781003221180-9

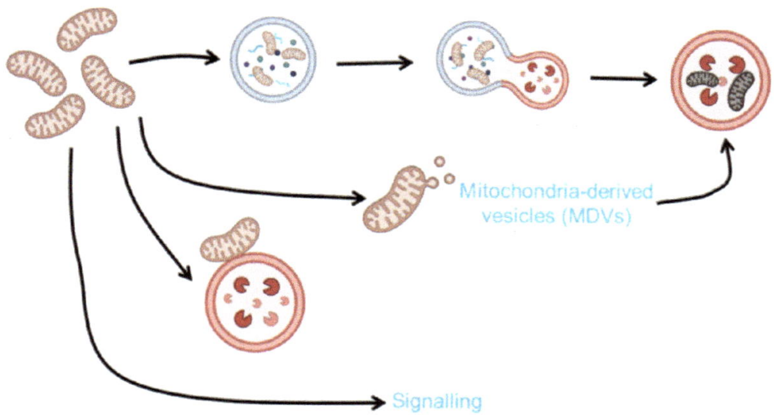

FIGURE 6.1 Pathways of mitochondria-lysosome communication. Mitophagy and mitochondria-derived vesicles bring mitochondrial contents to the lysosomes where they can be sensed by specific receptors. Furthermore, mitochondria and lysosomes form membrane contact sites. Finally, mitochondria and lysosome can engage signalling pathways to remotely communicate with each other.

the coordination between mitochondria, lysosomes, and metabolism is an area of active research that just started being explored, and at this point there are more questions than answers.

MITOPHAGY

The process of mitophagy consists of the selective or non-selective labelling of mitochondria and their enclosure inside autophagosomes. These then fuse with lysosomes and the catabolic machinery of the lysosomes degrades the contents of the autophagosomes and recycles building blocks back to the cytoplasm and other cellular organelles (Figure 6.2). Mitophagy is one pathway involved in mitochondrial quality control but can also occur as part of developmental programs, such as the elimination of mitochondria in reticulocytes to allow erythrocyte maturation or iron deficiency-induced mitophagy. The distinction between these determines if the labelling of mitochondria is selective (for those mitochondria that are dysfunctional) or non-selective (all mitochondria are labelled).

There are multiple regulatory steps on mitophagy. The first consists in the labelling of the mitochondria. This is carried out by proteins that have an LIR domain, which allows them to bind both to LC3 and ubiquitin-labelled cargo, thus ensuring that the phagophore expands around the cargo. While the PINK1-Parkin mitophagy pathway is widely studied because of the genetic links of these genes with Parkinson's disease, there are many other mitophagy receptors described, such as Optineurin (OPTN), NDP52, AMBRA1, BNIP3, NIX, among others (Birgisdottir et al., 2013). This subject has been object of several authoritative reviews (Pickles et al., 2018), and will not be discussed further in this chapter.

The second step of mitophagy is common to all other autophagy pathways, in that the autophagosome needs to form and is promoted by the activity of ULK1 and ULK2 (their relative roles are tissue-specific, with ULK2 being more important in brain while ULK1 is active in more tissues). These kinases (orthologues of yeast Atg1) promote the assembly of the machinery necessary to generate the autophagosomal membrane and eventually mature it. The role of autophagosome formation is discussed elsewhere (Vargas et al., 2023). ULK1/2 are regulated by both AMPK (activated) and mTORC1 (repressed) and thus contribute to coordinate autophagy with the status of cellular metabolism. This raises questions regarding what happens to selective mitophagy during periods of anabolism (in which mTORC1 is active and AMPK is repressed), which remain to be answered. Most studies on mitophagy have focused on acute insults to the mitochondrial network

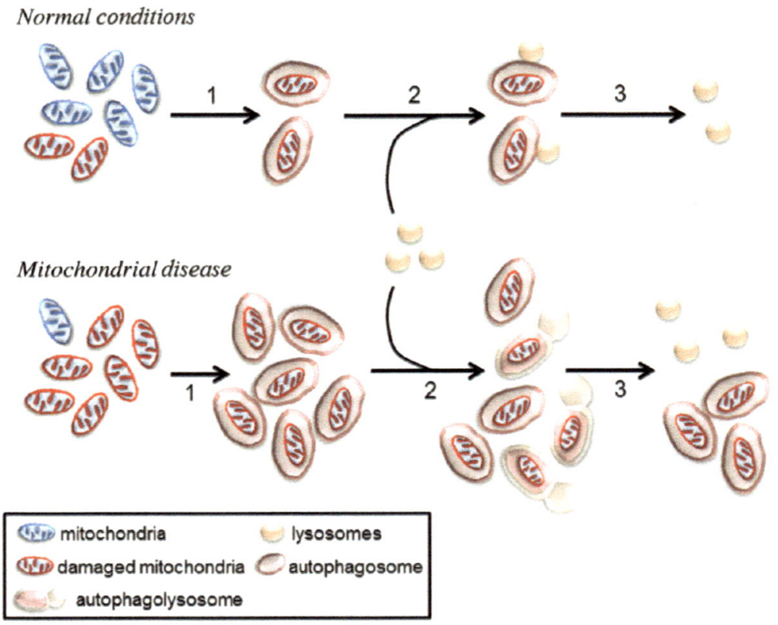

FIGURE 6.2 Graphic representation of mitophagy. Damaged mitochondria are labelled, resulting in the building of autophagosomes around them. Autophagosomes then fuse with lysosomes where the autophagosomal cargo is degraded by the lysosomal hydrolases.

in which AMPK is activated, thus promoting ULK1/2 activity. As the impacts of acute and chronic mitochondrial stress on AMPK signalling and lysosomal biogenesis and function are rather opposite, at least in some cell models and some tissues, it is plausible that acute and chronic mitophagy are also regulated differently, given that in an acute stress the logical response is to remove the damaged mitochondria, but in a chronic stress (e.g., mitochondrial disease or age-related mitochondrial malfunction) the cells must not eliminate their entire mitochondrial network. Future research will likely enlighten these aspects.

A third step of regulation deals with the stimulation of lysosomal capacity to cope with the arrival of autophagosomal cargo. This is handled by TFEB and other transcription factors of the microphtalmia family, which promote a transcriptional program of lysosomal biogenesis. Interestingly, this is also under the control of AMPK, which is needed both for TFEB activity (Young et al., 2016) and lysosomal function (Fernandez-Mosquera et al., 2019). Notably, by stimulating autophagy, lysosomal biogenesis/function, and mitochondrial biogenesis (Malik et al., 2023), AMPK acts as a coordinator of the whole process of removal of damaged mitochondria, their degradation, and their replacement by new mitochondria.

Another aspect that remains to be understood is the role of lysosomal fitness in mitophagy. Most acute mitophagy studies are made in cells that have intact, functional lysosomes, which can cope well with an increase in autophagy (specifically mitophagy) cargo delivery. However, in cells that already have defective lysosomes, it is still unclear if they can cope with an up-regulation of autophagic caused, for example, by excessive mitophagy. This is particularly important when taking into account that chronic mitochondrial malfunction also impairs lysosomes (Deus et al., 2020). It is often observed in tissues from patients and models of mitochondrial disease that there is accumulation of swollen lysosomes loaded with undigested material, as well as autophagosomes that were not delivered to the lysosomes (represented on Figure 6.2, bottom panel). How these events contribute to the onset or progression of disease remains to be determined, and will be discussed further below.

MDVs

Mitochondrial-derived vesicles (MDVs) were first described a decade ago by the McBride group as highly uniform structures formed from the mitochondrial membrane carrying selected mitochondrial cargo and target to the peroxisomes (Neuspiel et al., 2008). In their study, MAPL (mitochondria-anchored protein ligase) enriched structures with 70–100 nm in diameter containing one or both mitochondrial membranes were released from the mitochondria in a DRP1-independent manner and targeted to fuse with a subpopulation of peroxisomes (Neuspiel et al., 2008). Further studies demonstrated that Vps35, Vps26A, Vps29, and MAPL form a retromer complex that regulates the transport of MDVs to peroxisomes (Braschi et al., 2010). More recently, another fate was identified for MDVs targeting the late endosomal/multivesicular body for degradation into lysosomes (Soubannier, McLelland et al., 2012). These MDVs are stimulated during oxidative stress and their contents are enriched in oxidized proteins (Soubannier, McLelland et al., 2012; Soubannier, Rippstein et al., 2012). The selectivity of the incorporated cargo depends on the oxidative damage (Soubannier, Rippstein et al., 2012). Treatment with non-mitochondrial stressors, such as xanthine oxidase/xanthine (XO/X), resulted in single-membraned MDVs enriched in outer mitochondrial membrane proteins (TOM20 or VDAC) and lacking matrix or inner mitochondrial membrane markers (pyruvate dehydrogenase [PDH] or Complex III subunit core2). Contrarily, mitochondria treated with Antimycin A (Complex III inhibitor) generated MDVs with double membranes containing Complex III subunit Core2 (UQCRC2) without VDAC and TOM20 (Soubannier, McLelland et al., 2012; Soubannier, Rippstein et al., 2012). Oxidized mitochondrial components, including cardiolipin, seem to trigger membrane conformational changes from inside to incorporate them into the vesicles (Soubannier, Rippstein et al., 2012).

MDVs are part of an ancient mitochondrial quality control that regulates the clearance of defective mitochondrial proteins and lipids to avoid the degradation of the entire organelle by mitophagy. In addition, MDVs are important mediators involved in organelle crosstalk that have been associated to peroxisomal biogenesis and antigen presentation (Matheoud et al., 2016; Sugiura et al., 2017). The mitochondria bacterial origin can explain its ability to generate and load their content in vesicles (Popov, 2022). Indeed, bacteria can generate membrane vesicles as bioactive carriers to communicate with other bacteria and host cell (Brown et al., 2015; Jeong et al., 2022). The MDVs formation can be mapped to the ancestral archaebacteria vesicle formation, although the existence of diverse molecular mechanisms makes it difficult to understand the precise machinery involved.

The double-membrane MDVs targeted to lysosomes are dependent on the protein kinase PINK1 and the cytosolic ubiquitin E3 ligase Parkin (McLelland et al., 2014). Parkin and PINK1 are part of the conserved machinery that catalyses the polyubiquitination of several substrates at depolarized mitochondria and triggers degradation through the autophagy-lysosome pathway, by the process called mitophagy (Geisler et al., 2010; Okatsu et al., 2010; Sarraf et al., 2013).

In mildly oxidative stress conditions, MDV biogenesis requires both Parkin and PINK1 independently of activation of the canonical autophagic machinery, as it occurs in the absence of Atg5, Rab9, or Beclin-1 (McLelland et al., 2014; Soubannier, McLelland et al., 2012).

Recent studies have uncovered the proteins behind the shuttle to lysosomes via the endosomal system of PINK1/Parkin-dependent MDVs. Syntaxin-17 (Stx17), a Qa-SNARE (soluble N-ethylmaleimide-sensitive factor-attachment protein (SNAP) receptor), is loaded into vesicles and assembles into a ternary SNARE complex with SNAP29 and VAMP7 to mediate MDV–endolysosome fusion (McLelland et al., 2016). Another protein involved in TOM20^{+} MDVs traffic is the endosomal adaptor Tollip (Toll-interacting protein) (Ryan et al., 2020). Tollip mediates cargo trafficking through the association with the endosomal phosphatidylinositol 3-phosphate (PtdIns(3)P) and binding of ubiquitylated cargo (Xiao et al., 2015). Tollip is required for MDV acquisition of PtdIns(3)P-enriched membrane prior to Parkin recruitment (Ryan et al., 2020).

A recent Parkin-independent pathway targeting TOM20$^+$ MDV to lysosomes revealed an essential role for DRP1 in the scission sites (König et al., 2021). Prior to MDVs biogenesis, the pulling forces of microtubule-associated motor proteins, MIRO1 and MIRO2 (MIRO1/2), generate thin mitochondrial-membrane protrusions. The DRP1 receptors MID49, MID51, and MFF recruit DRP1 to the forming mitochondrial protrusions and assemble in structures close to the tip. Finally, MDVs are formed through a DRP1-mediated scission and transported to the endolysosomal pathway for degradation (König et al., 2021).

Interestingly, a recent study calculated that autophagy was responsible for only 35% of mitochondrial protein turnover and Parkin-dependent mitophagy accounted for 25%, suggesting that these processes preferentially eliminate the entire dysfunctional mitochondrial (Vincow et al., 2019). This work highlights the importance of non-autophagic mechanisms, including mitochondrial proteases, the ubiquitin proteasome system, and MDVs for mitochondrial protein turnover (Vincow et al., 2019). Indeed, most cells lacking the core autophagy genes struggle to survive. However, a few clones of cancer cells were able to endure and maintain mitochondrial homeostasis due to increased mitochondrial dynamics and MDVs generation (Towers et al., 2021). When LC3-mediated auto(mito)phagy is impaired, cells show a dependency on endocytic protein sorting nexin 9 (SNX9)-mediated MDVs to traffic damaged mitochondrial components to lysosomes, providing a mechanism for quality control (Towers et al., 2021). This highlights how different mechanisms of mitochondria-lysosome crosstalk are in place and can compensate for one another.

Our knowledge of tissue-specific cargoes incorporated in MDVs is scarce. A study using a tissue-specific approach to identify the full proteome of brain MDVs revealed an enrichment of OXPHOS and small TIM chaperone proteins. Curiously, the comparative MDVs proteome analysis of brains of wild-type mice vs PINK1$^{-/-}$ and Parkin$^{-/-}$ mice was unable to identify significative differences in the content, suggesting that Parkin/PINK1-dependent MDVs formation has low representativity in the brain (Roberts et al., 2021). Another approach has shown that MDVs generated by isolated cardiac mitochondria under oxidative stress are enriched in proteins associated with metabolic and redox processes, including OXPHOS complexes, complexes III and V, and the Fe-S cluster (Vasam et al., 2021). Interestingly, under oxidative stress the MDVs incorporate mainly components vulnerable to oxidative damage, including proteins containing hyper-reactive cysteine residues, metabolic redox switches, comprising redox enzymes and those that mediate iron metabolism probably to prevent mitochondrial iron overload (Vasam et al., 2021). In this sense, MDVs offer a route to eliminate irreversibly damaged proteins and prevent mitochondrial iron overload.

Remarkably, the TOM20$^+$ MDV cargo included all transmembrane β-barrel proteins and the TOM import complex along with assembled protein complexes (König et al., 2021). Besides the protein machinery, phospholipids are essential for MDVs formation. The lipidome of TOM$^+$ MDV showed a strong reduction in cardiolipin levels and a significant increase of phosphatidic acid compared to the outer mitochondrial membrane (OMM) (König et al., 2021). Moreover, this subset of vesicles showed increased content in GPAM and AGPAT5, two enzymes responsible for the *de novo* biogenesis of phosphatidic acid. Phosphatidic acid are glycerophospholipids that regulates mitochondrial division and fusion in the OMM through the inhibition of Drp1 and also regulates membrane curvature and dynamin-dependent vesicle formation (Adachi et al., 2018; Kameoka et al., 2018).

A new functional role for MDVs has emerged as regulators of *de novo* peroxisome biogenesis (Sugiura et al., 2017). Several peroxisomal membrane proteins, including Pex3 and Pex14, are released from mitochondria within vesicles to efficiently generate functional peroxisomes (Sugiura et al., 2017). Additionally, MAPL, a protein loaded into MDVs to be delivered to peroxisomes, seems to be involved in peroxisome elongation and possibly fission and biogenesis in a cell-type specific manner (Mohanty et al., 2021). MAPL is a mitochondrial SUMO (small ubiquitin-like modifier) E3 ligase that positively regulates mitochondrial fission through the SUMOylation of DRP1, and growing evidences suggest that it also modulates peroxisome fission (Braschi et al., 2009; Mohanty et al., 2021).

MITOCHONDRIA-LYSOSOME CONTACT SITES

Mitochondria and lysosomes form membrane contact sites with each other. In these membrane appositions, the mitochondrial outer membrane and the lysosomal membrane are less than 30 nm apart (Wong et al., 2018)

Contact sites between organelles are maintained by protein tethers. For mitochondria-lysosome contacts, the tethers currently known include Rab7 and GDAP1. Rab7 exists in Rab7-GTP and GDP-bound Rab7, and the Rab7-GTP form is the one that is able to tether mitochondria and lysosomes. The protein TBC1D15, which replaces GTP with GDP on Rab7, acts as a negative regulator of the mitochondria-lysosome contact sites (Wong et al., 2018). GDAP1 is an atypical glutathione-S-transferase localized to the mitochondrial outer membrane, which is also necessary for lysosomal function. It interacts directly with the lysosomal membrane protein LAMP1, and the loss of GDAP1 results in both lysosomal impairment and decreased number of mitochondria-lysosome contact sites (Cantarero et al., 2021). Mutations in GDAP1 are one of the causes of Charcot-Marie-Tooth neuropathy, which opens the question of whether the mitochondria-lysosome contacts are a component of the pathology in this disease.

The physiological role of these contact sites remains poorly understood. It has been shown that Ca^{2+} released from the lysosomes at these contacts can trigger DRP1-dependent mitochondrial fission. However, it is unclear if lysosome-induced mitochondrial fission is just another form of mitochondrial fission. In particular, it remains to be determined if it shares the same downstream mechanisms as from other mechanisms in which, for example, the endoplasmic reticulum wraps around mitochondria to mark the fission site.

Besides exchange of ions, contact sites are also optimized platforms for exchange of metabolites and lipids. It remains so far unknown if mitochondria-lysosome contacts are used for lipid or metabolite transfer, although the role of lysosomes in the recycling of amino acids and the occurrence of the citrate cycle in mitochondria raise the tantalizing possibility that these contacts may have a role in anaplerosis of the citrate cycle. It also remains to be determined if these contacts have any role in the transfer of mitochondrial contents to the lysosomes, for example, through microautophagy.

The discovery of the mitochondria-lysosome contacts is very recent, and at present there are more questions than answers. The near future will certainly shed more light on the existence of further tethers, their regulation, and which physiological roles these contacts play in different tissues and cell types.

MITOCHONDRIA-LYSOSOME COMMUNICATION AND INFLAMMATION

The occurrence of inflammatory signalling has been reported in mitochondrial diseases (Rodríguez-Nuevo et al., 2018), lysosomal storage diseases (LSDs) (Vitner et al., 2010), as well as in ageing (chronic inflammation or "inflammaging") and age-related diseases (Nichols et al., 2019). It also starts to be appreciated that chronic inflammation is not simply a side effect of the disease progression but an actual contributor (Hunter, 2012). Pro-inflammation mechanisms have been studied in great detail, however almost exclusively in immune cells, with little understanding of inflammatory signalling in non-immune tissues.

The lysosome is a starting point for several pro-inflammatory pathways, as several Toll-like receptors (TLRs) are located at the endo/lysosomal membrane (TLR3, TLR7/8, TLR9) (Lee & Barton, 2014). TLRs are sensors for diverse signals associated with pathogens or damage, and upon recognition of those signals triggers an inflammatory cascade (Lee & Barton, 2014). In functional lysosomes and cells, the recognition of a damage/pathogenic signal triggers an acute inflammatory reaction followed by lysosome-dependent degradation of the signal (Kondo et al., 2012).

However, if the lysosomes are impaired and unable to digest molecules efficiently, the signal may remain in the lysosome for a longer period (see Figure 6.3). The endo/lysosomal TLRs respond mostly to RNA or DNA, typically of viral or bacterial origin (Lee & Barton, 2014), although

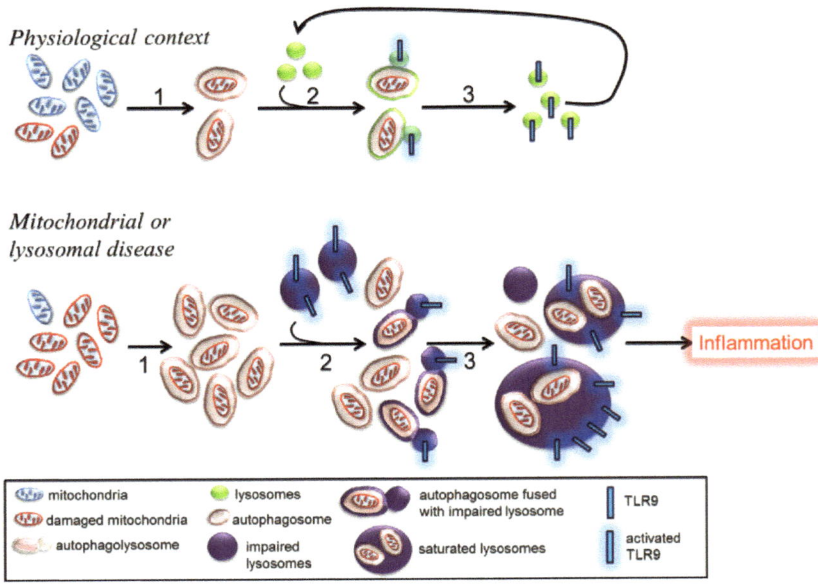

FIGURE 6.3 Mitochondria-lysosomal interaction in disease triggers inflammation. In physiological circumstances, damaged mitochondria are delivered by autophagy to lysosomes. New lysosomes form the autophagolysosome, compensating for the lysosomes consumed during the autophagic process. However, in mitochondrial or lysosomal diseases, lysosomes are impaired and filled with undigested contents. This includes mtDNA, which triggers chronic TLR9 activation and inflammation.

mitochondrial DNA (mtDNA) is also recognized by TLR9 (Lee & Barton, 2014). Mitochondrial contents are routinely delivered to lysosomes by autophagosomes, usually by selective autophagy of damaged or underperforming mitochondria (mitophagy) (Pickles et al., 2018). As pointed out above, mitochondrial contents can also reach the lysosomes via MDVs. In addition to the maintenance of mitochondrial turnover, MDVs were also suggested to be part of an exchange mechanism of mitochondrial components between damage and healthy organelle. Yeast MDVs were shown to have an intact membrane potential and a functional F_1F_O-ATP synthase complex with the capability to produce ATP and rescue respiratory deficient mitochondria (Hazan Ben-Menachem et al., 2023).

From the myriad of biological roles of MDVs, several have been described as having an impact in neurodegenerative disorders, cardiovascular defects, or adaptative immune response.

In the cardiac system, MDVs formation was shown to increase over cardioblast differentiation as a reflex of the increased dependence on the mitochondria metabolic function and also as a response to mitochondrial stressors, particularly doxorubicin (Cadete et al., 2016). Surprisingly, MDVs enriched in BCL-2 isolated from hypoxia-induced myocardial tissues can be uptaken by cardiomyocytes to inhibit hypoxia-induced apoptosis and alleviate myocardial ischemia (Li et al., 2020). MDVs presence was also described in human skeletal muscle disease caused by primary mitochondrial genetic disorders, reinforcing the role as mitochondrial quality control (Cadete et al., 2016; Picca et al., 2023). During osteogenesis, MDVs are secreted from mature osteoblasts to promote differentiation of osteoprogenitors. The osteogenic induction triggers mitochondrial fragmentation and formation of MDVs, which further release into the extracellular milieu that in turn can enhance bone regeneration (Suh et al., 2023).

MDVs mediate the mitochondrial antigen presentation (MitAP) from the mitochondria into the surface of the cell to trigger an immune response (Matheoud et al., 2016). This process is repressed by PINK1 and Parkin which, in Parkinson disease (PD), could be involved in immune dysregulation. PINK1 and Parkin loss of function, as seen in PD, induces the formation of MDVs carrying mitochondrial antigens and trafficking into lysosomes to be processed for antigen presentation at the cell surface and further recognition by the immune cells (Matheoud et al., 2016). Additionally, MDVs transport mtDNA that can be secreted within extracellular vesicles (EVs) and activate inflammatory pathways (Beatriz et al., 2022; Picca et al., 2020).

Syntaphilin (SNPH), a protein that interacts with microtubules to immobilize axonal mitochondria through its mitochondria-targeting sequence, was shown to be involved in the formation of a new class of MDVs independent of Parkin, Drp1, and autophagy (Cheng & Sheng, 2021; Lin et al., 2017). Axonal mitochondria under mild oxidative stress respond by upregulating the retrograde movement to favour the removal of damaged mitochondria. The stressed mitochondria can release SNPH cargo vesicles within seconds that ride along the late endosomes for lysosomal degradation at the soma (Lin et al., 2017). The dynamic budding process ensures a fast response of neurons to mitochondria damage before mitophagy activation in pathological conditions. Recently, mitochondrial membranous protrusions and vesicles were identified in the brain of distant species, indicating the traits are extremely phylogenetically conserved (Yao et al., 2020).

MITOCHONDRIAL-DERIVED VESICLES AND MITOVESICLES

EVs are double-layered lipid membrane-enclosed, nanosized particles secreted into the extracellular space by virtually all types of cells (Beatriz et al., 2021). EVs deliver bioactive cargoes among cells and recently were shown to be enriched in mitochondrial contents (Beatriz et al., 2022; D'Acunzo et al., 2021; Todkar et al., 2021). EV-mediated mitochondrial release seems to be driven by an autophagy/mitophagy impairment, suggesting that when a blockage in the intracellular mitochondrial quality control systems occurs, cells rely on EVs as an alternative route (Beatriz et al., 2022; Crewe et al., 2021; D'Acunzo et al., 2023; Phinney et al., 2015). MDVs and EVs share some endosomal compartments on route to the lysosome. The fusion of MDVs to multivesicular bodies (MVBs) to be released as EVs seems to be mediated by cluster of differentiation 38 (CD38)/cyclic ADP ribose (cADPR) signalling (Suh et al., 2023), SNX9 signalling, optic atrophy 1 (OPA1), and inhibited by Parkin (Todkar et al., 2021). The sorting of mitochondrial components was shown to be enhanced by some factors, such as oxidative stress, ionic homeostasis, mitochondrial proteases, metabolic status, and mitochondrial dynamics (Zhou et al., 2023).

Evidence suggests that the encapsulation of mitochondrial material in EVs is a coping mechanism to maintain cell homeostasis. Mitochondria are recognized as a source of mitochondrial damage-associated molecular patterns (mtDAMPs) that can be released by stressed or dying cells (Deus et al., 2022). Given the endosymbiotic origin of mitochondria, mitochondrial components share similarities with their bacterial equivalents which may trigger cellular immune responses in a process known as "sterile inflammation" (Deus et al., 2022; Grazioli & Pugin, 2018). Recently, a study proposed that cells prevent the secretion of oxidised proteins and consequent activation of the immune system by selectively targeting ROS-induced mitochondrial proteins containing EVs to lysosomes for degradation (Todkar et al., 2021). The selective incorporation of mtDAMPs into EVs depends on the mechanism of MDVs formation. OPA1 and Snx9-dependent MDVs favour the packaging into EVs for extracellular secretion while Parkin directs MDVs for lysosomal degradation to avoid the release of mtDAMPs involved in toxic responses in other organelles or cells (Todkar et al., 2021).

Mutations in fumarate hydratase (FH), an enzyme that catalyses the reversible conversion of fumarate to malate, promotes the generation of SNX9-dependent MDVs and the transport of mtDNA to the cytosol driving an inflammatory response (Zecchini et al., 2023).

Contrarily, mitovesicles can boost immune cell function in specific conditions. Mesenchymal stem cells (MSC) can enhance the phagocytic activity of macrophages through the mitochondria-containing extracellular vesicles and promote an anti-inflammatory phenotype (Ko et al., 2020). Similarly, MSC mitochondria-enriched EVs can induce nuclear translocation of the transcription factor NF-κB, resulting in the inhibition of TLR signalling in macrophages. This mechanism allows the desensitization of macrophages to the transfer of damage mitochondrial components from the MSC (Phinney et al., 2015).

All things considered, EVs mitochondrial cargo effects on the immune system can be ambivalent – from triggering sterile inflammation to enhancing macrophage phagocytic activity. The donor cell is a key determinant, but the nature of the mitochondrial perturbation and the conditions that induce cargo loading and EVs release are the major factors contributing to the beneficial or detrimental outcome of the response.

CONCLUDING REMARKS

Our understanding of mitochondria has evolved extraordinarily from the deep biophysical and biochemical aspects of mitochondrial metabolism towards the way mitochondria interact with the cytoplasm and other cellular organelles. Mitochondrial dynamics is no longer only about fission and fusion but also takes into account the whereabouts of the cell where mitochondria are and whether they are in contact with other organelles. The recent discoveries on how mitochondria and lysosomes are dependent on one another, how they communicate with each other, and the existence of physical membrane contacts between mitochondria and lysosomes raise a wide range of new questions. The near future will certainly illuminate the role of these contact sites in the regulation of metabolism and the signalling hubs that coordinate anabolism and catabolism. Furthermore, a broader understanding of intercellular communication between mitochondria, lysosomes, and other organelles, mediated by vesicle trafficking, is bound to unveil new systemic roles of mitochondrial signaling in inflammation, immunity, and neurodegeneration.

REFERENCES

Adachi, Y., Iijima, M., & Sesaki, H. (2018). An unstructured loop that is critical for interactions of the stalk domain of Drp1 with saturated phosphatidic acid. *Small GTPases*, *9*(6), 472–479. https://doi.org/10.1080/21541248.2017.1321614

Ballabio, A. (2016). The awesome lysosome. *EMBO Molecular Medicine*, *8*(2), 73–76. https://doi.org/10.15252/emmm.201505966

Beatriz, M., Vilaça, R., Anjo, S. I., Manadas, B., Januário, C., Rego, A. C., & Lopes, C. (2022). Defective mitochondria-lysosomal axis enhances the release of extracellular vesicles containing mitochondrial DNA and proteins in Huntington's disease. *Journal of Extracellular Biology*, *1*(10), e65.

Beatriz, M., Vilaça, R., & Lopes, C. (2021). Exosomes: Innocent bystanders or critical culprits in neurodegenerative diseases. *Frontiers in Cell and Developmental Biology*, *9*, 1047.

Birgisdottir, Å., Lamark, T., & Johansen, T. (2013). The LIR motif – crucial for selective autophagy. *Journal of Cell Science*, *126*(Pt 15), 3237–3247. https://doi.org/10.1242/jcs.126128

Braschi, E., Goyon, V., Zunino, R., Mohanty, A., Xu, L., & McBride, H. M. (2010). Vps35 mediates vesicle transport between the mitochondria and peroxisomes. *Current Biology*, *20*(14), 1310–1315. https://doi.org/10.1016/j.cub.2010.05.066

Braschi, E., Zunino, R., & McBride, H. M. (2009). MAPL is a new mitochondrial SUMO E3 ligase that regulates mitochondrial fission. *EMBO Reports*, *10*(7), 748–754. https://doi.org/10.1038/embor.2009.86

Brown, L., Wolf, J. M., Prados-Rosales, R., & Casadevall, A. (2015). Through the wall: extracellular vesicles in Gram-positive bacteria, mycobacteria and fungi. *Nature Reviews Microbiology*, *13*(10), 620–630. https://doi.org/10.1038/nrmicro3480

Cadete, V. J., Deschênes, S., Cuillerier, A., Brisebois, F., Sugiura, A., Vincent, A., … Burelle, Y. (2016). Formation of mitochondrial-derived vesicles is an active and physiologically relevant mitochondrial quality control process in the cardiac system. *Journal of Physiology*, *594*(18), 5343–5362. https://doi.org/10.1113/JP272703

Cantarero, L., Juárez-Escoto, E., Civera-Tregón, A., Rodríguez-Sanz, M., Roldán, M., Benítez, R., … Palau, F. (2021). Mitochondria–lysosome membrane contacts are defective in GDAP1-related Charcot-Marie-Tooth disease. *Human Molecular Genetics*, *29*(22), 3589–3605. https://doi.org/10.1093/hmg/ddaa243

Carmona-Gutierrez, D., Hughes, A. L., Madeo, F., & Ruckenstuhl, C. (2016). The crucial impact of lysosomes in aging and longevity. *Ageing Research Review*, *32*, 2–12. https://doi.org/10.1016/j.arr.2016.04.009

Cheng, X. T., & Sheng, Z. H. (2021). Developmental regulation of microtubule-based trafficking and anchoring of axonal mitochondria in health and diseases. *Developmental Neurobiology*, *81*(3), 284–299. https://doi.org/10.1002/dneu.22748

Crewe, C., Funcke, J. B., Li, S., Joffin, N., Gliniak, C. M., Ghaben, A. L., ... Scherer, P. E. (2021). Extracellular vesicle-based interorgan transport of mitochondria from energetically stressed adipocytes. *Cell Metabolism*, *33*(9), 1853–1868.e1811. https://doi.org/10.1016/j.cmet.2021.08.002

D'Acunzo, P., Pérez-González, R., Kim, Y., Hargash, T., Miller, C., Alldred, M. J., ... Levy, E. (2021). Mitovesicles are a novel population of extracellular vesicles of mitochondrial origin altered in Down syndrome. *Science Advances*, *7*(7). https://doi.org/10.1126/sciadv.abe5085

D'Acunzo, P., Ungania, J. M., Kim, Y., Barreto, B. R., DeRosa, S., Pawlik, M., ... Levy, E. (2023). Cocaine perturbs mitovesicle biology in the brain. *Journal of Extracellular Vesicles*, *12*(1), e12301. https://doi.org/10.1002/jev2.12301

Deus, C. M., Tavares, H., Beatriz, M., Mota, S., & Lopes, C. (2022). Mitochondrial damage-associated molecular patterns content in extracellular vesicles promotes early inflammation in neurodegenerative disorders. *Cells*, *11*(15). https://doi.org/10.3390/cells11152364

Deus, C. M., Yambire, K. F., Oliveira, P. J., & Raimundo, N. (2020). Mitochondria-lysosome crosstalk: from physiology to neurodegeneration. *Trends in Molecular Medicine*, *26*(1), 71–88. https://doi.org/10.1016/j.molmed.2019.10.009

Fernandez-Mosquera, L., Yambire, K. F., Couto, R., Pereyra, L., Pabis, K., Ponsford, A. H., ... Raimundo, N. (2019). Mitochondrial respiratory chain deficiency inhibits lysosomal hydrolysis. *Autophagy*, *15*(9), 1572–1591. https://doi.org/10.1080/15548627.2019.1586256

Fernández-Mosquera, L., Diogo, C. V., Yambire, K. F., Santos, G. L., Luna Sánchez, M., Bénit, P., ... Raimundo, N. (2017). Acute and chronic mitochondrial respiratory chain deficiency differentially regulate lysosomal biogenesis. *Scientific Report*, *7*, 45076. https://doi.org/10.1038/srep45076

Geisler, S., Holmström, K. M., Skujat, D., Fiesel, F. C., Rothfuss, O. C., Kahle, P. J., & Springer, W. (2010). PINK1/Parkin-mediated mitophagy is dependent on VDAC1 and p62/SQSTM1. *Nature Cell Biology*, *12*(2), 119–131. https://doi.org/10.1038/ncb2012

Grazioli, S., & Pugin, J. (2018). Mitochondrial damage-associated molecular patterns: from inflammatory signaling to human diseases. *Frontiers in Immunology*, *9*, 832. https://doi.org/10.3389/fimmu.2018.00832

Hazan Ben-Menachem, R., Lintzer, D., Ziv, T., Das, K., Rosenhek-Goldian, I., Porat, Z., ... Pines, O. (2023). Mitochondrial-derived vesicles retain membrane potential and contain a functional ATP synthase. *EMBO Reports*, *24*(5), e56114. https://doi.org/10.15252/embr.202256114

Hunter, P. (2012). The inflammation theory of disease. The growing realization that chronic inflammation is crucial in many diseases opens new avenues for treatment. *EMBO Reports*, *13*(11), 968–970. https://doi.org/10.1038/embor.2012.142

Jeong, D., Kim, M. J., Park, Y., Chung, J., Kweon, H. S., Kang, N. G., ... Kim, D. (2022). Visualizing extracellular vesicle biogenesis in gram-positive bacteria using super-resolution microscopy. *BMC Biology*, *20*(1), 270. https://doi.org/10.1186/s12915-022-01472-3

Kameoka, S., Adachi, Y., Okamoto, K., Iijima, M., & Sesaki, H. (2018). Phosphatidic acid and cardiolipin coordinate mitochondrial dynamics. *Trends in Cell Biology*, *28*(1), 67–76. https://doi.org/10.1016/j.tcb.2017.08.011

Ko, J. H., Kim, H. J., Jeong, H. J., Lee, H. J., & Oh, J. Y. (2020). Mesenchymal stem and stromal cells harness macrophage-derived amphiregulin to maintain tissue homeostasis. *Cell Reports*, *30*(11), 3806–3820.e3806. https://doi.org/10.1016/j.celrep.2020.02.062

Kondo, T., Kawai, T., & Akira, S. (2012). Dissecting negative regulation of Toll-like receptor signaling. *Trends in Immunology*, *33*(9), 449–458. https://doi.org/10.1016/j.it.2012.05.002

König, T., Nolte, H., Aaltonen, M. J., Tatsuta, T., Krols, M., Stroh, T., ... McBride, H. M. (2021). MIROs and DRP1 drive mitochondrial-derived vesicle biogenesis and promote quality control. *Nature Cell Biology*, *23*(12), 1271–1286. https://doi.org/10.1038/s41556-021-00798-4

Lawrence, R. E., & Zoncu, R. (2019). The lysosome as a cellular centre for signalling, metabolism and quality control. *Nature Cell Biology*, *21*(2), 133–142. https://doi.org/10.1038/s41556-018-0244-7

Lee, B. L., & Barton, G. M. (2014). Trafficking of endosomal Toll-like receptors. *Trends in Cell Biology*, *24*(6), 360–369. https://doi.org/10.1016/j.tcb.2013.12.002

Li, B., Zhao, H., Wu, Y., Zhu, Y., Zhang, J., Yang, G., ... Liu, L. (2020). Mitochondrial-derived vesicles protect cardiomyocytes against hypoxic damage. *Frontiers in Cell and Developmental Biology*, *8*, 214. https://doi.org/10.3389/fcell.2020.00214

Lin, M. Y., Cheng, X. T., Tammineni, P., Xie, Y., Zhou, B., Cai, Q., & Sheng, Z. H. (2017). Releasing syntaphilin removes stressed mitochondria from Axons independent of mitophagy under pathophysiological conditions. *Neuron*, *94*(3), 595–610.e596. https://doi.org/10.1016/j.neuron.2017.04.004

López-Otín, C., Blasco, M. A., Partridge, L., Serrano, M., & Kroemer, G. (2013). The hallmarks of aging. *Cell*, *153*(6), 1194–1217. https://doi.org/10.1016/j.cell.2013.05.039

Malik, N., Ferreira, B. I., Hollstein, P. E., Curtis, S. D., Trefts, E., Weiser Novak, S., ... Shaw, R. J. (2023). Induction of lysosomal and mitochondrial biogenesis by AMPK phosphorylation of FNIP1. *Science, 380*(6642), eabj5559. https://doi.org/10.1126/science.abj5559

Matheoud, D., Sugiura, A., Bellemare-Pelletier, A., Laplante, A., Rondeau, C., Chemali, M., ... Desjardins, M. (2016). Parkinson's disease-related proteins PINK1 and Parkin repress mitochondrial antigen presentation. *Cell, 166*(2), 314–327. https://doi.org/10.1016/j.cell.2016.05.039

McLelland, G. L., Lee, S. A., McBride, H. M., & Fon, E. A. (2016). Syntaxin-17 delivers PINK1/Parkin-dependent mitochondrial vesicles to the endolysosomal system. *Journal of Cell Biology, 214*(3), 275–291. https://doi.org/10.1083/jcb.201603105

McLelland, G. L., Soubannier, V., Chen, C. X., McBride, H. M., & Fon, E. A. (2014). Parkin and PINK1 function in a vesicular trafficking pathway regulating mitochondrial quality control. *EMBO Journal, 33*(4), 282–295. https://doi.org/10.1002/embj.201385902

Mohanty, A., Zunino, R., Soubannier, V., & Dilipkumar, S. (2021). A new functional role of mitochondria-anchored protein ligase in peroxisome morphology in mammalian cells. *Journal of Cellular Biochemistry, 122*(11), 1686–1700. https://doi.org/10.1002/jcb.30114

Neuspiel, M., Schauss, A. C., Braschi, E., Zunino, R., Rippstein, P., Rachubinski, R. A., ... McBride, H. M. (2008). Cargo-selected transport from the mitochondria to peroxisomes is mediated by vesicular carriers. *Current Biology, 18*(2), 102–108. https://doi.org/10.1016/j.cub.2007.12.038

Nichols, M. R., St-Pierre, M. K., Wendeln, A. C., Makoni, N. J., Gouwens, L. K., Garrad, E. C., ... Combs, C. K. (2019). Inflammatory mechanisms in neurodegeneration. *Journal of Neurochemistry, 149*(5), 562–581. https://doi.org/10.1111/jnc.14674

Okatsu, K., Saisho, K., Shimanuki, M., Nakada, K., Shitara, H., Sou, Y. S., ... Matsuda, N. (2010). p62/SQSTM1 cooperates with Parkin for perinuclear clustering of depolarized mitochondria. *Genes Cells, 15*(8), 887–900. https://doi.org/10.1111/j.1365-2443.2010.01426.x

Pagliarini, D. J., & Rutter, J. (2013). Hallmarks of a new era in mitochondrial biochemistry. *Genes & Development, 27*(24), 2615–2627. https://doi.org/10.1101/gad.229724.113

Phinney, D. G., Di Giuseppe, M., Njah, J., Sala, E., Shiva, S., St Croix, C. M., ... Ortiz, L. A. (2015). Mesenchymal stem cells use extracellular vesicles to outsource mitophagy and shuttle microRNAs. *Nature Communications, 6*, 8472. https://doi.org/10.1038/ncomms9472

Picca, A., Beli, R., Calvani, R., Coelho-Júnior, H. J., Landi, F., Bernabei, R., ... Marzetti, E. (2020). Older adults with physical frailty and sarcopenia show increased levels of circulating small extracellular vesicles with a specific mitochondrial signature. *Cells, 9*(4). https://doi.org/10.3390/cells9040973

Picca, A., Guerra, F., Calvani, R., Romano, R., Coelho-Junior, H. J., Bucci, C., ... Marzetti, E. (2023). Mitochondrial-derived vesicles in skeletal muscle remodeling and adaptation. *Seminars in Cell & Developmental Biology, 143*, 37–45. https://doi.org/10.1016/j.semcdb.2022.03.023

Pickles, S., Vigié, P., & Youle, R. J. (2018). Mitophagy and quality control mechanisms in mitochondrial maintenance. *Current Biology, 28*(4), R170–R185. https://doi.org/10.1016/j.cub.2018.01.004

Plotegher, N., & Duchen, M. R. (2017). Mitochondrial dysfunction and neurodegeneration in lysosomal storage disorders. *Trends in Molecular Medicine, 23*(2), 116–134. https://doi.org/10.1016/j.molmed.2016.12.003

Popov, L. D. (2022). Mitochondrial-derived vesicles: recent insights. *Journal of Cellular and Molecular Medicine, 26*(12), 3323–3328. https://doi.org/10.1111/jcmm.17391

Raimundo, N. (2014). Mitochondrial pathology: stress signals from the energy factory. *Trends in Molecular Medicine, 20*(5), 282–292. https://doi.org/10.1016/j.molmed.2014.01.005

Roberts, R. F., Bayne, A. N., Goiran, T., Lévesque, D., Boisvert, F. M., Trempe, J. F., & Fon, E. A. (2021). Proteomic profiling of mitochondrial-derived vesicles in brain reveals enrichment of respiratory complex sub-assemblies and small TIM chaperones. *Journal of Proteome Research, 20*(1), 506–517. https://doi.org/10.1021/acs.jproteome.0c00506

Rodríguez-Nuevo, A., Díaz-Ramos, A., Noguera, E., Díaz-Sáez, F., Duran, X., Muñoz, J. P., ... Zorzano, A. (2018). Mitochondrial DNA and TLR9 drive muscle inflammation upon Opa1 deficiency. *EMBO Journal, 37*(10). https://doi.org/10.15252/embj.201796553

Ryan, T. A., Phillips, E. O., Collier, C. L., Jb Robinson, A., Routledge, D., Wood, R. E., ... Tumbarello, D. A. (2020). Tollip coordinates Parkin-dependent trafficking of mitochondrial-derived vesicles. *EMBO Journal, 39*(11), e102539. https://doi.org/10.15252/embj.2019102539

Sarraf, S. A., Raman, M., Guarani-Pereira, V., Sowa, M. E., Huttlin, E. L., Gygi, S. P., & Harper, J. W. (2013). Landscape of the PARKIN-dependent ubiquitylome in response to mitochondrial depolarization. *Nature, 496*(7445), 372–376. https://doi.org/10.1038/nature12043

Soubannier, V., McLelland, G. L., Zunino, R., Braschi, E., Rippstein, P., Fon, E. A., & McBride, H. M. (2012). A vesicular transport pathway shuttles cargo from mitochondria to lysosomes. *Current Biology, 22*(2), 135–141. https://doi.org/10.1016/j.cub.2011.11.057

Soubannier, V., Rippstein, P., Kaufman, B. A., Shoubridge, E. A., & McBride, H. M. (2012). Reconstitution of mitochondria derived vesicle formation demonstrates selective enrichment of oxidized cargo. *PLOS ONE*, *7*(12), e52830. https://doi.org/10.1371/journal.pone.0052830

Sugiura, A., Mattie, S., Prudent, J., & McBride, H. M. (2017). Newly born peroxisomes are a hybrid of mitochondrial and ER-derived pre-peroxisomes. *Nature*, *542*(7640), 251–254. https://doi.org/10.1038/nature21375

Suh, J., Kim, N. K., Shim, W., Lee, S. H., Kim, H. J., Moon, E., ... Lee, Y. S. (2023). Mitochondrial fragmentation and donut formation enhance mitochondrial secretion to promote osteogenesis. *Cell Metabolism*, *35*(2), 345–360.e347. https://doi.org/10.1016/j.cmet.2023.01.003

Todkar, K., Chikhi, L., Desjardins, V., El-Mortada, F., Pépin, G., & Germain, M. (2021). Selective packaging of mitochondrial proteins into extracellular vesicles prevents the release of mitochondrial DAMPs. *Nature Communications*, *12*(1), 1971. https://doi.org/10.1038/s41467-021-21984-w

Towers, C. G., Wodetzki, D. K., Thorburn, J., Smith, K. R., Caino, M. C., & Thorburn, A. (2021). Mitochondrial-derived vesicles compensate for loss of LC3-mediated mitophagy. *Developmental Cell*, *56*(14), 2029–2042.e2025. https://doi.org/10.1016/j.devcel.2021.06.003

Vargas, J. N. S., Hamasaki, M., Kawabata, T., Youle, R. J., & Yoshimori, T. (2023). The mechanisms and roles of selective autophagy in mammals. *Nature Reviews Molecular Cell Biology*, *24*(3), 167–185. https://doi.org/10.1038/s41580-022-00542-2

Vasam, G., Nadeau, R., Cadete, V. J. J., Lavallée-Adam, M., Menzies, K. J., & Burelle, Y. (2021). Proteomics characterization of mitochondrial-derived vesicles under oxidative stress. *FASEB Journal*, *35*(4), e21278. https://doi.org/10.1096/fj.202002151R

Vincow, E. S., Thomas, R. E., Merrihew, G. E., Shulman, N. J., Bammler, T. K., MacDonald, J. W., ... Pallanck, L. J. (2019). Autophagy accounts for approximately one-third of mitochondrial protein turnover and is protein selective. *Autophagy*, *15*(9), 1592–1605. https://doi.org/10.1080/15548627.2019.1586258

Vitner, E. B., Platt, F. M., & Futerman, A. H. (2010). Common and uncommon pathogenic cascades in lysosomal storage diseases. *Journal of Biological Chemistry*, *285*(27), 20423–20427. https://doi.org/10.1074/jbc.R110.134452

Wong, Y. C., Ysselstein, D., & Krainc, D. (2018). Mitochondria-lysosome contacts regulate mitochondrial fission via RAB7 GTP hydrolysis. *Nature*, *554*(7692), 382–386. https://doi.org/10.1038/nature25486

Xiao, S., Brannon, M. K., Zhao, X., Fread, K. I., Ellena, J. F., Bushweller, J. H., ... Capelluto, D. G. S. (2015). Tom1 modulates binding of Tollip to phosphatidylinositol 3-phosphate via a coupled folding and binding mechanism. *Structure*, *23*(10), 1910–1920. https://doi.org/10.1016/j.str.2015.07.017

Yambire, K. F., Fernandez-Mosquera, L., Steinfeld, R., Mühle, C., Ikonen, E., Milosevic, I., & Raimundo, N. (2019). Mitochondrial biogenesis is transcriptionally repressed in lysosomal lipid storage diseases. *eLife*, *8*. https://doi.org/10.7554/eLife.39598

Yambire, K. F., Rostosky, C., Watanabe, T., Pacheu-Grau, D., Torres-Odio, S., Sanchez-Guerrero, A., ... Raimundo, N. (2019). Impaired lysosomal acidification triggers iron deficiency and inflammation in vivo. *eLife*, *8*. https://doi.org/10.7554/eLife.51031

Yao, P. J., Eren, E., Petralia, R. S., Gu, J. W., Wang, Y. X., & Kapogiannis, D. (2020). Mitochondrial protrusions in neuronal cells. *iScience*, *23*(9), 101514. https://doi.org/10.1016/j.isci.2020.101514

Young, N. P., Kamireddy, A., Van Nostrand, J. L., Eichner, L. J., Shokhirev, M. N., Dayn, Y., & Shaw, R. J. (2016). AMPK governs lineage specification through Tfeb-dependent regulation of lysosomes. *Genes & Development*, *30*(5), 535–552. https://doi.org/10.1101/gad.274142.115

Zecchini, V., Paupe, V., Herranz-Montoya, I., Janssen, J., Wortel, I. M. N., Morris, J. L., ... Frezza, C. (2023). Fumarate induces vesicular release of mtDNA to drive innate immunity. *Nature*, *615*(7952), 499–506. https://doi.org/10.1038/s41586-023-05770-w

Zhou, X., Liu, S., Lu, Y., Wan, M., Cheng, J., & Liu, J. (2023). MitoEVs: a new player in multiple disease pathology and treatment. *Journal of Extracellular Vesicles*, *12*(4), e12320. https://doi.org/10.1002/jev2.12320

Part IV

Mitochondria in Health and Disease

From mtDNA Release to Complex I Assembly

7 mtDNA In and Out

On the Tightrope of Cellular Health

Andrea Irazoki and David Pla-Martín

INTRODUCTION

Inflammation is a tissue response to different stimuli that pursues the elimination of the initial damage or insult. The five signs of inflammation considered as an innate immune response include heat, pain, redness, swelling and loss of function [1]. Because too little inflammation can result in the destruction of the affected tissue and too much inflammation leads to chronic tissue damage, fine-tuned regulation of the inflammatory response is essential for the successful elimination of the initial insult.

Given its evolutionary origin from an ancient endosymbiotic process, many mitochondrial components can be recognized by immune sensors as damage-associated molecular patterns or DAMPs. These include extracellular ATP, succinate and N-formyl peptides, as well as mtDNA. The mitochondrial genome stills harbors remnants of its bacterial ancestor DNA that are potentially immunostimulatory. Indeed, its methylation profile is different to the nuclear DNA, making it appear more as a foreign molecule. Besides, many unique immunogenic nucleic acid species such as uncapped mRNAs, double-stranded RNAs (dsRNAs) and RNA-DNA hybrids generated during normal mtDNA maintenance can be recognized as foreign molecules. When located outside the mitochondrial network they activate cellular immune sensors [2]. Thus, these molecules are also considered as DAMPs.

Although mitochondria have mtDNA repair mechanisms activated upon replication errors, mtDNA is subjected to persistent oxidative damage. 8-Hydroxy-2′-deoxyguanosine (8-OHdG) is produced upon oxidation of the mtDNA, altering base-pairing properties of the molecule and affecting faithful mtDNA replication and translation [3, 4]. Oxidized mtDNA is thus eliminated, as it represents a source for mutations which can be expanded over time during mitochondrial turnover, leading to cellular dysfunction [5]. The release of mtDNA, either in an oxidized or non-oxidized form, or other mitochondrial components can occur as a reaction to cellular and mitochondrial stress, leading to the activation of immune sensors. These events tune the response directed to recruit immunological mechanisms and eliminate the associated damage.

Several pattern recognition receptors (PRRs), including nucleic acid sensors, are located in different subcellular compartments surveilling bacterial or viral infection, ready to generate a rapid primary inflammatory response if necessary. Given that mtDNA harbors similar features to foreign DNA, it can be recognized as a DAMP, resulting in the activation of inflammatory responses. Such sensors include TLR9, inflammasomes and cyclic GMP-AMP synthase (cGAS). Due to the diversity and different subcellular location of the DNA sensors that can recognize mtDNA as an insult, both in immune and non-immune cells, their activation results in a plethora of inflammatory responses (Table 7.1).

In summary, mtDNA mislocation within the cell and activation of the immune pathways represent a crossroad between mitochondrial dysfunction and pro-survival mechanisms (Figure 7.1). Early detection of mitochondrial dysfunction by cellular sensors exemplifies a cellular instrument directed to localize the cellular damage and avoid its expansion. This adds a new level of control required to avoid the amplification of the immunological signal which might lead to catastrophic consequences for tissue homeostasis.

DOI: 10.1201/9781003221180-11

TABLE 7.1
mtDNA as a DAMP for DNA-Sensing PRRs: Cellular DNA Sensors and Associated Inflammatory Responses Triggered by the Engagement of mtDNA

DNA-Sensing PRR	Stimulatory Features	Intracellular Location	Signaling Pathway	Immune Response
TLR9	• Hypomethylated mtDNA • Oxidized mtDNA • mtDNA-protein complex	Endosomal compartment	• MYD88-NFκB axis • MYD88-IRF3 axis	• Cytokine release • Type I IFN response
cGAS	• mtDNA (20–500 bp) • Oxidized mtDNA • mtDNA-mtRNA hybrid • mtDNA-protein complex	Cytosol	• STING-IRF3 axis • STING-NFκB axis	• Type I IFN response • Cytokine release
NLRP3	• Oxidized mtDNA	Cytosol	• ASC-caspase 1 axis	• IL1β and IL18 release
NLRC4	• mtDNA • Oxidized mtDNA	Cytosol	• Caspase 1	• IL1β release
AIM2	• mtDNA	Cytosol	• ASC-caspase 1 axis • Caspase 8	• IL1β and IL18 release • Pyroptosis • Apoptosis
IFI16	• mtDNA	Nucleus Cytosol	• ASC-caspase 1 axis • STING-IRF3 axis	• IL1β and IL18 release • Type I IFN response

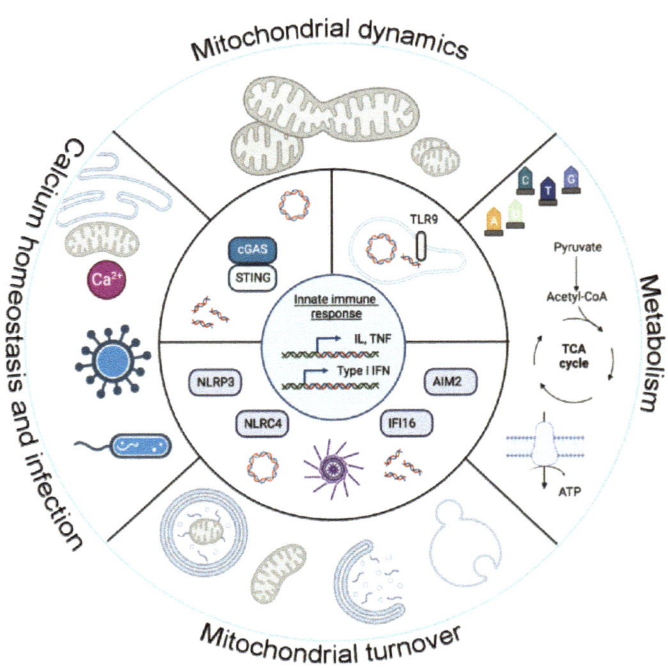

FIGURE 7.1 Mitochondria engagement of the innate immune response. Dysfunctional mitochondria or ill-functioning mitochondrial processes lead to mtDNA release. Detection from the different cellular DNA sensors modulate the activation of signaling cascades from the innate immune response. The expression of pro-inflammatory cytokines and interferon-related genes are directed to resolve the cellular damage.

mtDNA: GENERAL CONSIDERATIONS

The mitochondrial genome (mtDNA) is a small, double-stranded circular DNA molecule that encodes for 37 genes: 13 subunits of the oxidative phosphorylation (OXPHOS) system, 2 ribosomal subunits (12S and 16S) and 22 tRNAs. The encoded proteins are essential subunits of Complex I, III, IV and V, with Complex II being the only one which is completely encoded in the nuclear genome.

In contrast to the nuclear genome, the number of mtDNA copies per cell varies from hundreds to thousands, depending on the cell type. mtDNA copy number is precisely regulated by the mTORC1/PGC-1a axis [6, 7], adapting mitochondrial function to the cellular metabolic needs. Thus, the mtDNA copy number is determined by the dynamic equilibrium between synthesis and degradation, but usually varies between 10 [3] and 10 [4] copies per cell [8].

The mtDNA is not free and, as it occurs for the nuclear genome, is organized and compacted into a nucleoprotein complex within the mitochondrial matrix, named the mitochondrial nucleoid. The organization of the mitochondrial genome in nucleoids provides mtDNA a certain degree of protection against oxidative and other type of base damage [9]. Different approaches, based on quantification of the number of nucleoids and mtDNA within human cells, showed that the number of mitochondrial genomes ranges from 2 to 10 molecules per nucleoid. Nevertheless, next-generation imaging tools have narrowed the average to 1.4 mitochondrial genomes per nucleoid, suggesting that a single nucleoid is more likely to be composed by one or two molecules of mtDNA [10, 11]. Proteins involved in maintaining mtDNA copy number by protecting or replicating the mitochondrial genome are associated with nucleoids. The Transcription Factor A (TFAM) has a double role. On the one hand, it packs and protects the mtDNA, whereas, on the other hand, it is essential for mtDNA transcription and replication [12, 13]. mtDNA replication is carried out by the mitochondrial replisome, formed by the heterotrimeric mitochondrial DNA polymerase γ (POLG1), the mitochondrial single-stranded DNA-binding protein (mtSSB) and the hexameric replicative helicase Twinkle [14]. Twinkle, is required for the strand separation at the replication fork of mtDNA from 5′ to 3′ end [15]. Mitochondrial RNA polymerase (POLRMT) is necessary to form the primers for initiation of DNA synthesis [16, 17].

Recent techniques allowing the detection of transient and weak interactions have identified other proteins associated with mitochondrial nucleoids. This group of proteins, related to mtDNA organization and distribution within the mitochondrial matrix, enable the compartmentalization of the mitochondrial genome and provide an unexpected dynamic behavior, adding another level of control and adapting mtDNA transcription, translation and turnover to cellular needs. Among them, the scaffolding protein ATAD3A directs mtDNA movement, bringing nucleoids in close contact to the inner membrane, where respiratory chain subunits need to be formed [18, 19]. Others such as FASTKD1, identified by proximity proteomics, localizes in RNA granules and controls mtDNA expression by *ND3* mRNA translation [20]. RNA granules, in the vicinity of nucleoids, regulate mitochondrial ribosome assembly and, in general, different aspects of mRNA processing, maturation and translation [21]. Indeed, they also serve as storage for components of mtDNA replication [22], highlighting its position for tight regulation of mtDNA functions.

Thus, mitochondrial nucleoid proteins, together with peripheral factors, enable crosstalk between the mitochondrial genome, mitochondrial function and cellular processes, adapting oxidative phosphorylation, mitochondrial biogenesis and nuclear gene expression and transcription to cellular needs [9, 23].

CYTOSOLIC DETECTION OF mtDNA

Cytosolic detection of DNA is the ability of cells to detect DNA molecules present in the cytosol, originated either from cellular damage, stress or other pathogen infection. The ability of the cytosolic sensors to detect mtDNA relies on the similarities that this molecule has to foreign DNA. mtDNA release responds to a cellular stress which impairs mitochondrial function, leading to instability of the organelle and followed by translocation of the mtDNA. Hence, the mitochondrial control of the inflammatory response must be understood as a modulation of the cascade more than participating in an active form.

The mechanisms underlying cytosolic DNA detection involve several specialized receptors such as cGAS and inflammasomes, which recognize different forms of DNA and initiate downstream signaling pathways. Each sensor is localized in a specific subcellular compartment, and thus, reacts to different types of mitochondrial stress (Table 7.1).

cGAS

cGAS is a cytosolic dsDNA sensor which synthetizes cyclic GMP-AMP (cGAMP) upon activation. Upon stimulation, cGAS forms 2:2 complexes with cytosolic DNA [24], inducing conformational changes to host ATP and GTP and synthesize 2′3′cGAMP [25]. Both oxidized and non-oxidized mtDNA fragments, as well as mtDNA-mtRNA hybrids, activate cGAS in the cytosol [26]. Remarkably, the mitochondrial transcription factor A (TFAM), as well as other DNA-protein complexes inducing U-turns in DNA molecules strongly stimulate cGAS [27].

Activation of cGAS is dependent on DNA length. Although low DNA concentrations are sufficient for cGAS dimerization, it requires DNA fragments raging between 20 bp and 500 bp [28]. Newly generated cGAMP activates STING (the stimulator of interferon genes) located in the ER and, more specifically, in the mitochondria-associated membranes (MAMs) [29]. These regions are specialized subcellular areas in charge of Ca^{2+} buffering, mitochondrial dynamics and lipid transfer, but also immune signaling [30, 31], connecting mitochondrial and cellular fitness to the immunological response. Consequently, STING activation results in the recruitment of TRAF6 and TBK1 kinases, to promote activation of NFκB or interferon response factor 3 (IRF3)-mediated immune response. This leads to NFκB-dependent cytokine release or the type I interferon (IFN) response [32].

Maintenance of Ca^{2+} homeostasis is essential for cellular function. The main organelle storing Ca^{2+} ions is the endoplasmic reticulum (ER) and mitochondria exert regulatory functions. By shaping Ca^{2+} waves, mitochondria buffer Ca^{2+} ions between the two organelles and modulate downstream responses. Ca^{2+} transients are generated upon multiple stimuli, including normal cellular function but also after infection with pathogens. Upon ER-Ca^{2+} depletion, ER-Ca^{2+} stores are replenished through a mechanism called store-operated calcium entry (SOCE) [33]. The activation of this mechanism requires the dimerization in the ER membrane of the Ca^{2+} sensor protein STIM1, which translocates to the plasma membrane and contacts with the calcium release activation channel ORAI1. This interaction triggers extracellular Ca^{2+} influx from the extracellular space. Importantly, STING is also able to bind Ca^{2+} ions, which are required for its own dimerization [34]. Furthermore, STING interacts with STIM1 in the membrane of the ER [35]. By this interaction, STIM1 inhibits activation of STING in resting conditions. On the contrary, STING activation permits free diffusion of STIM1 to the plasma membrane, opening of ORAI1 and extracellular Ca^{2+} influx [36]. In addition, STING is also regulated by multiple Ca^{2+} channels including ATP2A2, ATP2B1, VDAC1 and VDAC3. However, their role might be secondary and activated upon different stressors, linking again immune response to cellular homeostasis [37].

Given the above, the MAM system establishes itself as a crossroad between DAMPs recognition and signal amplification. It is not surprising that the mechanisms related to mtDNA release are also linked to MAMs function. Thus, upon Ca^{2+} overload, the long-lasting mPTP opening, execution of apoptosis through BAK/BAX pores, opening of VDAC pores upon oxidative stress, mitophagy impairment and alterations in proteins involved in mitochondrial dynamics are all processes regulated up to some point in MAMs. Dysfunction of any of these processes results in mtDNA release and its detection by cytosolic DNA sensors.

Inflammasomes

Inflammasomes are protein super-complexes, whose activation leads to the production and secretion of IL1β and IL18 via the action of inflammatory caspases. Two signals are required for inflammasome activation: a priming signal mediated by NFκB, and a secondary signal mediated by the recognition of a DAMP or PAMP (pathogen-associated molecular pattern) facilitated by the PRRs of the inflammasomes. The combination of these two signals favors the oligomerization, and thus activation of the inflammasome super-complexes. The recognizable DAMP or PAMPs include bacterial or viral components, potassium efflux, ROS, cathepsins, ATP or DNA [38]. mtDNA recognition by inflammasomes induces IL1β- and IL18-mediated inflammation. However, only NLRP3, NLRC4, AIM2 and IFI16 inflammasomes can recognize mtDNA as a DAMP (Table 7.1).

NLRP3 is the most characterized type of inflammasome which can identify cytosolic-oxidized mtDNA molecules. Because of this, a certain type of mitochondrial dysfunction characterized by ROS overproduction and mtDNA leakage to the cytosol is a potential activator of the NLRP3 inflammasome. Besides playing an essential role as a priming signal molecule on NLRP3 inflammasome activation, NFκB negatively regulates NLRP3. NLRP3 oligomerization takes place in the interface between mitochondria and ER. Upon NFκB priming signal, p62, which is also an NFκB target gene, is recruited onto the surface of mitochondria. p62 targets mitochondria for mitophagy and, at the same time, initiates the degradation of the NLRP3 super-complexes, in order to shut down the cascade again [39].

NLRC4, AIM2 and IFI16 inflammasomes are less known. The NLRC4 inflammasome can recognize both oxidized and non-oxidized mtDNA, nevertheless it has a preference for oxidized mtDNA. Thus, upon *Pseudomonas aeruginosa* infection, NLRC4 activation by oxidized mtDNA leak is responsible for IL1β production in bone marrow-derived macrophages [40]. AIM2 and IFI16 inflammasomes show similar structural features, allowing them to recognize both viral and bacterial DNA. The IFI16 inflammasome is believed to be a nucleus-associated inflammasome component, given its role as a DNA sensor for viruses replicating within the nucleus [41, 42]. Despite this, the IFI16 inflammasome is also able to recognize mtDNA leaking to the cytosol upon infection by norovirus [43]. Similarly, the AIM2 inflammasome drives anti-viral responses against some RNA viruses [44], but in the presence of cytosolic mtDNA, its activation can result in pyroptosis or apoptosis through the engagement of caspase 1 or 8, respectively, depending on the cytosolic DNA concentration [45].

ENDOSOMAL DETECTION OF mtDNA

Endosomes are membrane-bound organelles involved in sorting, processing and transport of intracellular components. Endosomes can mature into various subtypes, such as early endosomes, recycling endosomes, lysosomes or late endosomes and multivesicular bodies. Each subtype is characterized by a different proteome and specific biochemical features [46]. Upon different stressors, mtDNA can be found in the endosomal compartment. Its detection leads to the activation of various signaling pathways that trigger innate immune responses and promote the clearance of cellular damage. The mechanisms underlying endosomal mtDNA detection involve specialized TLRs, which recognize different forms of nucleic acids and initiate downstream immunostimulatory pathways.

The Toll-like receptor protein 9 (TLR9) is a receptor for unmethylated CpG DNA present in bacterial DNA. Prior to stimulation, TLR9 is localized in the ER, shuttled to the plasma membrane, endocytosed and directed to the endosomal system and finally, to the lysosomal compartment to the lysosomal compartment [47]. In general, upon bacterial endocytosis by circulating macrophages and neutrophils, endocytic vesicles in traffic through the endosome system are internalized in phagosomes and delivered to lysosomes for enzymatic degradation. Lysosomal degradation of microbes facilitates DNA recognition by TLR9s receptors.

How mtDNA reaches the endosomal compartment remains unclear. Endocytosis of extracellular circulating mtDNA delivered upon cellular death could have a role in amplifying immunological response. Nevertheless, disruption of the mitochondrial quality control mechanisms also supports the activation of sterile immune responses, suggesting a direct link between both mechanisms. In fact, mitochondrial-derived vesicles (MDVs) mediate trafficking of mitochondrial components. mtDNA-containing MDVs can be exocytosed as extracellular vesicles (EVs), which can be later endocytosed by recipient cells. This process enables the encounter of the recipient cells' TLR9 with foreign mtDNA, which results in the engagement of the cognate inflammatory response [48]. dsDNA binding to TLR9 results in conformational changes in the cytosolic domains of the receptor, rising the recruitment of the signaling proteins of the innate immune response, including the myeloid differentiation primary response protein 88 (MYD88) [49]. MYD88 recruits IRAK-1 which, through activation of the inflammatory kinases TRAF6 and TBK1, follows with the activation of the NFκB and IRF3, resulting in cytokines and interferon Type I release [50, 51].

mtDNA recognition by TLR stimulates activation of inflammasomes, suggesting a crosstalk between both inflammatory responses [52]. Thus, TLR-mediated MYD88 activation induces deoxyribonucleotide entry to mitochondria, enhancing mtDNA synthesis, which results in aberrant mtDNA accumulation, likely in an environment with ROS overproduction. This stimulates the leakage of oxidized mtDNA to the cytosol, allowing its recognition by NLRP3 inflammasome, therefore connecting aberrant mtDNA synthesis induced by TLR activation with NLRP3 inflammasome [52].

MECHANISMS OF mtDNA ESCAPE

Given the variety of DNA sensors present in the intracellular space, maintenance of mtDNA within mitochondria is key to prevent sterile activation of the immune responses. mtDNA release is not an active process and is only triggered by mitochondrial dysfunction. A number of events have been directly linked to the mislocalization of mtDNA, including loss of mitochondrial integrity, altered bioenergetics and disrupted clearance of damaged mitochondria.

Membrane Pores

Disturbance of mitochondrial integrity triggers mtDNA release. The inner mitochondrial membrane (IMM) is highly impermeable to all molecules, since it establishes a proton gradient and a membrane potential ($\Delta\psi_m$), generated by the electron transport chain. Variation in the biochemical properties of the mitochondria leads to mtDNA translocation acting as an initial signal of mitochondrial dysfunction.

Opening of the mPTP was originally described as a consequence of Ca^{2+} overload. The transient opening of the permeability transition pore (mPTP) leads to mitochondrial membrane potential loss, mitochondrial swelling and mtDNA release [53]. Ca^{2+} release from the endoplasmic reticulum occurs in a variety of situations, including neurotransmitter release, stimulation by a plethora of hormones, glucose uptake in pancreatic cells, muscle contraction, and ER stress induced by pathogen infection [54]. Upon viral infection, distinct cellular sensors orchestrate a signaling cascade leading to the release of pro-inflammatory cytokines. For instance, the mitochondrial antiviral signaling protein (MAVS), essential for detecting RNA viruses, is located in MAMs and peroxisomes [55]. The mitochondrial calcium uniporter (MCU) interacts with MAVS, inducing Ca^{2+} release upon ER stress and amplifying antiviral signal [56].

Under oxidative stress, activation of the mitochondrial permeability transition pore results in mtDNA escape [57]. Transient opening of the mPTP is responsible for the homeostasis of solutes such as ions, small molecules, metabolites and proteins of the matrix. In contrast, long-lasting opening of the mPTP promotes irreversible loss of the mitochondrial membrane potential, reduced ATP synthesis, cytochrome *c* release, mitochondrial swelling and cell death [58]. Under these conditions, mtDNA is processed into fragments smaller than 700 bp that are released through the mPTP [59].

Cellular stress can result in the outer mitochondrial membrane (OMM) permeabilization mediated by the formation of BAK/BAX pores, resulting in the well-established intrinsic mitochondrial pathway of apoptosis [60]. In normal conditions, BAK (BCL2 antagonist/killer) is primarily localized in an inactive form at the OMM as monomer, or in a monomer-dimer equilibrium. In contrast, BAX (BCL2-associated X protein) is primarily localized as a monomer in the cytosol. The oligomerization and pore formation by BAK and BAX proteins is regulated by a complex interplay of pro-apoptotic and anti-apoptotic factors [61]. The pro-apoptotic BCL2 family members, such as BIM and PUMA, promote BAK and BAX activation [62], whereas the anti-apoptotic BCL2 family members, such as BCL2 and BCL-XL, inhibit their activation by preventing BAX translocation to the mitochondria [63, 64].

Upon a specific stimulus or imbalance between pro-apoptotic and anti-apoptotic factors, BAX undergoes a conformational change, exposing its hydrophobic domain and mitochondrial targeting signal, enabling BAX translocation to the mitochondrial outer membrane. Once in its mitochondrial localization, BAX undergoes further conformational changes, oligomerizes with other BAX molecules and forms BAK/BAX pores [65]. These pores allow the release of cytochrome *c*, mtDNA and other mitochondrial components. mtDNA escape through BAK/BAX pores to the extracellular space during apoptosis can trigger different inflammatory pathways [61, 66, 67]. It is not clear, however, if during this process mtDNA is released in fragments, similar to the release through mtPTP, or if the complete nucleoid can be also expelled.

Similarly, VDAC (voltage-dependent anion channel) oligomers play a role in the exit of mtDNA [67]. VDAC pores are large protein channels located in the OMM responsible for the exchange of metabolites and ions between the cytosol and the mitochondria. The opening of VDAC pores is regulated by various factors, including voltage, metabolites, and interactions with other proteins. In addition, oxidative stress can modify VDAC activities, linking mitochondrial functionality to pore formation [68]. Additionally, VDAC pores are able to interact with mtDNA and stabilize its oligomeric state, thus facilitating its extrusion from the mitochondria [67]. However, in this case, fragmentation of the mtDNA in 110 bp fragments by the nuclease FEN1 is pivotal to allow DNA to pass through VDAC pores [57].

Although the formation of mtPTP, BAK/BAX pores and VDAC oligomers has been identified as the key molecular mechanism by which mtDNA exits mitochondria, there is still not sufficient knowledge addressing the question how mtDNA can initially travel across the inner membrane. It has been reported that deficiency in the IMM protein Prohibitin 1 (PHB1) enables mtDNA to cross the internal membrane, and subsequently exit to the cytosol through VDAC oligomers or the mtPTP [69]. This mechanism could be an important downstream effect of a primary mitochondrial dysfunction, allowing the mtDNA release to signalize cellular stress.

Regardless of the type of mitochondrial pore by which mtDNA exits to the cytosol upon conditions of stress, this results in the activation of cytosolic DNA sensors. The length of the mtDNA fragments that are released, as well as their oxidation status, would determine whether cGAS, the inflammasomes or a combination of many other pathways, are activated.

Mitochondrial Dynamics

Proper mitochondrial function and cellular homeostasis significantly relies on the dynamic nature of mitochondria. Indeed, the mitochondrial network within a cell is in constant change of shape and form. Mitochondrial dynamics is characterized by constant mitochondrial fission and fusion. Tuned balance between these two events permits the fast adaptation of the mitochondrial metabolism to the cellular needs [70]. Mitochondrial dynamics controls not only the intracellular distribution of mitochondria but also the elimination of ill-functioning organelles, thus mitigating cellular damage. Mitochondrial dynamics is also directly related to mitochondrial transport and turnover, a crucial process in large and branched cells, such as neuronal cells [31]. Besides, mitochondrial dynamics is also fundamental for the proper distribution of mitochondrial nucleoids within the network [71].

Thus, it is not surprising that interfering with the molecular mechanisms involved in fusion and fission of mitochondrial membranes also leads to mtDNA release and activation of the innate immune response [72]. These responses, however, differ depending on the initial insult that modulates mitochondrial morphology. For instance, mitochondrial fragmentation, induced either by reduced fusion or increased fission, is linked to mtDNA arrival in the endosomal compartment and engagement of TLR9-dependent inflammatory responses [66, 73]. On the contrary, mitochondrial fragmentation induced by the deficiency of the OPA1 protease YME1L, promotes cytosolic mtDNA leakage, potentially through VDAC pores and cGAS-STING activation [74]. Similarly, reduced mitochondrial fission triggers mtDNA release to the cytosol and activation of the cGAS-STING axis [73].

The mitochondrial antiviral response is also deeply influenced by mitochondrial dynamics. Thus, activation of MAVS by dsRNA triggers mitochondrial membrane remodeling, mitochondrial fission and BAK/BAX-dependent apoptosis [75]. In agreement, the mitochondrial fission factor MFF facilitates MAVS clustering to activate the innate immune response [76]. In contrast, Mitofusin 2 (MFN2), another resident protein of MAMs involved in mitochondrial fusion, has also been identified as an interacting partner for MAVS, acting as an inhibitor of the antiviral response [77]. Upon activation, MAVS forms large polymers that activate the cytosolic kinases IKK and TBK1, which in turn activate NFκB and IRF3, respectively, to induce interferons [78].

Mitochondrial Clearance

Disrupted mitochondrial clearance by mitophagy can cause damaged mtDNA to accumulate, followed by mitochondrial dysfunction and mtDNA release. Several studies have demonstrated that impairment of mitophagy can result in mtDNA release, which activates diverse inflammatory responses. These studies include downregulation of auto/mitophagy proteins such as LC3, FUNDC1, BNIP3, PINK1 and Parkin [79–82]. In these cases, defective or delayed removal of mitochondrial components can cause the accumulation of DAMPs, leading to the activation of immune sensors.

The initiation of bulk mitochondrial quality control responds to the presence of severe insults disturbing mitochondrial homeostasis. The discovery of selective mitochondrial quality control pathways added a new level of control, preserving mitochondrial function and eliminating only ill-functioning parts of the network. The endosomes, membrane-bound organelles involved in the cellular sorting machinery, participate in selective removal of mitochondrial components. The contribution of endosomes to mitochondrial quality control was shown to be both dependent as well as independent of mitochondrial macroautophagy. In this context, a concert of RAB proteins, small GTPases in charge of vesicle sorting, fission, tethering, docking and fusion, participate in mitochondrial quality control and determine the selectivity of the mitochondrial turnover. For instance, RAB10 accumulates on depolarized mitochondria and binds to the autophagy receptor optineurin, facilitating mitophagy [83]. RAB9, however, stimulates the recruitment of trans-Golgi membranes to damage mitochondria upon ULK1 phosphorylation, thus initiating macroautophagy of mitochondria [84]. RAB7 endosomes are necessary for autophagosome formation during macroautophagy, and they can also be recruited to damaged mitochondria [85, 86]. Finally, RAB35 controls bulk autophagy, including mitochondria, through binding of ubiquitinated targets such as NDP52 [87].

In a higher level of selectivity, selected pieces of mitochondria previously labeled by Parkin can be recognized by RAB5 [88]. RAB5 endosomes containing engulfed mitochondrial particles maturate toward RAB7 late endosomes and are degraded in the lysosomal compartment. Interestingly, the isoform RAB5C binds to Mitofusins 1 and 2, facilitating mitochondria-endosomes contact sites and potential detection of mtDNA by TLRs [72]. RAB7, which plays a critical role in membrane fusion and mitophagy [89], modulates also the apoptotic response by degrading inner mitochondrial components [90]. In this context, the protein complex involved in maturation of the endosomes, the retromer, also interferes with mitochondrial turnover. Thus, the retromer core component VPS35, promotes mitochondrial fission by degrading MFN2 and removing inactive DRP1 [91, 92]. Moreover, VPS35 has been associated with the formation of mitochondrial-derived vesicles directed to peroxisomes and the elimination of mitochondrial nucleoids upon mtDNA damage [93, 94].

Endosomes are also regulators of diverse mitochondrial functions. Transport of nuclear-encoded mRNAs of mitochondrial proteins is vital in highly polarized cells such as neurons, allowing spatiotemporal coupling of mitochondrial transport and local mRNA translation [95]. As explained before, the endosomes are essential for both adaptive and innate immune response, as they carry the Toll-like receptors TLR3, TLR7 and TLR9. TLRs are initially synthesized in the endoplasmic reticulum and traffic to the plasma membrane through the vesicular sorting machinery of the Golgi apparatus. TLRs are then endocytosed and internalized in the endosomal compartment [96]. While TLR3 and TLR7 detect double- and single-stranded RNA, respectively, TLR9 is able to recognize DNA [97]. Since mtDNA is hypomethylated, it acts as a TLR9 agonist. Although a particular oxidative status of mtDNA is not required for TLR9 activation, the presence of 2-OHdG enhances the immunostimulatory capacity of mtDNA toward TLR9 [98].

The contribution of the endosomes to the innate immune response elicited by mitochondrial material is linked to mitochondrial turnover pathways. In fact, MDVs mediate trafficking of mitochondrial components to endosomes in order to allow antigen presentation [99]. mtDNA-containing MDVs, in a specific pathological environment, have been reported to initiate a cGAS-dependent inflammatory response [100]. Finally, as observed by impairment of initial steps of macroautophagy, disturbance of mitochondrial clearance by dysfunctional lysosomes leads either to detection of mtDNA by vesicle sensors [96], or release of mitochondrial components into the cytoplasm and further detection by cGAS or inflammasomes [101]. In addition, elimination of dysfunctional mitochondria by RAB5 endosomes limits the activation of NLRP3 inflammasome upon mitochondrial damage [102]. Therefore, the endosomes assure proper elimination of damaged mitochondria, including mtDNA, avoiding pleiotropic effects produced by its mislocalization, establishing as the organelles between mitochondria and the innate immune response.

Metabolic Imbalance

As the main metabolic organelle, mitochondria also contain proteins involved in bioenergetic pathways, coupling the electron transport chain to the synthesis of various substrates as carbon source. While mitochondria are commonly viewed primarily as catabolic organelles, responsible for the oxidation of various substrates such as pyruvate from glycolysis, amino acids and lipids to generate ATP, it is crucial to underscore their role in biosynthesis. Notably, mitochondria play a key part in the production of lipids and pyrimidines, serving as essential building blocks necessary for the synthesis of cellular compounds that facilitate cell growth and division. Disturbance of mitochondrial metabolism triggers mtDNA release and the innate immune response [74, 100]. However, the specific molecular mechanisms underlying mtDNA release are different. For instance, pyrimidine imbalance results in a mitochondrial metabolism rewiring dependent on YME1L. Thus, mtDNA release can be observed in absence of YME1L, caused by nucleotide imbalance and hence mtDNA replication defects [74]. Interestingly, accumulation of fumarate, a process associated with the pathogenesis of leiomyoatosis and renal cell carcinoma, leads to encapsulation of mtDNA in MDVs, which are later released into the cytosol, activating the innate immune response [100].

Dysregulated lipid metabolism starts similar inflammatory responses. Alterations in lipid biogenesis in adipose tissue results in ROS production and reduced OXPHOS, leading to mtDNA leakage and a subsequent cGAS-dependent inflammatory response [103]. Similarly, in regulatory T cells (Tregs), blockage of the fatty acid binding protein FABP5 reduces fatty acid oxidation, inducing mtDNA exit to the cytosol and cGAS activation [104]. Nevertheless, the molecular mechanisms linking dysregulation of lipid metabolism and leakage of mtDNA to the cytosol remain still largely unaddressed.

mtDNA RELEASE AND DISEASE

Inflammation is considered a hallmark of a wide variety of disorders, ranging from autoimmune to metabolic diseases, cancer, neurodegeneration and muscle dystrophies, but is also observed during normal aging. In fact, the recently born term "*inflammaging*" encompasses the pro-inflammatory aging associated phenotypes [105]. Tackling inflammation in disease and age-related disorders has been proposed as a potential universal therapeutic strategy. There is, however, a lack of knowledge on the triggers of chronic inflammation observed in those conditions. Given the recent and increasing evidence pointing toward the direct causality of mtDNA mislocation-inducing mitochondrial dysfunction to the activation of sterile inflammatory responses, it might be possible that mtDNA recognition by DNA sensors underlies *inflammaging* and represents a common feature in many mitochondria-related disorders.

Obesity-associated cardiovascular disease, muscle atrophy and neurodegeneration are defined hallmarks of aging [106]. These pathophysiological conditions are tightly related to tissue inflammation, which has been considered more as a driver rather than as an aging consequence [105]. Thus, age-advanced subjects with obesity-associated cardiovascular disease or current smokers carrying a mutation in *STING* (p.R293Q) showed a protective phenotype from combined aging-related alterations [107, 108]. Hence, it was proposed that obesity or smoking-induced mitochondrial dysfunction promotes mtDNA release to the cytosol and activation of the cGAS-STING axis.

Age-related muscle atrophy is related to decrease in the mitochondrial fusion protein MFN2 in the skeletal muscle. As a compensatory and survival mechanism, the mitophagy protein BNIP3 promotes mitochondrial turnover, ameliorating unhealthy muscle aging [109]. Accordingly, aged subjects can be stratified in regard to their BNIP3 protein expression in muscle biopsies, which strongly associates with muscle function and inflammation. In fact, individuals with high BNIP3 expression display more signs of healthy muscle and reduced muscle inflammation, whereas individuals with low BNIP3 levels presented an exacerbated phenotype. From the molecular point of view, BNIP3 deficiency leads to mitochondrial and lysosomal dysfunction, promoting the accumulation of damaged mitochondria in the endosomal compartment. There, mtDNA fragments can be recognized by TLR9 and engage an TLR9-NFκB axis-dependent inflammatory response [79].

Neurodegeneration

Neurodegeneration comprises a set of age-related or inherited disorders strongly linked to *inflammaging* of the brain or specific neuron subtypes. *In vivo* and *in vitro* models of amyotrophic lateral sclerosis (ALS) and Huntington's disease revealed that the neuroinflammatory phenotype is dependent on mtDNA release to the cytosol and cGAS-STING activation [110, 111]. Interestingly, administration of exogenous melatonin, an endogenous free radical scavenger, improves mitochondrial function, attenuates mtDNA-cGAS-STING inflammatory response and hence, decelerates neurodegeneration [110].

Increased cell-free mtDNA (ccf-mtDNA) levels have been detected in acute phases of relapsing-remitting multiple sclerosis, exactly when acute inflammation occurs, anticipating the neurodegenerative process [112]. In addition, neurodegenerative conditions, such as Alzheimer's and Parkinson's disease (PD) have shown decreased cerebrospinal fluid and increased ccf-mtDNA, suggesting a potential impairment in mtDNA homeostasis in the affected brain areas [112].

A chronic and ongoing neuroinflammatory state has been proposed to facilitate PD development [113]. Many genes leading to familiar forms of PD are related to cellular quality control mechanisms. Thus, mutations in genes encoding for mitophagy receptors PINK1 and Parkin [114, 115], lysosomal proteins such as ATP13A2 [116], the retromer VPS35 [117] and even the mitochondrial helicase Twinkle have been associated with PD [118]. In all these cases, the turnover of catecholamines in dopaminergic neurons increases mtDNA mutation rate [119], which activates quality control mechanisms [18]. Malfunctioning of quality control pathways might lead to mtDNA release and activation of pro-inflammatory cascades.

Peripheral neuropathies commonly affect the elderly population albeit some inherited forms exist. Chronic inflammatory demyelinating polyneuropathy (CIDP) and progressive or acute inflammatory neuropathy involve progressive weakness and reduced sensory capacity. The role of mitochondria in such cases is unclear, although some case reports for inflammatory neuropathies describe patients with reduced activity of the mitochondrial enzyme thymidine phosphorylase [120] or Complex I deficiency, leading to myopathy and CIDP [121]. The involvement of inflammation in other peripheral neuropathies such as Charcot-Marie Tooth (CMT) is also debated. CMT patients suffer more frequently from inflammatory neuropathy than the healthy population [122], although the mechanisms underlying this correlation are not clear. An inflammatory component has been found in a knock-out mouse model for the mitochondrial protein GDAP1 [123], associated with demyelinating and axonal CMT. Although the exact role of this protein is still unclear, it has been linked to Ca^{2+} homeostasis [124, 125], mitochondrial dynamics [126] and mitochondria-lysosomes interplay [127]. *In vitro* studies of MFN2 deficiency, related to axonal CMT, also suggest an inflammatory component [128, 129], but this has not been verified in human patients.

Cancer

Although cancer can occur at any age, the risk of developing cancer also increases with age. Chronic inflammation plays a major role in cancer progression. Thus, tissue inflammation is the precursor state for many cancer types, including lung and liver cancer. In this regard, mtDNA release through VDAC oligomers, disruption of mitophagy and presence of endosomal mtDNA underlies tissue inflammation [130, 131]. *In vitro* studies reported a link between cancer cell growth and mtDNA-dependent inflammation. For instance, in esophageal squamous cell carcinoma (ESCC), tumor growth is promoted in conditions of TFAM deficiency and cytosolic mtDNA-mediated inflammation [132]. Similarly, in leukemia cell lines exposed to autophagy stimulants, mtDNA is released to the cytosol via the oligomerization of BAK/BAX pores, followed by cGAS-STING activation [133].

The use of cytosolic mtDNA-mediated inflammation has been proposed as an anti-tumor immunotherapeutic strategy. Indeed, induction of ferroptosis in both localized and metastatic tumor models promote mtDNA release and cGAS-STING axis engagement, inducing a primarily systemic anti-tumor immunity that suppresses tumor growth [134]. Supporting these observations, another study described a molecular mechanism by which radiation-induced anti-tumor immunity relied on the role of the E3 ubiquitin ligase TIMR21, the formation of VDAC oligomers and mtDNA release [135].

In general, these evidences support the notion that promoting and enhancing mitochondrial fitness is key to prevent mtDNA-dependent *inflammaging* in several tissues and pathological conditions. Increased efforts should be devoted to identify more age-related pathologies characterized by mitochondrial dysfunction and inflammation that are linked to mtDNA mislocation, and to design novel therapeutic strategies targeting mtDNA-dependent sterile inflammation.

REFERENCES

1. Turvey, S. E. & Broide, D. H. Innate immunity. *J Allergy Clin Immunol* **125**, S24–S32 (2010). https://doi.org/10.1016/j.jaci.2009.07.016
2. Schlee, M. & Hartmann, G. Discriminating self from non-self in nucleic acid sensing. *Nat Rev Immunol* **16**, 566–580 (2016). https://doi.org/10.1038/nri.2016.78
3. Qian, W. & Van Houten, B. Alterations in bioenergetics due to changes in mitochondrial DNA copy number. *Methods* **51**, 452–457 (2010). https://doi.org/10.1016/j.ymeth.2010.03.006
4. Torres-Gonzalez, M., Gawlowski, T., Kocalis, H., Scott, B. T. & Dillmann, W. H. Mitochondrial 8-oxoguanine glycosylase decreases mitochondrial fragmentation and improves mitochondrial function

in H9C2 cells under oxidative stress conditions. *Am J Physiol Cell Physiol* **306**, C221–C229 (2014). https://doi.org/10.1152/ajpcell.00140.2013
5. Wu, L. L., Chiou, C. C., Chang, P. Y. & Wu, J. T. Urinary 8-OHdG: A marker of oxidative stress to DNA and a risk factor for cancer, atherosclerosis and diabetics. *Clin Chim Acta* **339**, 1–9 (2004). https://doi.org/10.1016/j.cccn.2003.09.010
6. Wang, L. L. *et al.* mTORC1-PGC1 axis regulates mitochondrial remodeling during reprogramming. *FEBS J* **287**, 108–121 (2020). https://doi.org/10.1111/febs.15024
7. Tran, M. T. *et al.* PGC1 alpha drives NAD biosynthesis linking oxidative metabolism to renal protection. *Nature* **531**, 528–532 (2016). https://doi.org/10.1038/nature17184
8. Shadel, G. S. & Clayton, D. A. Mitochondrial DNA maintenance in vertebrates. *Annu Rev Biochem* **66**, 409–435 (1997). https://doi.org/10.1146/annurev.biochem.66.1.409
9. Gilkerson, R. *et al.* The mitochondrial nucleoid: Integrating mitochondrial DNA into cellular homeostasis. *Cold Spring Harb Perspect Biol* **5**, a011080 (2013). https://doi.org/10.1101/cshperspect.a011080
10. Legros, F., Malka, F., Frachon, P., Lombes, A. & Rojo, M. Organization and dynamics of human mitochondrial DNA. *J Cell Sci* **117**, 2653–2662 (2004). https://doi.org/10.1242/jcs.01134
11. Kukat, C. *et al.* Super-resolution microscopy reveals that mammalian mitochondrial nucleoids have a uniform size and frequently contain a single copy of mtDNA. *P Natl Acad Sci USA* **108**, 13534–13539 (2011). https://doi.org/10.1073/pnas.1109263108
12. Ekstrand, M. I. *et al.* Mitochondrial transcription factor A regulates mtDNA copy number in mammals. *Hum Mol Genet* **13**, 935–944 (2004). https://doi.org/10.1093/hmg/ddh109
13. Xu, S. C. *et al.* Overexpression of Tfam protects mitochondria against beta-amyloid-induced oxidative damage in SH-SY5Y cells. *FEBS J* **276**, 4224–4233 (2009). https://doi.org/10.1111/j.1742-4658.2009.07094.x
14. Korhonen, J. A., Pham, X. H., Pellegrini, M. & Falkenberg, M. Reconstitution of a minimal mtDNA replisome in vitro. *EMBO J* **23**, 2423–2429 (2004). https://doi.org/10.1038/sj.emboj.7600257
15. Korhonen, J. A., Gaspari, M. & Falkenberg, M. TWINKLE Has 5' -> 3' DNA helicase activity and is specifically stimulated by mitochondrial single-stranded DNA-binding protein. *J Biol Chem* **278**, 48627–48632 (2003). https://doi.org/10.1074/jbc.M306981200
16. Kuhl, I. *et al.* POLRMT regulates the switch between replication primer formation and gene expression of mammalian mtDNA. *Sci Adv* **2**, e1600963 (2016). https://doi.org/10.1126/sciadv.1600963
17. Young, M. J. & Copeland, W. C. Human mitochondrial DNA replication machinery and disease. *Curr Opin Genet Dev* **38**, 52–62 (2016). https://doi.org/10.1016/j.gde.2016.03.005
18. Sen, A. *et al.* Mitochondrial membrane proteins and VPS35 orchestrate selective removal of mtDNA. *Nature Communications* **13**, 6704 (2022). https://doi.org/10.1038/s41467-022-34205-9
19. Ishihara, T., Ban-Ishihara, R., Ota, A. & Ishihara, N. Mitochondrial nucleoid trafficking regulated by the inner-membrane AAA-ATPase ATAD3A modulates respiratory complex formation. *Proc Natl Acad Sci USA* **119**, e2210730119 (2022). https://doi.org/10.1073/pnas.2210730119
20. Han, S. *et al.* Proximity biotinylation as a method for mapping proteins associated with mtDNA in living cells. *Cell Chem Biol* **24**, 404–414 (2017). https://doi.org/10.1016/j.chembiol.2017.02.002
21. Jourdain, A. A., Boehm, E., Maundrell, K. & Martinou, J. C. Mitochondrial RNA granules: Compartmentalizing mitochondrial gene expression. *J Cell Biol* **212**, 611–614 (2016). https://doi.org/10.1083/jcb.201507125
22. Hensen, F. *et al.* Mitochondrial RNA granules are critically dependent on mtDNA replication factors Twinkle and mtSSB. *Nucleic Acids Research* **47**, 3680–3698 (2019). https://doi.org/10.1093/nar/gkz047
23. He, J. *et al.* Mitochondrial nucleoid interacting proteins support mitochondrial protein synthesis. *Nucleic Acids Res* **40**, 6109–6121 (2012). https://doi.org/10.1093/nar/gks266
24. Li, X. *et al.* Cyclic GMP-AMP synthase is activated by double-stranded DNA-induced oligomerization. *Immunity* **39**, 1019–1031 (2013). https://doi.org/10.1016/j.immuni.2013.10.019
25. Xiao, T. S. & Fitzgerald, K. A. The cGAS-STING pathway for DNA sensing. *Mol Cell* **51**, 135–139 (2013). https://doi.org/10.1016/j.molcel.2013.07.004
26. Gehrke, N. *et al.* Oxidative damage of DNA confers resistance to cytosolic nuclease TREX1 degradation and potentiates STING-dependent immune sensing. *Immunity* **39**, 482–495 (2013). https://doi.org/10.1016/j.immuni.2013.08.004
27. Andreeva, L. *et al.* cGAS senses long and HMGB/TFAM-bound U-turn DNA by forming protein-DNA ladders. *Nature* **549**, 394–398 (2017). https://doi.org/10.1038/nature23890
28. Zhang, X. *et al.* The cytosolic DNA sensor cGAS forms an oligomeric complex with DNA and undergoes switch-like conformational changes in the activation loop. *Cell Rep* **6**, 421–430 (2014). https://doi.org/10.1016/j.celrep.2014.01.003

29. Ishikawa, H., Ma, Z. & Barber, G. N. STING regulates intracellular DNA-mediated, type I interferon-dependent innate immunity. *Nature* **461**, 788–792 (2009). https://doi.org/10.1038/nature08476
30. Missiroli, S. *et al.* Mitochondria-associated membranes (MAMs) and inflammation. *Cell Death Dis* **9**, 329 (2018). https://doi.org/10.1038/s41419-017-0027-2
31. Pass, T., Wiesner, R. J. & Pla-Martin, D. Selective neuron vulnerability in common and rare diseases-mitochondria in the focus. *Front Mol Biosci* **8** (2021). https://doi.org/10.3389/fmolb.2021.676187
32. Cai, X., Chiu, Y. H. & Chen, Z. J. J. The cGAS-cGAMP-STING pathway of cytosolic DNA sensing and signaling. *Molecular Cell* **54**, 289–296 (2014). https://doi.org/10.1016/j.molcel.2014.03.040
33. Putney, J. W. Capacitative calcium entry: From concept to molecules. *Immunol Rev* **231**, 10–22 (2009). https://doi.org/10.1111/j.1600-065X.2009.00810.x
34. Shu, C., Yi, G. H., Watts, T., Kao, C. C. & Li, P. W. Structure of STING bound to cyclic di-GMP reveals the mechanism of cyclic dinucleotide recognition by the immune system. *Nat Struct Mol Biol* **19**, 722–724 (2012). https://doi.org/10.1038/nsmb.2331
35. Srikanth, S. *et al.* The Ca^{2+} sensor STIM1 regulates the type I interferon response by retaining the signaling adaptor STING at the endoplasmic reticulum. *Nat Immunol* **20**, 152–162 (2019). https://doi.org/10.1038/s41590-018-0287-8
36. Zhou, Y. D. *et al.* STIM1 dimers undergo unimolecular coupling to activate Orai1 channels. *Nat Commun* **6** (2015). https://doi.org/10.1038/ncomms9395
37. Mathavarajah, S., Salsman, J. & Dellaire, G. An emerging role for calcium signalling in innate and autoimmunity via the cGAS-STING axis. *Cytokine Growth F R* **50**, 43–51 (2019). https://doi.org/10.1016/j.cytogfr.2019.04.003
38. Zheng, D. P., Liwinski, T. & Elinav, E. Inflammasome activation and regulation: Toward a better understanding of complex mechanisms. *Cell Discov* **6** (2020). https://doi.org/10.1038/s41421-020-0167-x
39. Zhong, Z. Y. *et al.* NF-kappa B restricts inflammasome activation via elimination of damaged mitochondria. *Cell* **164**, 896–910 (2016). https://doi.org/10.1016/j.cell.2015.12.057
40. Jabir, M. S. *et al.* Mitochondrial damage contributes to Pseudomonas aeruginosa activation of the inflammasome and is downregulated by autophagy. *Autophagy* **11**, 166–182 (2015). https://doi.org/10.4161/15548627.2014.981915
41. Kerur, N. *et al.* IFI16 acts as a nuclear pathogen sensor to induce the inflammasome in response to Kaposi sarcoma-associated herpesvirus infection. *Cell Host Microbe* **9**, 363–375 (2011). https://doi.org/10.1016/j.chom.2011.04.008
42. Unterholzner, L. *et al.* IFI16 is an innate immune sensor for intracellular DNA. *Nat Immunol* **11**, 997–U942 (2010). https://doi.org/10.1038/ni.1932
43. Jahun, A. S. *et al.* Leaked genomic and mitochondrial DNA contribute to the host response to noroviruses in a STING-dependent manner. *Cell Rep* **42** (2023). https://doi.org/10.1016/j.celrep.2023.112179
44. Ekchariyawat, P. *et al.* Inflammasome signaling pathways exert antiviral effect against Chikungunya virus in human dermal fibroblasts. *Infect Genet Evol* **32**, 401–408 (2015). https://doi.org/10.1016/j.meegid.2015.03.025
45. Sagulenko, V. *et al.* AIM2 and NLRP3 inflammasomes activate both apoptotic and pyroptotic death pathways via ASC. *Cell Death Differ* **20**, 1149–1160 (2013). https://doi.org/10.1038/cdd.2013.37
46. Duclos, S. *et al.* The endosomal proteome of macrophage and dendritic cells. *Proteomics* **11**, 854–864 (2011). https://doi.org/10.1002/pmic.201000577
47. Latz, E. *et al.* TLR9 signals after translocating from the ER to CpG DNA in the lysosome. *Nat Immunol* **5**, 190–198 (2004). https://doi.org/10.1038/ni1028
48. Rabas, N. *et al.* PINK1 drives production of mtDNA-containing extracellular vesicles to promote invasiveness. *J Cell Biol* **220** (2021). https://doi.org/10.1083/jcb.202006049
49. Latz, E. *et al.* Ligand-induced conformational changes allosterically activate Toll-like receptor 9. *Nat Immunol* **8**, 772–779 (2007). https://doi.org/10.1038/ni1479
50. Wesche, H., Henzel, W. J., Shillinglaw, W., Li, S. Y. & Cao, Z. D. MyD88: An Adapter That Recruits IRAK to the IL-1 Receptor Complex (Reprinted from Immunity, vol 7, pg 837–847, 1997). *J Immunol* **190**, 5–15 (2013).
51. Muroi, M. & Tanamoto, K. I. TRAF6 distinctively mediates MyD88- and IRAK-1-induced activation of NF-kappa B. *J Leukocyte Biol* **83**, 702–707 (2008). https://doi.org/10.1189/jlb.0907629
52. Zhong, Z. Y. *et al.* New mitochondrial DNA synthesis enables NLRP3 inflammasome activation. *Nature* **560**, 198–203 (2018). https://doi.org/10.1038/s41586-018-0372-z
53. Bernardi, P., Rasola, A., Forte, M. & Lippe, G. The mitochondrial permeability transition pore: Channel formation by F-ATP synthase, integration in signal transduction, and role in pathophysiology. *Physiol Rev* **95**, 1111–1155 (2015). https://doi.org/10.1152/physrev.00001.2015

54. Giorgi, C., Marchi, S. & Pinton, P. The machineries, regulation and cellular functions of mitochondrial calcium (vol 19, pg 713, 2018). *Nat Rev Mol Cell Bio* **19**, 746–746 (2018). https://doi.org/10.1038/s41580-018-0066-2
55. Horner, S. M., Liu, H. M., Park, H. S., Briley, J. & Gale, M. Mitochondrial-associated endoplasmic reticulum membranes (MAM) form innate immune synapses and are targeted by hepatitis C virus. *P Natl Acad Sci USA* **108**, 14590–14595 (2011). https://doi.org/10.1073/pnas.1110133108
56. Cheng, J. B. *et al.* Amplified RLR signaling activation through an interferon-stimulated gene-endoplasmic reticulum stress-mitochondrial calcium uniporter protein loop. *Sci Rep* **6** (2016). https://doi.org/10.1038/srep20158
57. Xian, H. X. *et al.* Oxidized DNA fragments exit mitochondria via mPTP- and VDAC-dependent channels to activate NLRP3 inflammasome and interferon signaling. *Immunity* **55**, 1370–1385.e8. (2022). https://doi.org/10.1016/j.immuni.2022.06.007
58. Giorgio, V., Guo, L. S., Bassot, C., Petronilli, V. & Bernardi, P. Calcium and regulation of the mitochondrial permeability transition. *Cell Calcium* **70**, 56–63 (2018). https://doi.org/10.1016/j.ceca.2017.05.004
59. Garcia, N. & Chavez, E. Mitochondrial DNA fragments released through the permeability transition pore correspond to specific gene size. *Life Sci* **81**, 1160–1166 (2007). https://doi.org/10.1016/j.lfs.2007.08.019
60. Riley, J. S. *et al.* Mitochondrial inner membrane permeabilisation enables mtDNA release during apoptosis. *EMBO J* **37** (2018). https://doi.org/10.15252/embj.201899238
61. Cosentino, K. *et al.* The interplay between BAX and BAK tunes apoptotic pore growth to control mitochondrial-DNA-mediated inflammation. *Mol Cell* **82**, 933–949.e9 (2022). https://doi.org/10.1016/j.molcel.2022.01.008
62. Ren, D. C. *et al.* BID, BIM, and PUMA are essential for activation of the BAX- and BAK-dependent cell death program. *Science* **330**, 1390–1393 (2010). https://doi.org/10.1126/science.1190217
63. Edlich, F. *et al.* Bcl-x(L) retrotranslocates Bax from the mitochondria into the cytosol. *Cell* **145**, 104–116 (2011). https://doi.org/10.1016/j.cell.2011.02.034
64. Murphy, K. M., Ranganathan, V., Farnsworth, M. L., Kavallaris, M. & Lock, R. B. Bcl-2 inhibits Bax translocation from cytosol to mitochondria during drug-induced apoptosis of human tumor cells. *Cell Death Differ* **7**, 102–111 (2000). https://doi.org/10.1038/sj.cdd.4400597
65. Westphal, D., Kluck, R. M. & Dewson, G. Building blocks of the apoptotic pore: How Bax and Bak are activated and oligomerize during apoptosis. *Cell Death Differ* **21**, 196–205 (2014). https://doi.org/10.1038/cdd.2013.139
66. Bao, D. K. *et al.* Mitochondrial fission-induced mtDNA stress promotes tumor-associated macrophage infiltration and HCC progression. *Oncogene* **38**, 5007–5020 (2019). https://doi.org/10.1038/s41388-019-0772-z
67. Kim, J. *et al.* VDAC oligomers form mitochondrial pores to release mtDNA fragments and promote lupus-like disease. *Science* **366**, 1531–1536 (2019). https://doi.org/10.1126/science.aav4011
68. Madesh, M. & Hajnoczky, G. VDAC-dependent permeabilization of the outer mitochondrial membrane by superoxide induces rapid and massive cytochrome c release. *J Cell Biol* **155**, 1003–1015 (2001). https://doi.org/10.1083/jcb.200105057
69. Liu, H. *et al.* Prohibitin 1 regulates mtDNA release and downstream inflammatory responses. *EMBO J* **41** (2022). https://doi.org/10.15252/embj.2022111173
70. Mishra, P. & Chan, D. C. Metabolic regulation of mitochondrial dynamics. *J Cell Biol* **212**, 379–387 (2016). https://doi.org/10.1083/jcb.201511036
71. Lewis, S. C., Uchiyama, L. F. & Nunnari, J. ER-mitochondria contacts couple mtDNA synthesis with mitochondrial division in human cells. *Science* **353**, aaf5549 (2016). https://doi.org/10.1126/science.aaf5549
72. Irazoki, A. *et al.* Disruption of mitochondrial dynamics triggers muscle inflammation through interorganellar contacts and mitochondrial DNA mislocation. *Nat Commun* **14**, 108 (2023). https://doi.org/10.1038/s41467-022-35732-1
73. Rodriguez-Nuevo, A. *et al.* Mitochondrial DNA and TLR9 drive muscle inflammation upon Opa1 deficiency. *EMBO J* **37** (2018). https://doi.org/10.15252/embj.201796553
74. Sprenger, H. G. *et al.* Cellular pyrimidine imbalance triggers mitochondrial DNA-dependent innate immunity. *Nat Metab* **3**, 636–650 (2021). https://doi.org/10.1038/s42255-021-00385-9
75. Hwang, M. S. *et al.* MAVS polymers smaller than 80 nm induce mitochondrial membrane remodeling and interferon signaling. *FEBS J* **286**, 1543–1560 (2019). https://doi.org/10.1111/febs.14772
76. Hanada, Y. *et al.* MAVS is energized by MFF which senses mitochondrial metabolism via AMPK for acute antiviral immunity. *Nat Commun* **11** (2020). https://doi.org/10.1038/s41467-020-19287-7

77. Yasukawa, K. *et al.* Mitofusin 2 inhibits mitochondrial antiviral signaling. *Sci Signal* **2** (2009). https://doi.org/10.1126/scisignal.2000287
78. Seth, R. B., Sun, L. J., Ea, C. K. & Chen, Z. J. J. Identification and characterization of MAVS, a mitochondrial antiviral signaling protein that activates NF-kappa B and IRF3. *Cell* **122**, 669–682 (2005). https://doi.org/10.1016/j.cell.2005.08.012
79. Irazoki, A. *et al.* Coordination of mitochondrial and lysosomal homeostasis mitigates inflammation and muscle atrophy during aging. *Aging Cell* **21** (2022). https://doi.org/10.1111/acel.13583
80. Li, W. H. *et al.* FUN14 domain-containing 1-mediated mitophagy suppresses hepatocarcinogenesis by inhibition of inflammasome activation in mice. *Hepatology* **69**, 604–621 (2019). https://doi.org/10.1002/hep.30191
81. Nakahira, K. *et al.* Autophagy proteins regulate innate immune responses by inhibiting the release of mitochondrial DNA mediated by the NALP3 inflammasome. *Nat Immunol* **12** (3), 222–230 (2011). https://doi.org/10.1038/ni.1980
82. Sliter, D. A. *et al.* Parkin and PINK1 mitigate STING-induced inflammation. *Nature* **561**, 258–262 (2018). https://doi.org/10.1038/s41586-018-0448-9
83. Wauters, F. *et al.* LRRK2 mutations impair depolarization-induced mitophagy through inhibition of mitochondrial accumulation of RAB10. *Autophagy* **16**, 203–222 (2020). https://doi.org/10.1080/15548627.2019.1603548
84. Saito, T. *et al.* An alternative mitophagy pathway mediated by Rab9 protects the heart against ischemia. *J Clin Invest* **129**, 802–819 (2019). https://doi.org/10.1172/Jci122035
85. Gutierrez, M. G., Munafo, D. B., Beron, W. & Colombo, M. I. Rab7 is required for the normal progression of the autophagic pathway in mammalian cells. *J Cell Sci* **117**, 2687–2697 (2004). https://doi.org/10.1242/jcs.01114
86. Jimenez-Orgaz, A. *et al.* Control of RAB7 activity and localization through the retromer-TBC1D5 complex enables RAB7-dependent mitophagy. *EMBO J* **37**, 235–254 (2018). https://doi.org/10.15252/embj.201797128
87. Minowa-Nozawa, A., Nozawa, T., Okamoto-Furuta, K., Kohda, H. & Nakagawa, I. Rab35 GTPase recruits NDP52 to autophagy targets (vol 36, pg 2790, 2017). *EMBO J* **36**, 3405 (2017). https://doi.org/10.15252/embj.201798293
88. Hammerling, B. C. *et al.* A Rab5 endosomal pathway mediates Parkin-dependent mitochondrial clearance. *Nat Commun* **8**, 14050 (2017). https://doi.org/10.1038/ncomms14050
89. Heo, J. M. *et al.* RAB7A phosphorylation by TBK1 promotes mitophagy via the PINK-PARKIN pathway. *Sci Adv* **4** (2018). https://doi.org/10.1126/sciadv.aav0443
90. Hamacher-Brady, A., Choe, S. C., Krijnse-Locker, J. & Brady, N. R. Intramitochondrial recruitment of endolysosomes mediates Smac degradation and constitutes a novel intrinsic apoptosis antagonizing function of XIAP E3 ligase. *Cell Death Differ* **21**, 1862–1876 (2014). https://doi.org/10.1038/cdd.2014.101
91. Wang, W. Z., Ma, X. P., Zhou, L. P., Liu, J. & Zhu, X. W. A conserved retromer sorting motif is essential for mitochondrial DLP1 recycling by VPS35 in Parkinson's disease model. *Hum Mol Genet* **26**, 781–789 (2017). https://doi.org/10.1093/hmg/ddw430
92. Tang, F. L. *et al.* VPS35 deficiency or mutation causes dopaminergic neuronal loss by impairing mitochondrial fusion and function. *Cell Rep* **12**, 1631–1643 (2015). https://doi.org/10.1016/j.celrep.2015.08.001
93. Sen, A., Boix, J. & Pla-Martin, D. Endosomal-dependent mitophagy coordinates mitochondrial nucleoid and mtDNA elimination. *Autophagy*, 1–2 (2023). https://doi.org/10.1080/15548627.2023.2170959
94. Braschi, E. *et al.* Vps35 Mediates Vesicle Transport between the Mitochondria and Peroxisomes. *Curr Biol* **20**, 1310–1315 (2010). https://doi.org/10.1016/j.cub.2010.05.066
95. Muntjes, K., Devan, S. K., Reichert, A. S. & Feldbrugge, M. Linking transport and translation of mRNAs with endosomes and mitochondria. *EMBO Rep* **22** (2021). https://doi.org/10.15252/embr.202152445
96. Lee, B. L. & Barton, G. M. Trafficking of endosomal Toll-like receptors. *Trends Cell Biol* **24**, 360–369 (2014). https://doi.org/10.1016/j.tcb.2013.12.002
97. Majer, O., Liu, B. & Barton, G. M. Nucleic acid-sensing TLRs: Trafficking and regulation. *Curr Opin Immunol* **44**, 26–33 (2017). https://doi.org/10.1016/j.coi.2016.10.003
98. Pazmandi, K. *et al.* Oxidative modification enhances the immunostimulatory effects of extracellular mitochondrial DNA on plasmacytoid dendritic cells. *Free Radic Biol Med* **77**, 281–290 (2014). https://doi.org/10.1016/j.freeradbiomed.2014.09.028
99. Matheoud, D. *et al.* Parkinson's disease-related proteins PINK1 and parkin repress mitochondrial antigen presentation. *Cell* **166**, 314–327 (2016). https://doi.org/10.1016/j.cell.2016.05.039
100. Zecchini, V. *et al.* Fumarate induces vesicular release of mtDNA to drive innate immunity. *Nature* **615**, 499–506 (2023). https://doi.org/10.1038/s41586-023-05770-w

101. Yamazaki, T. *et al.* Mitochondrial DNA drives abscopal responses to radiation that are inhibited by autophagy. *Nat Immunol* **21**, 1160–1171 (2020). https://doi.org/10.1038/s41590-020-0751-0
102. Wu, K. K. L. *et al.* The APPL1-Rab5 axis restricts NLRP3 inflammasome activation through early endosomal-dependent mitophagy in macrophages. *Nat Commun* **12** (2021). https://doi.org/10.1038/s41467-021-26987-1
103. Bai, J. L. *et al.* DsbA-L prevents obesity-induced inflammation and insulin resistance by suppressing the mtDNA release-activated cGAS-cGAMP-STING pathway. *P Natl Acad Sci USA* **114**, 12196–12201 (2017). https://doi.org/10.1073/pnas.1708744114
104. Field, C. S. *et al.* Mitochondrial integrity regulated by lipid metabolism is a cell-intrinsic checkpoint for Treg suppressive function. *Cell Metab* **31**, 422-+ (2020). https://doi.org/10.1016/j.cmet.2019.11.021
105. Salminen, A., Kaarniranta, K. & Kauppinen, A. Inflammaging: Disturbed interplay between autophagy and inflammasomes. *Aging* **4**, 166–175 (2012). https://doi.org/10.18632/aging.100444
106. Lopez-Otin, C., Blasco, M. A., Partridge, L., Serrano, M. & Kroemer, G. The hallmarks of aging. *Cell* **153**, 1194–1217 (2013). https://doi.org/10.1016/j.cell.2013.05.039
107. Hamann, L. *et al.* STING SNP R293Q is associated with a decreased risk of aging-related diseases. *Gerontology* **65**, 145–154 (2019). https://doi.org/10.1159/000492972
108. Hamann, L., Szwed, M., Mossakowska, M., Chudek, J. & Puzianowska-Kuznicka, M. First evidence for STING SNP R293Q being protective regarding obesity-associated cardiovascular disease in age-advanced subjects - a cohort study. *Immun Ageing* **17** (2020). https://doi.org/10.1186/s12979-020-00176-y
109. Sebastian, D. *et al.* Mfn2 deficiency links age-related sarcopenia and impaired autophagy to activation of an adaptive mitophagy pathway. *EMBO J* **35**, 1677–1693 (2016). https://doi.org/10.15252/embj.201593084
110. Jauhari, A. *et al.* Melatonin inhibits cytosolic mitochondrial DNA-induced neuroinflammatory signaling in accelerated aging and neurodegeneration. *J Clin Invest* **130**, 3124–3136 (2020). https://doi.org/10.1172/Jci135026
111. Yu, C. H. *et al.* TDP-43 triggers mitochondrial DNA release via mPTP to activate cGAS/STING in ALS. *Cell* **183**, 636–649.e18 (2020). https://doi.org/10.1016/j.cell.2020.09.020
112. Gambardella, S. *et al.* ccf-mtDNA as a potential link between the brain and immune system in neuro-immunological disorders. *Front Immunol* **10** (2019). https://doi.org/10.3389/fimmu.2019.01064
113. Tansey, M. G. *et al.* Inflammation and immune dysfunction in Parkinson disease. *Nat Rev Immunol* **22**, 657–673 (2022). https://doi.org/10.1038/s41577-022-00684-6
114. Valente, E. M. *et al.* Localization of a novel locus for autosomal recessive early-onset parkinsonism, PARK6, on human chromosome 1p35-p36. *Am J Hum Genet* **68**, 895–900 (2001). https://doi.org/10.1086/319522
115. Valente, E. M. *et al.* Hereditary early-onset Parkinson's disease caused by mutations in PINK1. *Science* **304**, 1158–1160 (2004). https://doi.org/10.1126/science.1096284
116. Ramirez, A. *et al.* Hereditary parkinsonism with dementia is caused by mutations in ATP13A2, encoding a lysosomal type 5 P-type ATPase. *Nat Genet* **38**, 1184–1191 (2006). https://doi.org/10.1038/ng1884
117. Vilariño-Güell, C. et al. VPS35 mutations in Parkinson disease. American journal of human genetics 89, 162–167 (2011). https://doi.org/10.1016/j.ajhg.2011.06.001
118. Breen, D. P., Munoz, D. G. & Lang, A. E. Twinkle-associated familial parkinsonism with Lewy pathology: Cause or predisposition? *Neurology* **95**, 644–647 (2020). https://doi.org/10.1212/Wnl.0000000000010674
119. Neuhaus, J. F. *et al.* Catecholamine metabolism induces mitochondrial DNA deletions and leads to severe adrenal degeneration during aging. *Neuroendocrinology* **104**, 72–84 (2017). https://doi.org/10.1159/000444680
120. Bedlack, R. S. *et al.* MNGIE neuropathy: Five cases mimicking chronic inflammatory demyelinating polyneuropathy. *Muscle Nerve* **29**, 364–368 (2004). https://doi.org/10.1002/mus.10546
121. Bernsen, P. L. J. A. *et al.* Successful treatment of pure myopathy, associated with Complex-I deficiency, with riboflavin and carnitine. *Arch Neurol* **48**, 334–338 (1991). https://doi.org/10.1001/archneur.1991.00530150106028
122. Ginsberg, L. *et al.* Coexistent hereditary and inflammatory neuropathy. *Brain* **127**, 193–202 (2004). https://doi.org/10.1093/brain/awh017
123. Fernandez-Lizarbe, S. *et al.* Neuroinflammation in the pathogenesis of axonal Charcot-Marie-Tooth disease caused by lack of GDAP1. *Exp Neurol* **320** (2019). https://doi.org/10.1016/j.expneurol.2019.113004

124. Barneo-Munoz, M. *et al.* Lack of GDAP1 induces neuronal calcium and mitochondrial defects in a knockout mouse model of charcot-marie-tooth neuropathy. *PLOS Genet* **11**, e1005115 (2015). https://doi.org/10.1371/journal.pgen.1005115
125. Pla-Martin, D. *et al.* Silencing of the Charcot-Marie-Tooth disease-associated gene GDAP1 induces abnormal mitochondrial distribution and affects Ca^{2+} homeostasis by reducing store-operated Ca^{2+} entry. *Neurobiol Dis* **55**, 140–151 (2013). https://doi.org/10.1016/j.nbd.2013.03.010
126. Pedrola, L. *et al.* GDAP1, the protein causing Charcot-Marie-Tooth disease type 4A, is expressed in neurons and is associated with mitochondria. *Hum Mol Genet* **14**, 1087–1094 (2005). https://doi.org/10.1093/hmg/ddi121
127. Cantarero, L. *et al.* Mitochondria-lysosome membrane contacts are defective in GDAP1-related Charcot-Marie-Tooth disease. *Hum Mol Genet* **29**, 3589–3605 (2021). https://doi.org/10.1093/hmg/ddaa243
128. Hou, S., Wang, L. L. & Zhang, G. P. Mitofusin-2 regulates inflammation-mediated mouse neuroblastoma N2a cells dysfunction and endoplasmic reticulum stress via the Yap-Hippo pathway. *J Physiol Sci* **69**, 697–709 (2019). https://doi.org/10.1007/s12576-019-00685-6
129. Khodzhaeva, V. *et al.* Mitofusin 2 deficiency causes pro-inflammatory effects in human primary macrophages. *Front Immunol* **12** (2021). https://doi.org/10.3389/fimmu.2021.723683
130. Bueno, M. *et al.* PINK1 attenuates mtDNA release in alveolar epithelial cells and TLR9 mediated profibrotic responses. *PLOS ONE* **14** (2019). https://doi.org/10.1371/journal.pone.0218003
131. Wu, N. N. *et al.* Site-specific ubiquitination of VDAC1 restricts its oligomerization and mitochondrial DNA release in liver fibrosis. *Exp Mol Med* **55**, 269–280 (2023). https://doi.org/10.1038/s12276-022-00923-9
132. Li, Y. J. *et al.* TFAM downregulation promotes autophagy and ESCC survival through mtDNA stress-mediated STING pathway. *Oncogene* **41**, 3735–3746 (2022). https://doi.org/10.1038/s41388-022-02365-z
133. Jia, J., Li, M. Q., Li, Y. J., Xiao, J. & Dai, H. M. The mtDNA-STING pathway plays an important role in both navitoclax- and S63845-induced autophagy and enhances cell death. *Cell Biol Toxicol* (2023). https://doi.org/10.1007/s10565-023-09804-x
134. Liang, J. L. *et al.* Specific activation of cGAS-STING pathway by nanotherapeutics- mediated ferroptosis evoked endogenous signaling for boosting systemic tumor immunotherapy. *Sci Bull* **68**, 622–636 (2023). https://doi.org/10.1016/j.scib.2023.02.027
135. Li, J. Y. *et al.* TRIM21 inhibits irradiation-induced mitochondrial DNA release and impairs antitumour immunity in nasopharyngeal carcinoma tumour models. *Nat Commun* **14**, 865 (2023). https://doi.org/10.1038/s41467-023-36523-y

8 Mitochondrial Complex I Assembly and Implications for Mitochondrial Disease

Luke E. Formosa[*]

COMPLEX I STRUCTURE, FUNCTION AND SUBUNIT BIOGENESIS

The electron transport chain plays an important role in the final steps of primary metabolism, using electrons from reducing equivalents formed during glycolysis and the Krebs cycle, namely NADH and $FADH_2$, to generate the proton motive force, which is then used to power the production of ATP. The first enzyme in the electron transport chain is Complex I (NADH:Ubiquinone oxidoreductase), a ~1 MDa complex in the inner mitochondrial membrane that couples the oxidation of NADH to NAD^+ and reduction of ubiquinone to ubiquinol, to the translocation of protons across the inner membrane from the matrix to the intermembrane space [1–3].

In mammals, Complex I forms a characteristic boot-shaped structure in the membrane and is composed of 45 separate subunits encoded by two spatially separated genomes [4] (Figure 8.1). The mitochondrial genome (mtDNA) encodes seven highly hydrophobic subunits, which are all part of the membrane arm of Complex I, whereas the remaining 38 subunits are encoded by nuclear DNA and forms six distinct modules related to their function and assembly [5]. The subunits of Complex I are designated differently depending on whether they are encoded by mtDNA or the nucleus. Those encoded by mtDNA are labelled with the prefix 'MT-ND' for 'mtDNA-encoded NADH Dehydrogenase chain' and given numbers 1 to 6 and 4L. Nuclear-encoded subunits are labelled with 'NDUF' for NADH dehydrogenase (NDU) and a name that historically indicates the isolation of the subunit in a particular detergent-extracted subcomplex, such as 'alpha subcomplex' (FA) or 'beta subcomplex' (FB), or named after Complex I cofactors such as flavoprotein (FV) or iron-sulfur protein (FS) [6]. It is important to note that only NDUFV1 contains a covalently bound flavin mononucleotide (FMN) cofactor, whereas NDUFV2 and NDUFV3 do not [7, 8]. Additionally, not all subunits designated as 'NDUFS' contain iron-sulfur (FeS) clusters, such as NDUFS2-6 do not contain these cofactors [7, 8]. Complex I assembly factors are similarly labelled with 'NDUF' and the designation 'AF' for 'assembly factor,' followed by a number indicating the order of their discovery. It is worth noting that the labelling of 'F' for the presence of the 'alpha subcomplex' is a misnomer, as assembly factors are not present in the final Complex I structure where these designations were made.

The peripheral arm of Complex I extends into the mitochondrial matrix and is the region of the enzyme responsible for NADH oxidation at the N (NADH dehydrogenase) module followed by the transfer of electrons through a series of FeS clusters to ubiquinone (Q) at the Q (ubiquinone reduction) module. The membrane arm of Complex I is composed of four highly conserved modules that together form the ubiquinone entry site (ND1 module) and the proton-pumping modules (ND2, ND4 and ND5 modules). The membrane arm modules have also been named according to their role as proton-pumping (P) modules and their position relative to the peripheral arm of Complex I [9]. The ND1 and ND2 modules are proximal to the peripheral arm and are also referred to as the 'P_P-a' and 'P_P-b' modules, respectively, while the ND4 and ND5 modules are distal to the peripheral

[*] Corresponding author: Luke.Formosa@monash.edu

 DOI: 10.1201/9781003221180-12

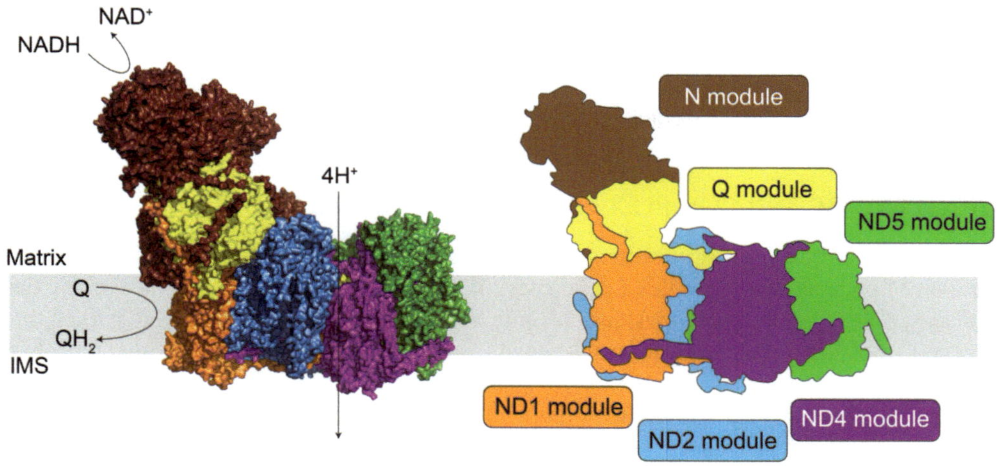

FIGURE 8.1 The structure and function of Complex I. Cryo-EM model of Complex I (PDB:6QA9) highlighting structural modules and functional regions where NADH oxidation, ubiquinone (Q) reduction and proton translocation occur. On the right is a schematic representation that will be used throughout this chapter.

arm and are referred to as the 'P_D-a' and 'P_D-b' modules, respectively [9]. The evolutionary roots of these Complex I modules are thought to be from the Multiple resistance and pH adaptation (Mrp) antiporter, a membrane-embedded multi-subunit Na^+/H^+ antiporter that resembles the antiporter-like subunits of the membrane arm of Complex I and membrane-bound hydrogenases that together coupled redox and ion translocation functions [10–13].

For Complex I to assemble correctly, the nuclear DNA-encoded subunits must be imported into the appropriate mitochondrial sub-compartments (matrix, inner membrane, intermembrane space) while the mtDNA subunits must be co-translationally inserted into the inner membrane as required. Subunits that are required in the matrix follow the canonical 'pre-sequence-mediated pathway' for mitochondrial protein import through the TOM and TIM23 motor complexes (Figure 8.2A), while transmembrane spanning subunits use the TIM23-sort or TIM22 translocases if they are encoded

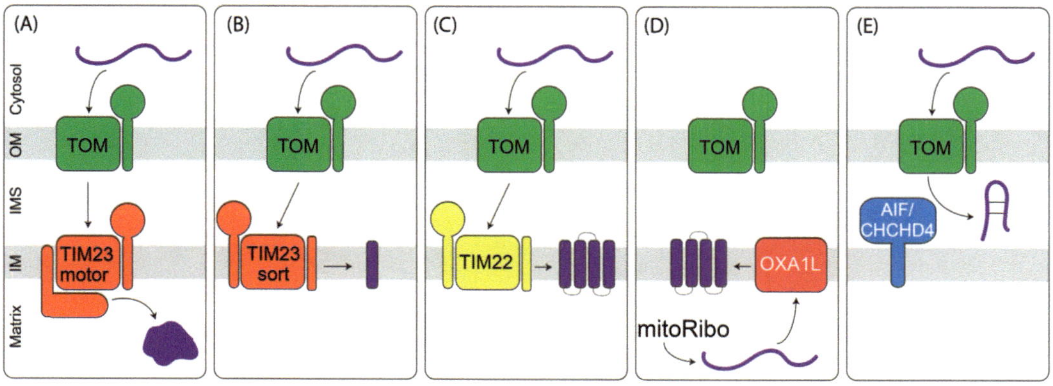

FIGURE 8.2 Biogenesis of Complex I subunits. To assemble Complex I, subunits and assembly factors must reach their correct destinations through the (A) TIM23 motor complex to reach the mitochondrial matrix, (B) the TIM23 sort complex or (C) TIM22 complex for nuclear-encoded membrane proteins, (D) the OXA1L insertase for mtDNA-encoded membrane proteins and (E) the AIF/MIA40/CHCHD4 import pathway for oxidatively folded intermembrane space proteins. *Abbreviations*: IM, inner membrane; IMS, intermembrane space; MitoRibo, MitoRibosome; OM, outer membrane; TOM, TOM complex.

by the nuclear genome (Figure 8.2B, C), or through the OXA1L insertase if encoded by the mtDNA (Figure 8.2D) [14–17]. A subset of subunits located on the intermembrane space (IMS) side of Complex I engage the AIF/CHCHD4 pathway, which oxidatively folds subunits by introducing disulfide bonds before their assembly (Figure 8.2E) [18, 19].

DISCOVERY OF THE COMPLEX I ASSEMBLY MACHINERY

The discovery and validation of proteins involved in the assembly of Complex I is crucial for the diagnosis of patients with Complex I deficiency, of which ~50% do not have a molecular cause for the disease [20, 21]. For many years, insights into how complex I is assembled were gained through the genetic diagnosis of patients with an isolated Complex I deficiency and subsequent mechanistic insights of the molecular basis of the mitochondrial dysfunction [22, 23]. A new age in mitochondrial biology research was started with the advent of genome editing, multi-omics and systems biology approaches, which have played an important role in mitochondrial biology more broadly by improving diagnostic rates for patients and identifying new components of the Complex I assembly machinery and understanding the consequences of their loss [24–27].

ASSEMBLY PATHWAY AND ASSEMBLY FACTORS

Assembly of the ND1/Q Module (Q/P_P-a)

De novo Complex I assembly starts with the integration of two distinct functional modules, the Q module and the ND1 module, that form the nexus between the membrane and peripheral arms of Complex I (Figure 8.3) [7, 28].

Assembly of the Q module begins with the subunits NDUFS2, NDUFS3 and NDUFA5 forming a soluble intermediate in the mitochondrial matrix [9]. The assembly factor NDUFAF7 is required at this stage for the symmetric dimethylation of NDUFS2 at Arginine-118 (Arg-85 following maturation) on the ω-N^G, $N^{G'}$ atoms of the side chain [29, 30]. The subunits NDUFS7 and NDUFS8 come together and undergo maturation with the delivery of FeS clusters (one [4Fe-4S] in NDUFS7 and two [4Fe-4S] in NDUFS8), thought to be mediated by the FeS cluster chaperone NUBPL/IND1 [31, 32]. Furthermore, during the early stages of NDUFS7 maturation, the assembly factor NDUFAF5, which belongs to the family of S-adenosylmethionine-dependent methyltransferases, hydroxylates

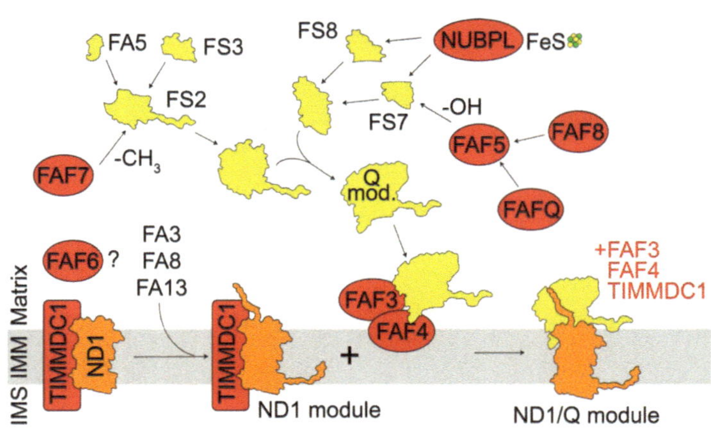

FIGURE 8.3 Assembly of the ND1/Q module. Subunits of the Q and ND1 module assemble separately by forming a soluble Q module that becomes attached to the membrane-embedded ND1 module. Q module subunits shown in yellow and ND1 module subunits indicated in orange. Assembly factors shown in red.

the subunit at Arginine-111 (Arg-73 following mitochondrial targeting sequence processing), which is required for proper Complex I assembly [33]. In addition, two assembly factors have been implicated in the assembly of Complex I by promoting the stability of NDUFAF5, namely NDUFAF8 and NDUFAFQ. The assembly factor NDUFAF8, a member of the twin CX_9C motif family, was identified as an orphan protein that interacts with NDUFAF5, and knockdown of NDUFAF8 led to a reduction in the steady-state levels of NDUFAF5 [34]. Since the initial discovery of NDUFAF8, these observations have been further supported by the identification of patients that harbour pathogenic mutations in the *NDUFAF8* gene, and the analysis of fibroblasts from these patients recapitulated the previous knockdown studies [34, 35]. More recently, the function of NDUFAF8 in Complex I was discovered that links Complex I assembly in the matrix to the activity of the disulfide relay system of the IMS [36]. NDUFAF8 was found to harbour a weak mitochondrial targeting signal that drives TIM23-mediated import into the matrix, and, in the process NDUFAF8 is also oxidized by the AIF/MIA40/CHCHD4 pathway (Figure 8.2A, E) [36]. Excess NDUFAF8 in the IMS was degraded by the i-AAA protease YME1L, while reduced NDUFAF8 in the matrix was degraded by the matrix resident protease CLPP, suggesting that NDUFAF8 can only participate in Complex I assembly if both import pathways are functional [36]. An additional Complex I assembly factor, NDUFAFQ/PYURF, has also been found to be important for the stability of NDUFAF5, with a reduction in the steady-state levels of NDUFAFQ leading to decreased levels of Complex I in knockout cells and patient-derived cells [37]. Interestingly, NDUFAFQ was also found to be important for the stability of COQ5, a protein important in the biosynthesis of ubiquinone, potentially linking these two distinct but related metabolic pathways [37].

Following the assembly of these subunits, the matured Q module interacts with the membrane-associated assembly factors NDUFAF3 and NDUFAF4 [38, 39]. Interestingly, both NDUFAF3 and NDUFAF4 remain strongly associated with the mitochondrial membrane [38], even though neither protein is predicted to contain a transmembrane anchor. It has since been discovered that NDUFAF4 is modified on its N terminus by N-myristoyltransferase (NMT) 1/2 with the addition of a C14 myristic acid on an essential conserved glycine and the loss of this modification (through NMT inhibition), leads to proteosome-mediated NDUFAF4 degradation [40, 41]. Indeed, a patient with Leigh syndrome was identified to harbour a homozygous p.Ala3Pro mutation in NDUFAF4 (located directly adjacent to the conserved glycine that is myristylated) and through *in vitro* protein myristylation analysis, the mutant protein expressed in patient cells could not be myristylated [42, 43]. This argues that myristylation of the NDUFAF4 N-terminus is crucial for NDUFAF4 function in the assembly of Complex I, uncovering the likely mechanism of mitochondrial disease. The close association of the NDUFAF3/NDUFAF4 assembly factor complex with the inner membrane and their interaction with a pre-assembled Q module hints that NDUFAFA3/NDUFAF4 may act as a platform for Q-module recruitment to the matrix side of the inner membrane for Complex I assembly to occur [9].

Early in the assembly process, the mtDNA-encoded subunit ND1 is synthesized in the matrix by the mitoribosome and is then thought to be inserted into the mitochondrial inner membrane by the general insertase OXA1L [16]. This is facilitated by the assembly factor TIMMDC1, which was identified to interact with ND1 [44, 45] and was also identified in the OXA1L interactome [16]. The ND1 subunit was also found to comigrate with the assembly factor TIMMDC1 [9] and loss of TIMMDC1 led to rapid ND1 degradation [5, 46]. Another protein implicated in the stability of ND1 is NDUFAF6, as loss of NDUFAF6 leads to rapid turnover of ND1 [47]. The formation of the mature ND1 module is completed with the addition of the Complex I subunits NDUFA3, NDUFA8 and NDUFA13 [9]. The assembly of Complex I continues with the formation of the ND1/Q module, where these two distinct assemblies come together to form an ~440 kDa complex.

The stability of the ND1 module is dependent on the integration of Q module during assembly, as loss of the Q module subunits or assembly factors results in rapid AFG3L2-mediated degradation of ND1 [30, 48]. Additionally, quantitative proteomics analysis of cells lacking NDUFAF4 showed strong decreases in both Q and ND1 module subunits, while the loss of TIMMDC1 had a far more drastic effect on the ND1 module compared to the Q module of Complex I, suggesting that this stability is not reciprocal in nature [5].

Assembly of the ND2 Module (P_P-b)

The ND2 module consists of several subunits encoded by mtDNA (ND2, ND3, ND4L and ND6), and requires at least seven assembly factors as well as the general insertase OXA1L (Figure 8.4) [16, 28]. These assembly factors have been termed the 'Mitochondrial Complex I Intermediate Assembly' (MCIA) complex [46, 49]. This complex consists of the assembly factors NDUFAF1, ECSIT, ACAD9, TMEM126B, TMEM186, COA1 and SFXN4 [9, 46, 49–55]. Assembly of this module begins with the translation of ND2, where COA1 is thought to form a nascent chain complex with the mitoribosome [54]. The newly synthesized ND2 is stabilized by direct contact with NDUFAF1, ECSIT and ACAD9 and the incorporation of the accessory subunits NDUFC1 and NDUFC2 [9, 46, 54]. Maturation of the ND2 module continues with the incorporation of the assembly factors TMEM126B and TMEM186, the latter interacting with newly synthesized mtDNA-encoded subunit ND3, presumably for delivery to the complex [9, 46]. The assembly of intermediates containing the MCIA complex rely strongly on NDUFAF1, ECSIT and ACAD9, as without any of these assembly factors higher order complexes are not formed [46]. In contrast, while TMEM126B is critical for the assembly of Complex I and mutations have been implicated in mitochondrial disease, the loss of TMEM126B does not preclude the formation of higher order complexes containing ACAD9, ECSIT and NDUFAF1 [46, 49, 56, 57] The final step in the assembly of the ND2 module is the incorporation of mtDNA-encoded ND6 and its assembly chaperone SFXN4, which can still maintain a binary interaction in the absence of the remainder of MCIA complexes, as well as the mtDNA-encoded subunit ND4L [9, 55].

Before being implicated in the assembly of Complex I, ECSIT was first identified as a signalling intermediate in innate immune signalling through Toll/IL-1 signal transduction pathway where it linked TRAF6 to MEKK-1 and NF-κB signalling [58]. It was later found that ECSIT plays a role in bone morphogenic protein (BMP) signalling during embryogenesis, through association with Smads to promote the expression of BMP-stimulated genes, playing a role of linking these signalling pathways [59]. Following the discovery of a mitochondrial pool of ECSIT that participates in Complex I assembly [52], it was found that ECSIT interacts with TRAF6 on the mitochondrial surface during lipopolysaccharide-induced TLR1/2/4 activation to promote mitochondrial reactive oxygen species (mROS) generation, which can clear the bacterial infection as part of the innate

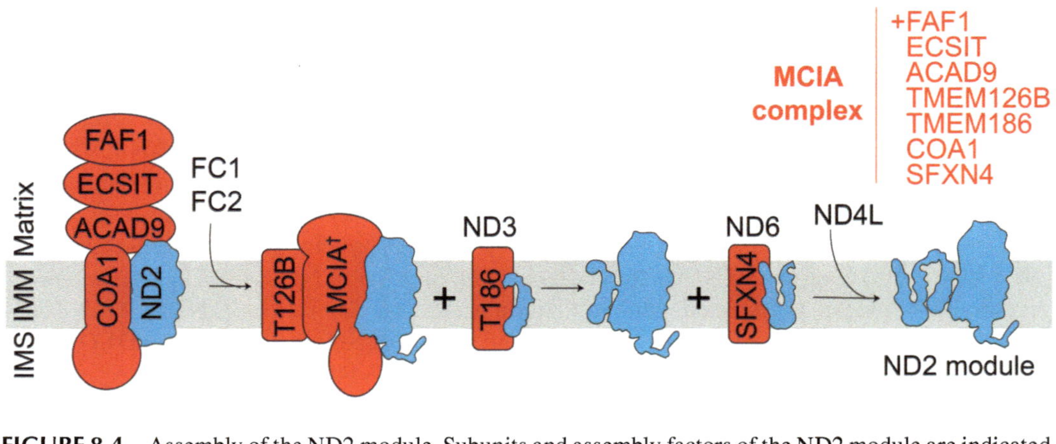

FIGURE 8.4 Assembly of the ND2 module. Subunits and assembly factors of the ND2 module are indicated. Assembly factors are shown in red. MCIA† indicates an initial complex of NDUFAF1, ECSIT, ACAD9 and COA1 during ND2 translation and assembly.

immune response [60]. While the role of ECSIT in Complex I assembly has been well established [9, 46, 52, 61], several open questions remain regarding its proposed cytosolic roles. For example, how is the distribution of ECSIT maintained between the cytosol and mitochondria, given that full-length ECSIT present in the cytosol contains an intact mitochondrial targeting signal, and how is stability of ECSIT maintained given its reliance on NDUFAF1 and ACAD9 in mitochondria? Further research is required to address these open questions.

ACAD9 was originally identified as a member of the Acyl-CoA dehydrogenase family, with *in vitro* activity towards long-chain unsaturated acyl-CoA substrates and an ortholog of the very long-chain acyl-CoA dehydrogenase (VLCAD), which is involved in the β-oxidation of long-chain fatty acids [62]. Indeed, in patients with mutations in the *VLCAD* gene, ACAD9 was found to be responsible for the generation of the diagnostic C14:1 acyl-carnitine species in a manner dependent on the catalytic activity of ACAD9 [63]. New molecular insights have been gained in understanding the interplay of ACAD9 in these two metabolic pathways with the discovery that ECSIT is involved in deflavinating holo-ACAD9, shifting its role from a catalytic role in β-oxidation to a chaperone role when participating in Complex I assembly [64, 65]. The deflavination of ACAD9 involved in Complex I assembly correlates with experimental data, indicating that catalytically active ACAD9 is not required for the function of ACAD9 in the assembly of Complex I [63].

Before being associated with the MCIA complex, COA1 was originally implicated in the assembly of the mitochondrial Complex IV as part of the **Mi**tochondrial **T**ranslation **R**egulation **A**ssembly Intermediate of **C**ytochrome *c* Oxidase (MITRAC), required for the COX1 translational regulation early in Complex IV assembly [66]. Later studies discovered that COA1 also comigrates with components of the MCIA complex [9]. Further analysis identified that the majority of COA1 was not present in the MITRAC complex but rather participated in the early stages of ND2 translation, and genetic ablation of COA1 did not appear to affect the assembly of Complex IV in cultured cells [46, 54]. COA1 is distantly related to TIM21, a component of the yeast and mammalian TIM23 complex, as well as part of MITRAC [66, 67] and has been lost or modified several times during evolution, including in rodent lineages [68]. It may be possible that COA1 plays a more prominent role in the assembly of Complex IV in a tissue-specific manner and further research will be important to clarify the multifaceted role of this protein in mammalian mitochondria.

Assembly of the ND4 Module (P_D-a)

The assembly of the ND4 module starts with the co-translational insertion of newly synthesized ND4 into the inner membrane, which is mediated by the OXA1L insertase and TMEM126A (Figure 8.5) [16, 69]. Indeed, isolation and proteomics analysis of the translating mitoribosome could enrich both OXA1L and TMEM126A, and co-immunoprecipitation of either epitope-tagged OXA1L or TMEM126A was able to enrich the other protein [16, 69, 70]. In addition, the assembly factor TMEM70 (also involved in the assembly of the ATP synthase c-ring [71, 72]) also participates in the early stage of assembly of the ND4 module and also interacts with TMEM126A [9, 69, 73].

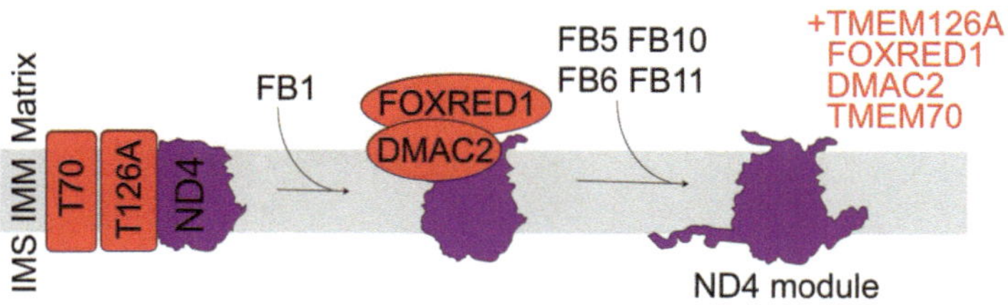

FIGURE 8.5 Assembly of the ND4 module. Subunits and assembly factors of the ND4 module are indicated. Assembly factors are shown in red.

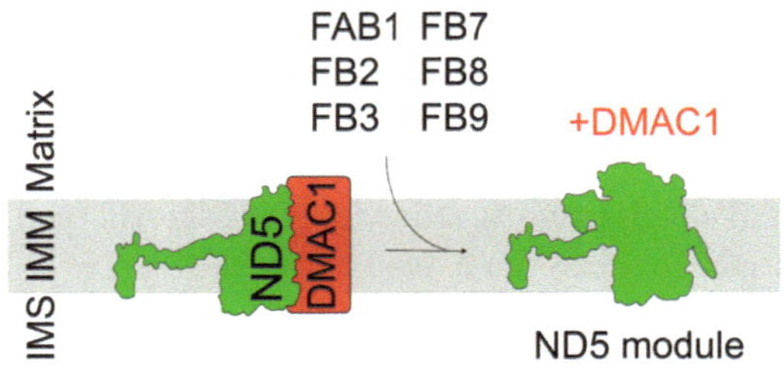

FIGURE 8.6 Assembly of the ND5 module. Subunits and the known assembly factor of the ND5 module (DMAC1) are indicated. Assembly factor DMAC1 is shown in red.

Following the incorporation of the accessory subunit NDUFB1, the assembly factors FOXRED1 and DMAC2 integrate into the assembly pathway [5, 9, 74]. Interestingly, in patient cells lacking FOXRED1, analysis of mtDNA-encoded proteins revealed a crippled ~475 kDa breakdown product that contained the mtDNA-encoded subunits ND1, ND2, ND3 and ND6, suggesting that in the absence of FOXRED1, the ND4 and ND5 subunits are unable to assemble [74]. The assembly of the ND4 module is completed with the addition of the accessory subunits NDUFB5, NDUFB6, NDUFB10 and NDUFB11 [9].

Assembly of the ND5 Module (P_D-b)

Compared to other Complex I assembly modules, relatively little is known about the assembly of the ND5 module (Figure 8.6). The assembly factor DMAC1 interacts with newly synthesized ND5 [5]. Following the synthesis and membrane integration of ND5, the remaining subunits of the module coalesce and include NDUFB2, NDUFB3, NDUFB7, NDUFB8, NDUFB9 and NDUFAB1 to form the completed module [9].

Assembly of the N Module

Assembly of the N-module begins in the mitochondrial matrix with the incorporation if FeS clusters into the core catalytic subunits NDUFS1 (one [2Fe-2S], two [4Fe-4S]), NDUFV1 (one [4Fe-4S]) and NDUFV2 (one [2Fe-2S]), which is thought to be mediated by NUBPL, as well as the incorporation of a non-covalent flavin mononucleotide (FMN) into NDUFV1 (Figure 8.7). Following this, NDUFV1 and NDUFV2 interact together, whereas NDUFS1 and NDUFA2 interact forming two intermediates [9]. The association of NDUFS1 and NDUFA2 may also involve the protein LYRM9, which was recently shown to interact with both proteins, as well as NDUFAB1 (as a member of the LYR family) [75, 76]. These two subassemblies then come together to form an immature N-module present in the mitochondrial matrix [9].

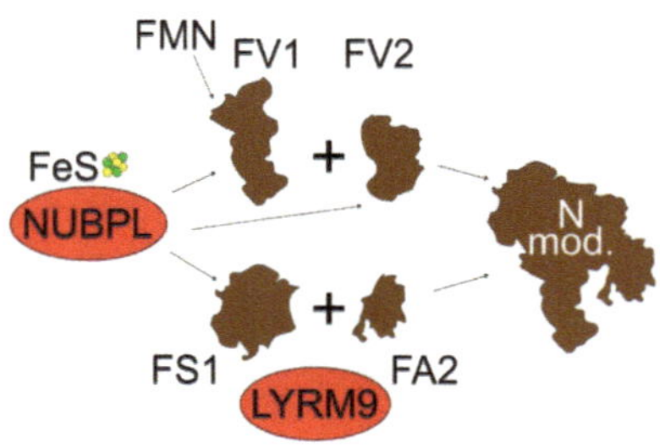

FIGURE 8.7 Assembly of the N module. Subunits, cofactors and assembly factors of the N module are indicated. Assembly factors are shown in red.

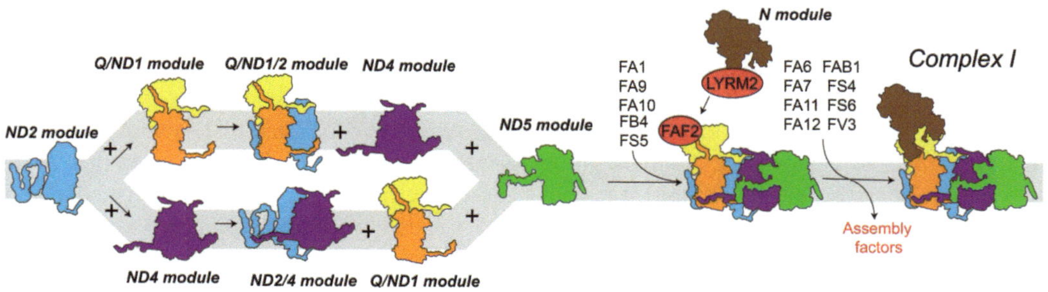

FIGURE 8.8 *De novo* assembly of Complex I from its modules. The major modules as starting points can assemble through two pathways to form the membrane arm of Complex I, followed by the addition of the N module as the last step in the assembly process. Assembly factors are shown in red.

Assembly of Complex I from Its Modules

With the preliminary assembly modules established, the membrane arm of Complex I continues to mature (Figure 8.8). The modular assembly diverges with the ND2 (P_P-b) module attaching to either the Q/ND1 (Q/P_P-a) module or the ND4 (P_D-a), which is then followed by the addition of the complementary module, followed by the ND5 module (P_D-b) to form the completed membrane arm [9]. At this stage, the assembly factor NDUFAF2 is associated with the late-stage ~815–830 kDa Complex I intermediate lacking the N module [5, 9, 77, 78]. The final step in the assembly of Complex I is thought to be the attachment of the N module to the assembled membrane arm. This is at least in part facilitated by the assembly factor LYRM2. As in the absence of LYRM2, the levels of mature Complex I were substantially reduced compared to control levels, with the accumulation of a Complex I intermediate lacking the N-module with NDUFAF2 still associated, which phenocopies the loss of the accessory subunit NDUFS4 [5, 76, 78]. During the attachment of the N-module, several other subunits associate (NDUFA6, NDUFA7, NDUFA11, NDUFA12, NDUFAB1, NDUFS4, NDUFS6 and NDUFV3), with the concomitant dissociation of Complex I assembly factors, leading to the formation of the fully assembled and enzymatically active Complex I [9]. Based on recent structural evidence of the fully assembled Complex I [7, 28, 79], many of the accessory subunits appear to extend and contact not only their neighbouring modules but could extend to interact with more distant modules also. It is possible that a role for some assembly factors may be to interact with these extensions to prevent unwanted interactions and promote their proper integration into the growing complex.

ASSEMBLY OF COMPLEX I WITH THE RESPIRATORY SUPERCOMPLEX

During the assembly of the respiratory chain complexes, the biogenesis of the respiratory chain supercomplex comprising complexes I, III and IV occurs. High-resolution cryogenic electron microscopy structures have illuminated the arrangement of the respiratory chain complexes within the respirasome [79–81]. Current models suggest that either each respiratory chain complex is assembled independently and subsequently Complex I combines with complexes III and IV to form a supercomplex [9] or that the supercomplex assembles cooperatively during the assembly of the individual enzymes [82]. There are several lines of evidence that support the model for the co-assembly of Complex I into the supercomplex. For example, primary assembly defects in complexes III or IV lead to reduced levels of Complex I [83–85]. Furthermore, cells lacking certain subunits of Complex I, which can still assemble the membrane arm, are able to associate with Complexes III and IV, indicating that fully assembled Complex I is not required for supercomplex formation [5, 86]. It has been shown that Complex I in human cells is almost exclusively located in the respirasome [57, 87], which suggests an important role for the supercomplex in the assembly or stability of Complex I.

REPAIR OF DAMAGED COMPLEX I

In order to maintain Complex I activity, Complex I subunits that are oxidatively damaged can be replaced by substituting with newly imported subunits without the need for *de novo* Complex I assembly [88]. This is also evident even when oxidative damage to Complex I subunits is not present. *In vitro* import analysis of Complex I subunits into isolated mitochondria revealed that, despite the import of only a single subunit, it is able to assemble into Complex I and intermediate subcomplexes without relying on other nuclear encoded proteins or *de novo* Complex I assembly [77]. More recently, the mechanism underlying a Complex I repair pathway has been uncovered. As the N-module is susceptible to damage, it was found that the matrix protease CLPXP can selectively degrade damaged subunits allowing for subunit replacement [88]. This 'degrade-and-replace' pathway for the N-module of Complex I is also apparent with N-module subunits being relatively over-synthesized compared to other Complex I module subunits, presumably to provide additional subunits for incorporation following turnover [89]. Furthermore, in cell models of Complex I dysfunction, in the absence of Complex I activity, the N-module was readily turned over, suggesting stability of N-module subunits is at least partially dependent on the availability of the membrane arm for the N module to associate with [5]. The inability to repair Complex I can also lead to mitochondrial disease, with the identification of several families with Leber's hereditary optic neuropathy discovered to harbour biallelic mutations in the chaperone DNAJC30, a nuclear-encoded protein that appears to play a role in efficient exchange of damaged N-module Complex I subunits [90]. How these repair pathways work together, and if other pathways exist to repair Complex I remains to be elucidated.

COMPLEX I PATHOLOGY ASSOCIATED WITH MUTATIONS IN ASSEMBLY FACTORS

Mutations in Complex I assembly factors can lead to Complex I deficiency with diverse clinical presentations and severities (Table 8.1). As with most mitochondrial diseases, Complex I deficiency tends to affect tissues of high energetic demand, in particular nervous tissue, the heart and skeletal muscle. Therefore, the loss of Complex I can result in several pathologies such as MELAS (mitochondrial encephalomyopathy, lactic acidosis and stroke-like episodes) and various neuropathies including LHON (Leber hereditary optic neuropathy) and optic atrophy [91, 92]. The relationship between mutations in Complex I assembly factors, which at the molecular level all contain a Complex I deficiency, and the highly variable clinical presentation of these patients is not well understood and remains to be fully explored. In this section, three phenotypes associated with Complex I deficiency due to mutations in assembly factors will be highlighted – Leigh syndrome, optic atrophy and cardiomyopathy [93].

Leigh Syndrome

Leigh syndrome, also known as subacute necrotizing encephalomyelopathy, was first reported by Denis Leigh in 1951 as a unique case with no prior report in the literature [94]. Leigh syndrome occurs in approximately 1 in 40,000 live births, and is characterized by the onset of symptoms between the ages of 3 months and 1 year, often following a metabolic challenge such as a viral infection [95]. Leigh syndrome causes a rapidly deteriorating disease that is often fatal, with death occurring by 3–4 years of age [43, 94, 96]. Several patients with primary Complex I deficiency harbouring mutations in assembly factors that build the N, Q and ND4 modules have presented with Leigh syndrome, either alone or in conjunction with other neurological or metabolic symptoms (Table 8.1). Indeed, Leigh syndrome is one of the most common clinical presentations of Complex I deficiency [97].

TABLE 8.1
Pathogenic Mutations Identified in Complex I Assembly Factors

Module	Gene	Mutation	Disease Presentation	Ref.
N	NDUFAF2/B17.2L	p.Arg45*	Progressive encephalopathy	[123]
		p.Met1?	Hypotonia, nystagmus and ataxia, interspersed with acute episodes of encephalopathy	[124]
		p.Trp3*	Lactic acidosis, leukodystrophy and symmetric brainstem lesions	[125]
Q	NDUFAF3/C3ORF60	p.Gly77Arg	Lactic acidemia	[38]
		p.Arg122Pro	Macrocephaly and severe muscle weakness and hypotonia	
		p.Met1? p.Arg122Pro	Diffuse leukomalacia	
		p.Ala165Val	Leigh syndrome	[126]
		p.117Valdup p.Pro169Thr	Cavitating leukoencephalopathy	[127]
	NDUFAF4/C6ORF66	p.Leu65Pro	Mitochondrial encephalomyopathy, severe encephalopathy, intractable acidosis, muscle hypotonia and bilateral optic atrophy. Severe cardiomyopathy in one patient	[39]
		p.Aal3Pro	Leigh syndrome	[43]
		p.Gly8Asp	Lactic acidosis, leukodystrophy, muscle hypotonia, developmental delay, symmetrical basal ganglia and brainstem lesions	[125]
	NDUFAF5/C20ORF7	p.Lys52Thr p.Met279Arg	Leigh syndrome	[96]
		p.Met279Arg	Leigh-like syndrome with onset following infection	
		p.Lys109Asn p.Gly250Val	Leigh syndrome	
		p.Glu142Ala p.Met279Arg	Bilateral striatal necrosis	[128]
		p.Leu229Pro	Lactic acidosis, abnormalities of the diaphragm and central nervous system and adrenal insufficiency	[129]
	NDUFAF7/MidA	p.Asp266Glu	Pathologic myopia	[130]
	NDUFAF8/C17ORF89	p.Phe18Serfs*32 (one intronic mutation)	Leigh syndrome	[35]
		p.Phe55Leu	Leigh syndrome	
	NDUFAFQ/PYURF	p.Gln97Hisfs*6	Muscle hypotonia, failure to thrive, developmental delay, optic atrophy and persistently elevated lactate levels	[37]
	NUBPL/IND1	p.Gly56Arg + p.Asp273Glnfs*31	Predominant abnormalities of the cerebellar cortex, deep cerebral white matter and corpus callosum	[131]
		p.Gly56Arg + p.Asp273Glnfs*31 p.Asp105Tyr		

(Continued)

TABLE 8.1 *(Continued)*
Pathogenic Mutations Identified in Complex I Assembly Factors

Module	Gene	Mutation	Disease Presentation	Ref.
		p.Gly56Arg + p.Asp273Glnfs*31		
		p.Leu193Phe		
		p.Gly56Arg + p.Asp273Glnfs*31	Mitochondrial encephalomyopathy	[132]
		Complex rearrangement		
		p.Leu104Pro p.Phe242Leu	Developmental delay, generalized hypotonia and myopathy	[133]
ND1	TIMMDC1/C3ORF1	p.Gly199_Thr200ins5* (intronic)	Muscular hypotonia, developmental delay and neurological deterioration	[134]
		p.Arg129* p.Gly199_Thr200ins5* (intronic)	Hypotonia and respiratory insufficiency	[135]
		p.Arg129* p.Gly199_Thr200ins5* (intronic)	Failure to thrive in the early postnatal period, poor feeding, hypotonia, peripheral neuropathy and drug-resistant epilepsy	[136]
ND2	NDUFAF1/CIA30	p.Thr207Pro p.Lys253Arg/ splice defect (ΔVal252ΔLys253)	Hypertrophic cardiomyopathy, developmental delay, hypotonia and lactic acidosis	[51]
		p.Arg211Cys p.Gly245Arg	Hypertrophic cardiomyopathy	[116]
		p.Asp83Asn p.His93Arg	Leukodystrophy	[137]
	ACAD9	p.Arg266Gln p.Arg417Cys	Hypertrophic cardiomyopathy, encephalomyopathy and lactic acidosis	[138]
		p.Phe44Ile p.Arg266Gln	Hypertrophic cardiomyopathy, mild exercise intolerance, encephalopathy and lactic acidosis	
		p.Ala326Pro p.Arg532Trp	Hypertrophic cardiomyopathy, encephalopathy and lactic acidosis	
		p.Arg518His p.Glu63* p.Glu413Lys	Exercise intolerance, hypertrophic cardiomyopathy, lactic acidosis and failure to thrive	[53]
		p.Arg532Trp p.Arg127Gln p.Arg469Trp	Fatigability, exercise intolerance and lactic acidosis	[139]
		p.Ala220Val	Hypertrophic cardiomyopathy	[140]
		p.Arg414Ser p.Leu558Profs*45	Pulmonary hypertension, dilated and hypertrophic cardiomyopathy	[120]
		p.Val546Leu	Hypertrophic cardiomyopathy	
	TMEM126B	p.Gly212Val	Exercise intolerance	[56]
		p.Asn134Ilefs*2 p.Gly212Val	Exercise intolerance, muscle weakness in lower limbs and pelvis and retinitis pigmentosa	
		p.Gly212Val	Multi-organ involvement including respiratory failure, cardiomyopathy and renal acidosis, severe growth failure, chronic renal insufficiency and lactic acidosis	

(Continued)

TABLE 8.1 *(Continued)*
Pathogenic Mutations Identified in Complex I Assembly Factors

Module	Gene	Mutation	Disease Presentation	Ref.
		p.Gly212Val; p.Asp133Asn	Progressive exercise-induced myalgia and fatigue	[57]
		p.Gly212Val p.Gln70*	Exercise intolerance, premature exertional muscle fatigue, myalgia, dyspnea and cardiac palpitations	
		p.Gly212Val p.Asp133Asn	Proximal muscle weakness and lifelong exercise intolerance marked by myalgia	
		c.82-2 A > G (intronic) p.Lys98Glnfs*2	Leigh-like syndrome and muscle weakness	[141]
	SFXN4	p.Pro78Leufs*26	Intrauterine growth retardation, macrocytic anaemia, microcephaly, hypotonia, vision impairment, speech delay and lactic acidosis	[142]
		p.Thr158Metfs*38 p.Arg247Lysfs*19	Intrauterine growth retardation, microcephaly, fine motor skills defects, visual-motor integration and speech delay	
		p.Gln323Hisfs*20 p.*388Glnext2	Intrauterine growth retardation, macrocytic anaemia, lactic acidosis, microcephaly and epilepsy. Peripheral neuropathy with mild ataxia and tremor. Moderate-to-severe intellectual disability. Non-progressive psychomotor delay	[143]
ND4	FOXRED1	p.Gln232* p.Asn430Ser	Leigh syndrome	[132]
		p.Arg352Trp	Infantile-onset encephalomyopathy	[144]
	TMEM126A	p.Arg55*	Optic atrophy, moderate hypertrophic cardiomyopathy and mild hearing loss	[110]
		p.Ser36Leu	Optic atrophy	[145]
		p. Gln166Arg	Optic atrophy and deafness	[146]
	TMEM70†	c.317-2A > G (intronic)	Cardiomyopathy	[147]‡
Unclear	NDUFAF6/ C8ORF38	p.Gln99Arg	Leigh syndrome	[148]
		c.298-768 T > C (intronic)	Fanconi syndrome (Acadian variant), proximal tubular dysfunction, slowly progressive chronic kidney disease and pulmonary interstitial fibrosis	[149]
		p.Tyr187Asnfs*65 p.Ile124Thr	Leigh syndrome, isolated bilateral striatal necrosis and progressive dystonia	[150]
Other	OXA1L	p.Ser170Glnfs*18 p.Cys207Phe	Severe encephalopathy, hypotonia and developmental delay. Combined OXPHOS deficiency (CI/IV/V)	[16]
	ATAD3 locus	ATAD3 duplication	Hypertrophic cardiomyopathy and neurological defects	[151]
	AIFM	p.Gly308Glu (hemizygous)	Ventriculomegaly, developmental delay and muscular dystrophy	[152]

† TMEM70 is also a Complex V assembly factor.

‡ Patients had a mild Complex I deficiency and stronger Complex V deficiency which may be the main cause of the cardiomyopathy.

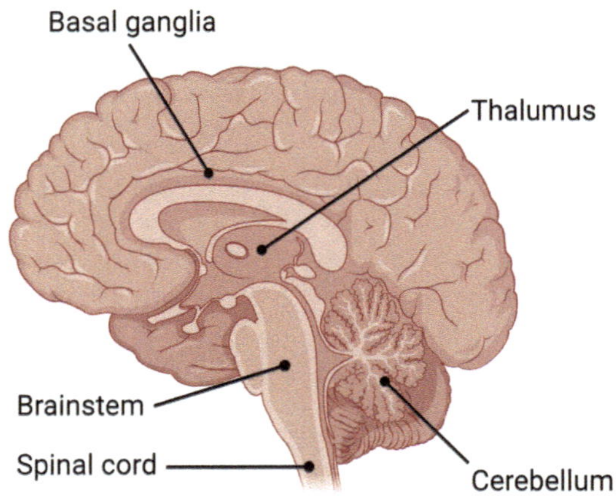

FIGURE 8.9 Regions of the brain commonly affected by Leigh syndrome. Symmetric lesions are often identified in the basal ganglia and thalamus, as well as the brainstem, cerebellum and spinal cord. (Created with BioRender.com.)

Leigh syndrome is defined now by the three most commonly described aspects [98]

1. neurodegenerative with variable symptoms
2. mitochondrial dysfunction caused by inherited genetic defects
3. bilateral central nervous system lesions that are associated with abnormalities in diagnostic imaging such as magnetic resonance imaging (MRI) or CT scan.

If these criteria are only partially fulfilled, or if patients present with atypical symptoms, yet are still indicative of Leigh syndrome, the term 'Leigh-like syndrome' is used [98]. Typically, patients with Leigh syndrome will most frequently show defects in the brainstem or the basal ganglia. However, other areas of the brain may also include the thalamus, cerebellum and the spinal cord (Figure 8.9) [99]. These defects are characterized by bilateral symmetric lesions visible as regions of strong intensity on T2-weighted MRI, which is able to detect paramagnetic deoxyhemoglobin, methemoglobin or hemosiderin [100, 101]. In addition, patients that present with Leigh syndrome may also present with additional clinical phenotypes, including increased blood and cerebrospinal fluid lactate levels, cardiomyopathy and ophthalmoplegia, with cardiomyopathy the most common non-neurological phenotype in these patients [102, 103]. It should be noted that mutations in Complex I assembly factors are not the only cause of Leigh syndrome. Indeed, mutations in Complex I mtDNA-encoded subunits, nuclear-encoded structural genes, as well as subunits and assembly factors of other OXPHOS enzymes and genes involved in other biosynthetic processes [104].

Optic Atrophy

Optic atrophy is characterized by damage to the optic nerve leading to reduced vision [91]. The vulnerability of retinal ganglion cells to mitochondrial dysfunction is attributed to their unmyelinated axons that are reliant on ATP for efficient propagation of the action potential and is widely acknowledged as a key factor underlying visual pathway impairment [105]. Optic atrophy is thought to be the most common form of hereditary optic neuropathy and is thought to affect 1 in 25,000 to 1 in 12,000 people, depending on the particular region [106].

There are several genetic causes of optic atrophy that are linked to various mitochondrial functions such as mitochondrial dynamics [91, 107, 108]. In addition, a role for Complex I deficiency leading to optic atrophy is becoming more apparent with the discovery of *TMEM126A*, where mutations in this gene led to optic atrophy yet the molecular function was unknown, and later being identified as an assembly factor for the ND4 module of Complex I [69, 109–111]. While mutations in the *TMEM126A* gene are synonymous with optic atrophy, other causes of Complex I assembly defects such as *NDUFAF4* and *NDUFAFQ* are also being identified that lead to optic atrophy [37, 39]. Why mutations in these Complex I assembly factor genes lead to optic atrophy and not others remains to be elucidated.

CARDIOMYOPATHY

Cardiomyopathy describes several different disorders that converge on dysfunction of the heart and is characterized by abnormal muscle structure, the ability of the heart to perform its function or both [112]. Cardiomyopathy is another common clinical presentation of Complex I deficiency and may be associated with other symptoms, including exercise intolerance, muscle myopathy and lactic acidosis. It is estimated that between 20% and 40% of patients with mitochondrial disease will have cardiomyopathy, with the mortality of patients diagnosed with cardiomyopathy at 71%, whereas of patients without cardiomyopathy at 26% [113–115]. In patients with Complex I deficiency, cardiomyopathy is commonly diagnosed in patients with mutations in the *NDUFAF1* and *ACAD9* genes, which are part of the MCIA complex involved in the biogenesis of the ND2 module (Table 8.1) [46, 116, 117].

Presentation of cardiomyopathy in mitochondrial disease is most often as hypertrophic cardiomyopathy. However, other mitochondrial cardiomyopathies may also present as dilated or restrictive cardiomyopathy (as well as others) [115, 118]. Hypertrophic cardiomyopathy is characterized by thickening (hypertrophy) of the left ventricle wall and ventricular septum without dilation in the absence of other factors. such as aortic stenosis (narrowing of the aortic valve), coronary artery disease or hypertension (Figure 8.10A, B) [112, 119]. This can lead to irregular heartbeat, heart failure and sudden cardiac death [112]. In contrast, dilated cardiomyopathy due to Complex I deficiency may be rarer, patients have been identified with dilated cardiomyopathy due to mutations in *ACAD9* [120, 121]. Dilated cardiomyopathy is caused by enlargement of the left ventricle, associated with thinning of the ventricle muscle leading to an inability to efficiently pump blood and reduced pumping capacity, known as a decreased ejection fraction in the absence of hypertension or heart disease (Figure 8.10A, C) [122].

CONCLUSION

The assembly of Complex I is a sophisticated and intricate process that involves numerous subunits from two genomes and more than 20 assembly factors. The assembly process occurs in a modular fashion, with the assembly of separate modules coming together to form the fully assembled enzyme that is facilitated by assembly factors as well as other respiratory chain enzymes during the formation of respiratory chain supercomplexes. Mutations in assembly factor genes can lead to isolated Complex I deficiency, which manifests as a range of pathologies, including neurological abnormalities and cardiomyopathy. However, many questions still remain about the assembly of Complex I, the clinical manifestations of Complex I deficiency and the genetic basis of the disease. Future research will undoubtedly uncover new assembly factors and help to elucidate the genotype-phenotype relationship in Complex I deficiency, ultimately leading to a better understanding of Complex I-related mitochondrial disease.

ACKNOWLEDGEMENTS

I would like to thank Prof. Mike Ryan (Monash University) for critically reading the manuscript. This work was supported by a National Health and Medical Research Council Investigator Grant (2010149) and funding from the Mito Foundation. Figures 8.9 and 8.10 were created with BioRender.com.

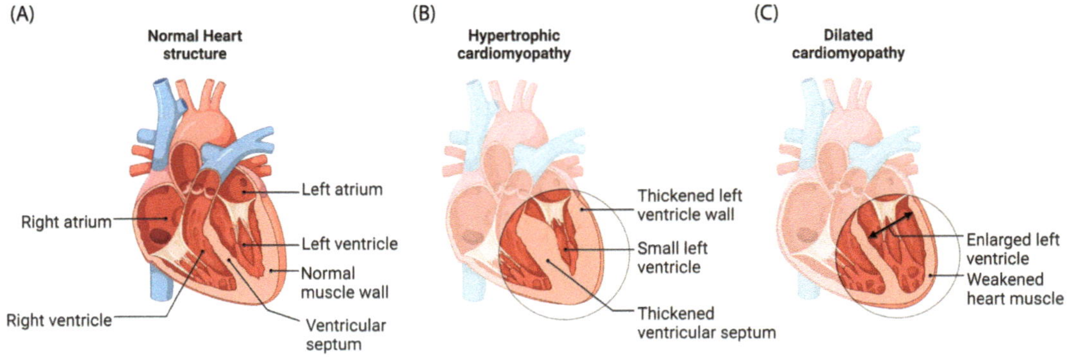

FIGURE 8.10 Cardiomyopathy in complex I deficiency and mitochondrial disease. (A) Healthy heart anatomy. (B) Hypertrophic cardiomyopathy results in thickening of the ventricular septum and left ventricle wall. (C) Dilated cardiomyopathy results in an enlarged left ventricle and weakened heart muscle. (Created with BioRender.com.)

REFERENCES

1. Sazanov, L.A., *From the 'black box' to 'domino effect' mechanism: what have we learned from the structures of respiratory complex I.* Biochemical Journal, 2023. **480**(5): p. 319–333.
2. Hirst, J., *Mitochondrial complex I.* Annual Review of Biochemistry, 2013. **82**(1): p. 551–575.
3. Brandt, U., *Energy converting NADH: quinone oxidoreductase (complex I).* Biochemistry, 2006. **75**(1): p. 69–92.
4. Letts, J.A., et al., *Structures of respiratory supercomplex I+III$_2$ reveal functional and conformational crosstalk.* Molecular Cell, 2019. **75**(6): p. 1131–1146.e6.
5. Stroud, D.A., et al., *Accessory subunits are integral for assembly and function of human mitochondrial complex I.* Nature, 2016. **538**(7623): p. 123–126.
6. Carroll, J., et al., *Analysis of the subunit composition of complex i from bovine heart mitochondria.* Molecular & Cellular Proteomics, 2003. **2**(2): p. 117–126.
7. Fiedorczuk, K., et al., *Atomic structure of the entire mammalian mitochondrial complex I.* Nature, 2016. **538**(7625): p. 406–410.
8. Vinothkumar, K.R., J. Zhu, and J. Hirst, *Architecture of mammalian respiratory complex I.* Nature, 2014. **515**(7525): p.80–84.
9. Guerrero-Castillo, S., et al., *The assembly pathway of mitochondrial respiratory chain complex I.* Cell Metabolism, 2017. **25**: p. 1–12.
10. Sazanov, L.A., *A giant molecular proton pump: structure and mechanism of respiratory complex I.* Nature Reviews Molecular Cell Biology, 2015. **16**(6): p. 375–388.
11. Steiner, J. and L. Sazanov, *Structure and mechanism of the Mrp complex, an ancient cation/proton antiporter.* eLife, 2020. **9**: p. e59407.
12. Yu, H., et al., *Structure of an ancient respiratory system.* Cell, 2018. **173**(7): p. 1636–1649.e16.
13. Parey, K., et al., *High-resolution cryo-EM structures of respiratory complex I: mechanism, assembly, and disease.* Science Advances, 2019. **5**(12): p. eaax9484.
14. Formosa, L.E., et al., *Building a complex: assembly of mitochondrial respiratory chain complex I.* Seminars in Cell & Developmental Biology, 2018. **76**: p. 154–162.
15. Jackson, T.D., et al., *The TIM22 complex mediates the import of sideroflexins and is required for efficient mitochondrial one-carbon metabolism.* Molecular Biology of the Cell, 2021. **32**(6): p. 475–491.
16. Thompson, K., et al., *OXA1L mutations cause mitochondrial encephalopathy and a combined oxidative phosphorylation defect.* EMBO Molecular Medicine, 2018. **10**(11). :e9060. doi: 10.15252/emmm.201809060.
17. Kang, Y., L.F. Fielden, and D. Stojanovski, *Mitochondrial protein transport in health and disease.* Seminars in Cell & Developmental Biology, 2018. **76**: p. 142–153.

18. Hangen, E., et al., *Interaction between AIF and CHCHD4 regulates respiratory chain biogenesis.* Molecular Cell, 2015. **58**(6): p. 1001–1014.
19. Salscheider, S.L., et al., *AIFM1 is a component of the mitochondrial disulfide relay that drives complex I assembly through efficient import of NDUFS5.* EMBO Journal, 2022. **41**(17): p. e110784.
20. Kirby, D., et al., *Respiratory chain complex I deficiency: an underdiagnosed energy generation disorder.* Neurology, 1999. **52**: p. 1255–1264.
21. Hoefs, S.J., et al., *Molecular base of biochemical complex I deficiency.* Mitochondrion, 2012. **12**(5): p. 520–532.
22. Frazier, A.E., D.R. Thorburn, and A.G. Compton, *Mitochondrial energy generation disorders: genes, mechanisms, and clues to pathology.* Journal of Biological Chemistry, 2019. **294**(14): p. 5386–5395.
23. Stenton, S.L. and H. Prokisch, *Genetics of mitochondrial diseases: identifying mutations to help diagnosis.* EBioMedicine, 2020. **56**: p. 102784.
24. Thompson, K., et al., *Investigation of oxidative phosphorylation activity and complex composition in mitochondrial disease*, in *Handbook of Clinical Neurology*, R. Horvath, M. Hirano, and P.F. Chinnery, Editors. 2023, Elsevier: Amsterdam, Netherlands. p. 127–139.
25. Labory, J., et al., *Multi-omics approaches to improve mitochondrial disease diagnosis: challenges, advances, and perspectives.* Frontiers in Molecular Biosciences, 2020. **7**. Article number (NCBI Pubmed) 590842.
26. Pagliarini, D.J. and J. Rutter, *Hallmarks of a new era in mitochondrial biochemistry.* Genes & Development, 2013. **27**(24): p. 2615–2627.
27. Sung, A.Y., B.J. Floyd, and D.J. Pagliarini, *Systems biochemistry approaches to defining mitochondrial protein function.* Cell Metabolism, 2020. **31**(4): p. 669–678.
28. Zhu, J., K.R. Vinothkumar, and J. Hirst, *Structure of mammalian respiratory complex I.* Nature, 2016. **536**(7616): p. 354–358.
29. Rhein, V.F., et al., *NDUFAF7 methylates arginine 85 in the NDUFS2 subunit of human complex I.* Journal of Biological Chemistry, 2013. **288**(46): p. 33016–33026.
30. Zurita Rendon, O., et al., *The arginine methyltransferase NDUFAF7 is essential for complex I assembly and early vertebrate embryogenesis.* Human Molecular Genetics, 2014. **23**(19): p. 5159–5170.
31. Sheftel, A.D., et al., *Human indl, an iron-sulfur cluster assembly factor for respiratory complex I.* Molecular and Cellular Biology, 2009. **29**(22): p. 6059–6073.
32. Bych, K., et al., *The iron-sulphur protein Indl is required for effective complex I assembly.* EMBO Journal, 2008. **27**(12): p. 1736–1746.
33. Rhein, V.F., et al., *NDUFAF5 hydroxylates NDUFS7 at an early stage in the assembly of human complex I.* Journal of Biological Chemistry, 2016. **291**(28): p. 14851–14860.
34. Floyd, B.J., et al., *Mitochondrial protein interaction mapping identifies regulators of respiratory chain function.* Molecular Cell, 2016. **63**(4): p. 621–632.
35. Alston, C.L., et al., *Pathogenic Bi-allelic mutations in NDUFAF8 cause leigh syndrome with an isolated complex I deficiency.* American Journal of Human Genetics, 2020. **106**(1): p. 92–101.
36. Peker, E., et al., *A two-step mitochondrial import pathway couples the disulfide relay with matrix complex I biogenesis.* The Journal of Cell Biology, 2023. **222**(7). e202210019. doi: 10.1083/jcb.202210019.
37. Rensvold, J.W., et al., *Defining mitochondrial protein functions through deep multiomic profiling.* Nature, 2022. **606**(7913): p. 382–388.
38. Saada, A., et al., *Mutations in NDUFAF3 (C3ORF60), encoding an NDUFAF4 (C6ORF66)-interacting complex I assembly protein, cause fatal neonatal mitochondrial disease.* American Journal of Human Genetics, 2009. **84**(6): p. 718–727.
39. Saada, A., et al., *C6ORF66 is an assembly factor of mitochondrial complex I.* American Journal of Human Genetics, 2008. **82**(1): p. 32–38.
40. Thinon, E., et al., *Global profiling of co- and post-translationally N-myristoylated proteomes in human cells.* Nature Communications, 2014. **5**(1): p. 4919.
41. Timms, R.T., et al., *A glycine-specific N-degron pathway mediates the quality control of protein* N-*myristoylation.* Science, 2019. **365**(6448): p. eaaw4912.
42. Lueg, G.A., et al., N-*myristoyltransferase inhibition is synthetic lethal in MYC-deregulated cancers.* bioRxiv, 2021: p. 2021.03.20.436222.
43. Baertling, F., et al., *Mutations in mitochondrial complex I assembly factor NDUFAF3 cause Leigh syndrome.* Molecular Genetics and Metabolism, 2017. **120**(3): p. 243–246.

44. Andrews, B., et al., *Assembly factors for the membrane arm of human complex I.* Proceedings of the National Academy of Sciences of the United States of America, 2013. **110**(47): p. 18934–18939.
45. Guarani, V., et al., *TIMMDC1/C3orf1 functions as a membrane-embedded mitochondrial complex I assembly factor through association with the MCIA complex.* Molecular and Cellular Biology, 2014. **34**(5): p. 847–861.
46. Formosa, L.E., et al., *Dissecting the Roles of Mitochondrial Complex I Intermediate Assembly Complex Factors in the Biogenesis of Complex I.* Cell Reports, 2020. **31**(3): p. 107541.
47. McKenzie, M., et al., *Mutations in the gene encoding C8orf38 block complex I assembly by inhibiting production of the mitochondria-encoded subunit ND1.* Journal of Molecular Biology, 2011. **414**(3): p. 413–426.
48. Rendón, O.Z. and E.A. Shoubridge, *Early complex I assembly defects result in rapid turnover of the ND1 subunit.* Human Molecular Genetics, 2012. **21**(17): p. 3815–3824.
49. Heide, H., et al., *Complexome profiling identifies TMEM126B as a component of the mitochondrial complex I assembly complex.* Cell Metabolism, 2012. **16**(4): p. 538–549.
50. Vogel, R.O., et al., *Human mitochondrial complex I assembly is mediated by NDUFAF1.* The FEBS Journal, 2005. **272**(20): p. 5317–5326.
51. Dunning, C.J., et al., *Human CIA30 is involved in the early assembly of mitochondrial complex I and mutations in its gene cause disease.* EMBO Journal, 2007. **26**(13): p. 3227–3237.
52. Vogel, R.O., et al., *Cytosolic signaling protein Ecsit also localizes to mitochondria where it interacts with chaperone NDUFAF1 and functions in complex I assembly.* Genes & Development, 2007. **21**(5): p. 615–624.
53. Nouws, J., et al., *Acyl-CoA dehydrogenase 9 is required for the biogenesis of oxidative phosphorylation complex I.* Cell Metabolism, 2010. **12**(3): p. 283–294.
54. Wang, C., et al., *MITRAC15/COA1 promotes mitochondrial translation in a ND2 ribosome-nascent chain complex.* EMBO Reports, 2020. **21**(1): p. e48833–e48833.
55. Jackson, T.D., et al., *Sideroflexin 4 is a complex I assembly factor that interacts with the MCIA complex and is required for the assembly of the ND2 module.* Proceedings of the National Academy of Sciences of the United States of America, 2022. **119**(13): p. e2115566119.
56. Alston, C.L., et al., *Biallelic mutations in TMEM126B cause severe complex I deficiency with a variable clinical phenotype.* American Journal of Human Genetics, 2016. **99**(1): p. 217–227.
57. Sánchez-Caballero, L., et al., *Mutations in complex I assembly factor TMEM126B result in muscle weakness and isolated complex I deficiency.* American Journal of Human Genetics, 2016. **99**(1): p. 208–216.
58. Kopp, E., et al., *ECSIT is an evolutionarily conserved intermediate in the Toll/IL-1 signal transduction pathway.* Genes & Development, 1999. **13**(16): p. 2059–2071.
59. Xiao, C., et al., *Ecsit is required for Bmp signaling and mesoderm formation during mouse embryogenesis.* Genes & Development, 2003. **17**(23): p. 2933–2949.
60. West, A.P., et al., *TLR signalling augments macrophage bactericidal activity through mitochondrial ROS.* Nature, 2011. **472**(7344): p. 476–480.
61. Carneiro, F.R.G., et al., *An essential role for ECSIT in mitochondrial complex I assembly and mitophagy in macrophages.* Cell Reports, 2018. **22**(10): p. 2654–2666.
62. Ensenauer, R., et al., *Human acyl-CoA dehydrogenase-9 plays a novel role in the mitochondrial beta-oxidation of unsaturated fatty acids.* The Journal of Biological Chemistry, 2005. **280**(37): p. 32309–32316.
63. Nouws, J., et al., *ACAD9, a complex I assembly factor with a moonlighting function in fatty acid oxidation deficiencies.* Human Molecular Genetics, 2013. **23**(5): p. 1311–1319.
64. Giachin, G., et al., *Assembly of the mitochondrial complex I assembly complex suggests a regulatory role for deflavination.* Angewandte Chemie, 2021. **60**(9): p. 4689–4697.
65. Xia, C., et al., *Molecular mechanism of interactions between ACAD9 and binding partners in mitochondrial respiratory complex I assembly.* iScience, 2021. **24**(10): p. 103153.
66. Mick, D.U., et al., *MITRAC links mitochondrial protein translocation to respiratory-chain assembly and translational regulation.* Cell, 2012. **151**(7): p. 1528–1541.
67. van der Laan, M., et al., *A role for Tim21 in membrane-potential-dependent preprotein sorting in mitochondria.* Current Biology, 2006. **16**(22): p. 2271–2276.
68. Shinde, S.S., et al., *Recurrent erosion of COA1/MITRAC15 exemplifies conditional gene dispensability in oxidative phosphorylation.* Scientific Reports, 2021. **11**(1): p. 24437.
69. Formosa, L.E., et al., *Optic atrophy–associated TMEM126A is an assembly factor for the ND4-module of mitochondrial complex I.* Proceedings of the National Academy of Sciences, 2021. **118**(17): p. e2019665118.

70. Itoh, Y., et al., *Mechanism of membrane-tethered mitochondrial protein synthesis.* Science, 2021. **371**(6531): p. 846–849.
71. Carroll, J., et al., *TMEM70 and TMEM242 help to assemble the rotor ring of human ATP synthase and interact with assembly factors for complex I.* Proceedings of the National Academy of Sciences, 2021. **118**(13): p. e2100558118.
72. Bahri, H., et al., *TMEM70 forms oligomeric scaffolds within mitochondrial cristae promoting in situ assembly of mammalian ATP synthase proton channel.* Biochimica et Biophysica Acta. Molecular Cell Research, 2021. **1868**(4): p. 118942.
73. Sánchez-Caballero, L., et al., *TMEM70 functions in the assembly of complexes I and V.* Biochimica et Biophysica Acta (BBA) — Bioenergetics, 2020. **1861**(8): p. 148202.
74. Formosa, L.E., et al., *Characterization of mitochondrial FOXRED1 in the assembly of respiratory chain complex I.* Human Molecular Genetics, 2015. **24**(10): p. 2952–2965.
75. Morgenstern, M., et al., *Quantitative high-confidence human mitochondrial proteome and its dynamics in cellular context.* Cell Metabolism, 2021. **33**(12): p. 2464–2483.e18.
76. Dibley, M.G., et al., *The mitochondrial acyl-carrier protein interaction network highlights important roles for LYRM family members in complex I and mitoribosome assembly.* Molecular & Cellular Proteomics, 2020. **19**(1): p. 65–77.
77. Lazarou, M., et al., *Analysis of the assembly profiles for mitochondrial- and nuclear-DNA-encoded subunits into complex I.* Molecular and Cellular Biology, 2007. **27**(12): p. 4228–4237.
78. Adjobo-Hermans, M.J.W., et al., *NDUFS4 deletion triggers loss of NDUFA12 in Ndufs4−/− mice and Leigh syndrome patients: a stabilizing role for NDUFAF2.* Biochimica et Biophysica Acta (BBA) — Bioenergetics, 2020. **1861**(8): p. 148213.
79. Gu, J., et al., *The architecture of the mammalian respirasome.* Nature, 2016. **537**(7622): p. 639–643.
80. Letts, J.A., K. Fiedorczuk, and L.A. Sazanov, *The architecture of respiratory supercomplexes.* Nature, 2016. **537**(7622): p. 644–648. doi: 10.7554/eLife.21290.
81. Sousa, J.S., et al., *Functional asymmetry and electron flow in the bovine respirasome.* eLife, 2016. **5**: p. e21290.
82. Protasoni, M., et al., *Respiratory supercomplexes act as a platform for complex III-mediated maturation of human mitochondrial complexes I and IV.* EMBO Journal, 2020. **39**(3): p. e102817–e102817.
83. Acín-Pérez, R., et al., *Respiratory complex III is required to maintain complex I in mammalian mitochondria.* Molecular Cell, 2004. **13**(6): p. 805–815.
84. Diaz, F., et al., *Cytochrome c oxidase is required for the assembly/stability of respiratory complex I in mouse fibroblasts.* Molecular and Cellular Biology, 2006. **26**(13): p. 4872–4881.
85. Li, Y., et al., *An assembled complex IV maintains the stability and activity of complex I in mammalian mitochondria.* The Journal of Biological Chemistry, 2007. **282**(24): p. 17557–17562.
86. Stroud, D.A., et al., *Gene knockout using transcription activator-like effector nucleases (TALENs) reveals that human NDUFA9 protein is essential for stabilizing the junction between membrane and matrix arms of complex I.* The Journal of biological chemistry, 2013. **288**(3): p. 1685–1690.
87. Fernández-Vizarra, E., et al., *Two independent respiratory chains adapt OXPHOS performance to glycolytic switch.* Cell Metabolism, 2022. **34**(11): p. 1792–1808.
88. Szczepanowska, K., et al., *A salvage pathway maintains highly functional respiratory complex I.* Nature Communications, 2020. **11**(1): p. 1643.
89. Bogenhagen, D.F. and J.D. Haley, *Pulse-chase SILAC-based analyses reveal selective oversynthesis and rapid turnover of mitochondrial protein components of respiratory complexes.* The Journal of Biological Chemistry, 2020. **295**(9): p. 2544–2554.
90. Stenton, S.L., et al., *Impaired complex I repair causes recessive Leber's hereditary optic neuropathy.* The Journal of Clinical Investigation, 2021. **131**(6): p. e138267. doi: 10.1172/JCI138267.
91. Carelli, V., C. La Morgia, and P. Yu-Wai-Man, *Mitochondrial optic neuropathies.* Handbook of Clinical Neurology, 2023. **194**: p. 23–42.
92. El-Hattab, A.W., et al., *MELAS syndrome: clinical manifestations, pathogenesis, and treatment options.* Molecular Genetics and Metabolism, 2015. **116**(1): p. 4–12.
93. Mavraki, E., et al., *Genetic testing for mitochondrial disease: the United Kingdom best practice guidelines.* European Journal of Human Genetics, 2023. **31**(2): p. 148–163.
94. Leigh, D., *Subacute necrotizing encephalomyelopathy in an infant.* Journal of Neurology, Neurosurgery, and Psychiatry, 1951. **14**(3): p. 216–221.
95. Thorburn, D.R., J. Rahman, and S. Rahman, *Mitochondrial DNA-Associated Leigh Syndrome and NARP*, in *GeneReviews(®)*, M.P. Adam, et al., Editors. 1993, University of Washington: Seattle (WA).

96. Simon, M.T., et al., *Novel mutations in the mitochondrial complex I assembly gene NDUFAF5 reveal heterogeneous phenotypes.* Molecular Genetics and Metabolism, 2019. **126**(1): p. 53–63.
97. Fassone, E. and S. Rahman, *Complex I deficiency: clinical features, biochemistry and molecular genetics.* Journal of Medical Genetics, 2012. **49**(9): p. 578–590.
98. Baertling, F., et al., *A guide to diagnosis and treatment of Leigh syndrome.* Journal of Neurology, Neurosurgery & Psychiatry, 2014. **85**(3): p. 257–265.
99. Lake, N.J., et al., *Leigh syndrome: neuropathology and pathogenesis.* Journal of Neuropathology & Experimental Neurology, 2015. **74**(6): p. 482–492.
100. Lake, N.J., et al., *Leigh syndrome: one disorder, more than 75 monogenic causes.* Annals of Neurology, 2016. **79**(2): p. 190–203.
101. Chavhan, G.B., et al., *Principles, techniques, and applications of T2*-based MR imaging and its special applications.* Radiographics, 2009. **29**(5): p. 1433–1449.
102. Chang, X., et al., *A meta-analysis and systematic review of Leigh syndrome: clinical manifestations, respiratory chain enzyme complex deficiency, and gene mutations.* Medicine (Baltimore), 2020. **99**(5): p. e18634.
103. Lee, J.S., et al., *Leigh syndrome in childhood: neurologic progression and functional outcome.* Journal of Clinical Neurology, 2016. **12**(2): p. 181–187.
104. Schubert Baldo, M. and L. Vilarinho, *Molecular basis of Leigh syndrome: a current look.* Orphanet Journal of Rare Diseases, 2020. **15**(1): p. 31.
105. Carelli, V., et al., *Retinal ganglion cell neurodegeneration in mitochondrial inherited disorders.* Biochim Biophys Acta, 2009. **1787**(5): p. 518–528.
106. Strachan, E.L., et al., *The role of mitochondria in optic atrophy with autosomal inheritance.* Frontiers in Neuroscience, 2021. **15**: p. 784987.
107. Hartmann, B., et al., *Homozygous YME1L1 mutation causes mitochondriopathy with optic atrophy and mitochondrial network fragmentation.* Elife, 2016. **5**: p. e16078. doi: 10.7554/eLife.16078.
108. Yapa, N.M.B., et al., *Mitochondrial dynamics in health and disease.* FEBS Letters, 2021. **595**(8): p. 1184–1204.
109. D'Angelo, L., et al., *NDUFS3 depletion permits complex I maturation and reveals TMEM126A/OPA7 as an assembly factor binding the ND4-module intermediate.* Cell Reports, 2021. **35**(3): p. e109002. doi: 10.1016/j.celrep.2021.109002.
110. Hanein, S., et al., *TMEM126A, encoding a mitochondrial protein, is mutated in autosomal-recessive nonsyndromic optic atrophy.* American Journal of Human Genetics, 2009. **84**(4): p. 493–498.
111. Desir, J., et al., *TMEM126A mutation in a Moroccan family with autosomal recessive optic atrophy.* Molecular Vision, 2012. **18**: p. 1849–1857.
112. Meyers, D.E., H.I. Basha, and M.K. Koenig, *Mitochondrial cardiomyopathy: pathophysiology, diagnosis, and management.* Texas Heart Institute Journal, 2013. **40**(4): p. 385–394.
113. Holmgren, D., et al., *Cardiomyopathy in children with mitochondrial disease: clinical course and cardiological findings.* European Heart Journal, 2003. **24**(3): p. 280–288.
114. Scaglia, F., et al., *Clinical spectrum, morbidity, and mortality in 113 pediatric patients with mitochondrial disease.* Pediatrics, 2004. **114**(4): p. 925–931.
115. El-Hattab, A.W. and F. Scaglia, *Mitochondrial cardiomyopathies.* Frontiers in Cardiovascular Medicine, 2016. **3**: p. 25–25.
116. Fassone, E., et al., *Mutations in the mitochondrial complex I assembly factor NDUFAF1 cause fatal infantile hypertrophic cardiomyopathy.* Journal of Medical Genetics, 2011. **48**(10): p. 691–697.
117. Repp, B.M., et al., *Clinical, biochemical and genetic spectrum of 70 patients with ACAD9 deficiency: is riboflavin supplementation effective?* Orphanet Journal of Rare Diseases, 2018. **13**(1): p. 120.
118. Finsterer, J. and S. Kothari, *Cardiac manifestations of primary mitochondrial disorders.* International Journal of Cardiology, 2014. **177**(3): p. 754–763.
119. Faber, L., *Hypertrophic cardiomyopathy*, in *Cardiology Secrets* (Fifth Edition), G.N. Levine, Editor. 2018, Elsevier: Philadelphia, PA. p. 216–223.
120. Dewulf, J.P., et al., *Evidence of a wide spectrum of cardiac involvement due to ACAD9 mutations: Report on nine patients.* Molecular genetics and metabolism, 2016. **118**(3): p. 185–189.
121. Lagoutte-Renosi, J., et al., *Lethal neonatal progression of fetal cardiomegaly associated to ACAD9 deficiency*, in *JIMD Reports, Volume 28*, E. Morava, et al., Editors. 2016, Springer, Berlin, Heidelberg. p. 1–10.
122. Bayona-Bafaluy, M.P., et al., *Genetic aspects of the oxidative phosphorylation dysfunction in dilated cardiomyopathy.* Mutation Research/Reviews in Mutation Research, 2020. **786**: p. 108334.

123. Ogilvie, I., N.G. Kennaway, and E.A. Shoubridge, *A molecular chaperone for mitochondrial complex I assembly is mutated in a progressive encephalopathy.* Journal of Clinical Investigation, 2005. **115**(10): p. 2784–2792.
124. Barghuti, F., et al., *The unique neuroradiology of complex I deficiency due to NDUFA12L defect.* Molecular Genetics and Metabolism, 2008. **94**(1): p. 78–82.
125. Haack, T.B., et al., *Mutation screening of 75 candidate genes in 152 complex I deficiency cases identifies pathogenic variants in 16 genes including NDUFB9.* Journal of Medical Genetics, 2012. **49**(2): p. 83–89.
126. Baertling, F., et al., *Mutations in mitochondrial complex I assembly factor NDUFAF3 cause Leigh syndrome.* Molecular Genetics and Metabolism, 2017. **120**(3): p. 243–246.
127. Ishiyama, A., et al., *NDUFAF3 variants that disrupt mitochondrial complex I assembly may associate with cavitating leukoencephalopathy.* Clinical Genetics, 2018. **93**(5): p. 1103–1106.
128. Bi, H., et al., *A novel variation in the mitochondrial complex I assembly factor NDUFAF5 causes isolated bilateral striatal necrosis in childhood.* Frontiers in Neurology, 2021. **12**: p. 675616.
129. Sugiana, C., et al., *Mutation of C20orf7 disrupts complex I assembly and causes lethal neonatal mitochondrial disease.* American Journal of Human Genetics, 2008. **83**(4): p. 468–78.
130. Wang, B., et al., *A novel potentially causative variant of NDUFAF7 revealed by mutation screening in a Chinese family with pathologic myopia.* Investigative Ophthalmology & Visual Science, 2017. **58**(10): p. 4182–4192.
131. Kevelam, S.H., et al., *NUBPL mutations in patients with complex I deficiency and a distinct MRI pattern.* Neurology, 2013. **80**(17): p. 1577–1583.
132. Calvo, S.E., et al., *High-throughput, pooled sequencing identifies mutations in NUBPL and FOXRED1 in human complex I deficiency.* Nature Genetics, 2010. **42**: p. 851.
133. Protasoni, M., et al., *Novel compound heterozygous pathogenic variants in nucleotide-binding protein like protein (NUBPL) cause leukoencephalopathy with multi-systemic involvement.* Molecular Genetics and Metabolism, 2020. **129**(1): p. 26–34.
134. Kremer, L.S., et al., *Genetic diagnosis of Mendelian disorders via RNA sequencing.* Nature Communications, 2017. **8**: p. 15824.
135. Naber, M., et al., *Deep intronic TIMMDC1 variant delays diagnosis of rapidly progressive complex I deficiency.* European Journal of Medical Genetics, 2021. **64**(1): p. 104120.
136. Kumar, R., et al., *Oligonucleotide correction of an intronic TIMMDC1 variant in cells of patients with severe neurodegenerative disorder.* NPJ Genomic Medicine, 2022. **7**(1): p. 9.
137. Wu, L., et al., *Leukodystrophy associated with mitochondrial complex I deficiency due to a novel mutation in the NDUFAF1 gene.* Mitochondrial DNA. Part A, DNA Mapping, Sequencing, and ANALYSIs, 2016. **27**(2): p. 1034–1037.
138. Haack, T.B., et al., *Exome sequencing identifies ACAD9 mutations as a cause of complex I deficiency.* Nature Genetics, 2010. **42**(12): p. 1131–1134.
139. Gerards, M., et al., *Riboflavin-responsive oxidative phosphorylation complex I deficiency caused by defective ACAD9: new function for an old gene.* Brain, 2011. **134**(Pt 1): p. 210–219.
140. Nouws, J., et al., *A patient with complex I deficiency caused by a novel ACAD9 mutation not responding to riboflavin treatment.* JIMD Reports, 2014. **12**: p. 37–45.
141. Zhou, X., et al., *Novel biallelic mutations in TMEM126B cause splicing defects and lead to Leigh-like syndrome with severe complex I deficiency.* Journal of Human Genetics, 2023. **68**(4): p. 239–246.
142. Hildick-Smith, G.J., et al., *Macrocytic anemia and mitochondriopathy resulting from a defect in sideroflexin 4.* American Journal of Human Genetics, 2013. **93**(5): p. 906–914.
143. Sofou, K., et al., *Prenatal onset of mitochondrial disease is associated with sideroflexin 4 deficiency.* Mitochondrion, 2019. **47**: p. 76–81.
144. Fassone, E., et al., *FOXRED1, encoding an FAD-dependent oxidoreductase complex-I-specific molecular chaperone, is mutated in infantile-onset mitochondrial encephalopathy.* Human Molecular Genetics, 2010. **19**(24): p. 4837–4847.
145. Kloth, K., et al., *Novel likely pathogenic variants in TMEM126A identified in non-syndromic autosomal recessive optic atrophy: two case reports.* BMC Medical Genetics, 2019. **20**(1): p. 62.
146. La Morgia, C., et al., *First TMEM126A missense mutation in an Italian proband with optic atrophy and deafness.* Neurology. Genetics, 2019. **5**(3): p. e329.
147. Wortmann, S.B., et al., *Biochemical and genetic analysis of 3-methylglutaconic aciduria type IV: a diagnostic strategy.* Brain, 2009. **132**(Pt 1): p. 136–146.
148. Pagliarini, D.J., et al., *A mitochondrial protein compendium elucidates complex I disease biology.* Cell, 2008. **134**(1): p. 112–123.

149. Hartmannová, H., et al., *Acadian variant of Fanconi syndrome is caused by mitochondrial respiratory chain complex I deficiency due to a non-coding mutation in complex I assembly factor NDUFAF6.* Human Molecular Genetics, 2016. **25**(18): p. 4062–4079.
150. Baide-Mairena, H., et al., *Mutations in the mitochondrial complex I assembly factor NDUFAF6 cause isolated bilateral striatal necrosis and progressive dystonia in childhood.* Molecular Genetics and Metabolism, 2019. **126**(3): p. 250–258.
151. Frazier, A.E., et al., *Fatal perinatal mitochondrial cardiac failure caused by recurrent de novo duplications in the ATAD3 locus.* Med, 2021. **2**(1): p. 49–73.
152. Berger, I., et al., *Early prenatal ventriculomegaly due to an AIFM1 mutation identified by linkage analysis and whole exome sequencing.* Molecular Genetics and Metabolism, 2011. **104**(4): p. 517–520.

Part V

Advanced Methods in Mitochondrial Biology and Metabolism Research

9 The Optogenetic Landscape of the Mitochondrion

Odette Deen-Rozalen[§], *Marius Nieke*[§], *Sara Shumka*[§], *Hannes M. Beyer*[*], *and Matias D. Zurbriggen*[*]

INTRODUCTION

Optogenetics – a collection of methods to interface biological systems with optical stimulation – has considerably evolved during the past decade. Originating from the use of opsins for controlling neuron activity through light-gated ion channels and pumps, optogenetics transitioned into a discipline with broad application in cell biology across kingdoms. Deisseroth defined optogenetics as "*the combination of genetic and optical methods to achieve gain or loss of function of well-defined events in specific cells of living tissue*"[1]. Nowadays, the term "optogenetics" refers to any genetically encoded system or product thereof that performs specialized light-mediated functions[2]. The past decade has witnessed ambitious progress in the development of optogenetic tools and photochemical solutions to control the excitability and signaling of cells[3–7]. These tools increasingly employ natural and engineered photosensitive switches, rewired to diverse biological functions within cells or the extracellular space, giving rise to non-opsin-based optogenetics. A broad scope of innovation emerged to control diverse cellular functions spatially and temporally. Some of these advances have prompted the discovery of photosensory signaling mechanisms[8–11], the development of photoreceptor engineering strategies[12–15], and the emergence of high-resolution structures of natural and engineered light-activated proteins[16–19]. Additionally, major advances in the development of required hardware, i.e., light sources and spatially resolved illumination technologies, accelerated the implementation of optogenetic approaches into biological systems[20,21]. Light-controlled protein-protein interactions and the use of photosensors that respond to photoexcitation with pronounced conformational changes, as well as photoregulated enzymes, constitute next-generation optogenetic components. These tools enable accurate regulation and analysis of dynamic protein activities in living cells combined with precise sample illumination strategies. All these efforts aim to inform how to construct photo-inducible signaling solutions from natural proteins, resulting in modern biotechnological tools that will enhance our understanding of diverse and dynamic (sub)cellular events[22].

Recently, optogenetics has further yielded strategies to specifically control cell organelle function, providing new insights into fundamental processes occurring within a cell. The study of mitochondrial biology, for example, would much benefit from the precision provided by optogenetic methods in order to disentangle their role as endosymbiotic cellular constituents and the balanced crosstalk with the cytosol. Mitochondria are crucial cell organelles, given their pivotal role in adenosine triphosphate (ATP) generation, regulation of cell fate decisions, calcium homeostasis, and cellular metabolism[23,24]. Several optogenetic techniques have been applied to mitochondria for regulating their membrane potential, protein localization, as well as to control apoptosis[15,25–31] and mitophagy[32,33]. Due to the spatiotemporal precision intrinsic to optogenetic approaches, such applications will likely grow in number shortly and assist in exploring diverse functions within the cellular mitochondrial network. In this review, we summarize recent advancements and applications of optogenetic approaches in mitochondria, structured in two main parts: (1) non-opsin-based optogenetics and (2) opsin-based

[*] Corresponding authors: hannes.beyer@uni-duesseldorf.de; matias.zurbriggen@uni-duesseldorf.de
[§] Equally contributing authors

DOI: 10.1201/9781003221180-14

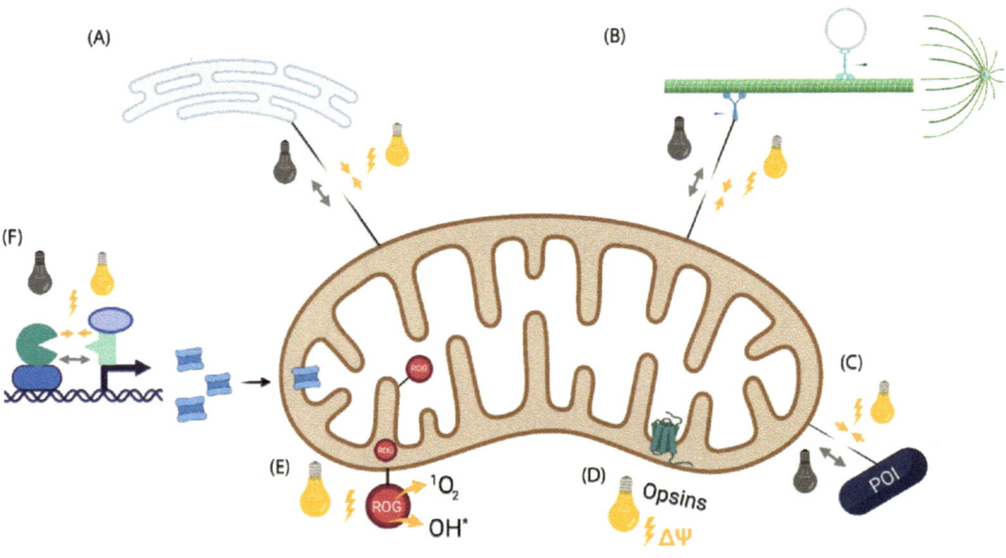

FIGURE 9.1 Overview of optogenetic systems used in mitochondria research. (A) Light-activated attachment of the outer mitochondrial membrane (OMM) to the ER. (B) Light-activated attachment of the OMM to mitotic motor proteins. (C) Light activated recruitment of target proteins (POI, proteins of interest), for example, BAX, AMBRA-1 to the OMM. (D) Insertion of light-activated ion channels and pumps into the inner mitochondrial membrane (IMM). (E) Targeting of light-activated reactive oxygen generating (ROG) proteins to the OMM, IMM, or intermembrane space (IMS). (F) Light-activated gene expression of an ion channel targeted to the IMM. (Created with BioRender.com.)

optogenetics. We provide a comprehensive picture of the current optogenetic toolkit for use in mitochondria and discuss possible avenues for future optogenetic investigations (Figure 9.1).

NON-OPSIN-BASED OPTOGENETICS

Non-opsin-based optogenetics (also referred to as next-generation optogenetics or optogenetics 2.0) involves both naturally occurring and synthetically engineered photosensitive switches that can influence various biological functions within cells or the extracellular space. Existing optogenetic tools to control cellular processes can be classified in a variety of ways, such as by the origin of the utilized photoreceptors, the wavelength of induction, signaling mechanisms, or the scope of application. In this review, we primarily discuss applications involving membrane-mediated signaling events, constituting the major field of optogenetic applications pursued in mitochondria to date. In eukaryotic cells, some approaches for optically inducing intracellular signaling mediated *via* integral membrane and membrane-associated proteins include membrane receptors (OPN4, OptoXR, JellyOp)[34–36], direct membrane lipid binding (*Bc*LOV4)[9], clustering (LINC, CLICR, LARIAT)[37–39], single-chain photoswitching (Dronpa, PA-Rac, Lumitox)[40–42], and hetero-dimer association and dissociation (CRY2:CIBN, iLID, TULIPs, Magnets, FKF1:Gigantea, BphP1:Pps-R2, LOVTRAP, PhyB:PIF6)[13–15,22,43–48]. As many of these systems have previously been thoroughly reviewed[2,22,24,49], we wish to briefly introduce the most impactful assets to the optogenetic toolbox, before shifting the focus toward the potential of optogenetics for the study of mitochondria.

Photoswitches Employed as Optogenetic Tools and Their Mechanisms

The targeted interaction of proteins represents a major mechanistic principle involved in a wide range of biological pathways. Consequently, molecular switches regulating protein-protein interactions upon light stimulation form the basis of various developed optogenetic principles employing a range of photosensitive proteins originating from plants, bacteria, cyanobacteria, and fungi. The most common of these include light-oxygen-voltage-sensing (LOV) domains, cryptochromes, phytochromes, and cyanobacteriochromes, which have been extensively exploited for modulating signaling pathways and cell membrane dynamics[2,22,24,49]. The comprehensive online resource "Optobase" (www.optobase.org) hosts an ever-growing resource bank of optogenetic switches, applications, and related literature, guiding interested researchers on the choice of the optogenetic components for implementation into various biological systems[50]. In the following, we describe the mode of function of common optogenetic switches and their applications.

LOV Domains

LOV domains are constituents of blue light photoreceptors found in organisms from various kingdoms of life including plants, bacteria, and fungi[51,52]. The photoactivity of LOV domains is mediated by light-dependent interactions with a flavin such as flavin adenine dinucleotide (FAD), a cofactor abundant in eukaryotic cells. The LOV2 domain of *Avena sativa* phototropin 1 is widely used for designing synthetic single- and dual-chain photoswitches. In its natural context, it regulates the activity of a serine/threonine protein kinase. *As*LOV2 possesses a C-terminal Jα-helix which undocks from the core domain. A conformational shift in the overall protein tertiary structure initiated by blue light absorption causes the reversible generation of a flavin-cysteinyl adduct in the hydrophobic core of the LOV sensor, which ultimately converts the photosensory input into biochemical signaling outputs[8,53]. This conformational shift has been utilized as the molecular switching event in engineered photoswitches, for example, used to control numerous signaling outputs including signaling cascades, the activity of DNA binding proteins, small GTPases, membrane-bound Ca^{2+} channels, and cell death[9,41,54–56]. Figure 9.2A and Table 9.1 describe a multitude of LOV domain-based tools that are also employed in optogenetic applications related to mitochondria. The most common ones (iLID, TULIP, LOVTRAP) are all based on the *As*LOV2 domain (Figure 9.2, Table 9.1).

The improved light-inducible dimer (iLID), for example, constitutes a pair of optical heterodimers based on an engineered *As*LOV2 variant caging SspA and its adaptor protein SspB[13]. The iLID/SspB combination has been designed to allow for interactions with a tandem-dimer SspB construct for improved photodimerization and functional utility.[13,49,56] Similarly, the tunable light-inducible dimerization tags (TULIPs) are based on the interaction of the *As*LOV2 domain-fused tag and an engineered cognate PDZ domain (ePDZ), which can recruit proteins to a variety of cellular structures with high spatial precision[47]. Further, LOV domains, such as the one from aureochrome 1 of *Vaucheria frigida* (VfAU1-LOV), which homodimerize when exposed to blue light and can activate various distantly related receptor tyrosine kinases, have found various optogenetic applications like controlling signaling pathways or targeted gene expression[57–59]. *Bc*LOV4, a *Botrytis cinerea* photoreceptor, directly binds plasma membrane (PM) lipids in response to blue light stimulation *via* an amphipathic helix-mediated electrostatic interaction[9,60]. This LOV-domain-based single-component system has fast membrane association/dissociation kinetics and is ideal for high-performance control. Vivid (VVD) is another blue-light LOV domain-containing photoreceptor derived from *Neurospora crassa*, that can be very useful for guiding protein-protein interactions as it switches from a monomeric to a homodimeric state when exposed to blue light[14,61]. Magnets are additional versatile tools for optogenetically influencing molecular processes. These are re-engineered variants of VVD with better dimerization properties, enabling the formation of heterodimers initiated by engineered electrostatic interactions[14].

Other optogenetic tools involving the *As*LOV2 domain include LOVTRAP[15,62], LINuS[63], and LEXY[64]. LOVTRAP enables precise and reversible control of protein activity, by using the protein Zdark (Zdk) which selectively binds to the dark state of *As*LOV2. A protein of interest (POI) can be sequestered away from the region of action in the dark by being fused to either *As*LOV2 or Zdk. However, Zdk separates from *As*LOV2 when exposed to light, enabling the POI to return to its initial location[62]. Another LOV domain-based tool that uses Zdk is called Z-lock. This tool allows for reversible, light-controlled steric inhibition of protein active sites. When fused to the termini of

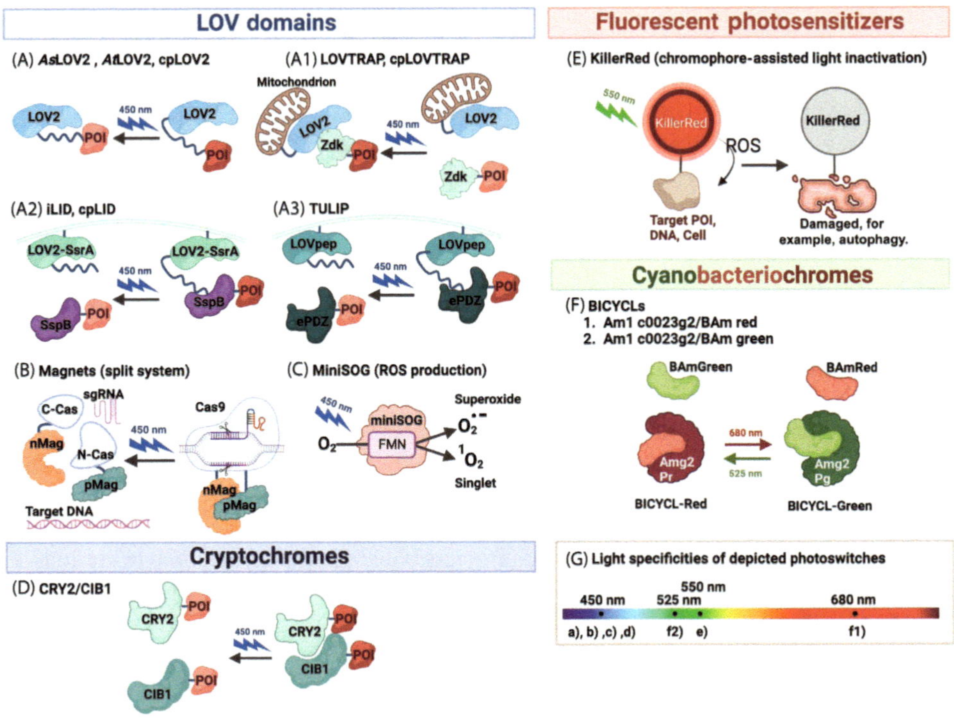

FIGURE 9.2 Schematic modes of function of the most common photoswitches with possible applications in mitochondria. (A) *Avena sativa* (*As*LOV2), *Arabidopsis thaliana* (*At*LOV2) light-oxygen-voltage domain, and the circularly permuted *As*LOV2 domain (cpLOV2) undergo blue light-induced conformational changes which have been used to recruit a protein of interest (POI) to the outer mitochondrial membrane (OMM). (A1) LOVTRAP and cpLOVTRAP use Zdark (Zdk) which selectively binds to the dark state of the *As*LOV2 domain protein enabling precise and reversible control of protein activity and have been used to recruit a POI to the OMM. (A2) iLID and cpLID constitute pairs of optical heterodimers based on an engineered *As*LOV2 variant caging SspA and its adaptor protein SspB. They have been used for endoplasmic reticulum (ER)-mitochondria tethering. (A3) Tunable, light-controlled interacting protein tags (TULIPs) are based on the interaction of the *As*LOV2 domain and an engineered cognate PDZ domain (ePDZ), which can recruit proteins to a variety of structures, like the OMM, with high spatial precision. (B) Magnets are re-engineered variants of Vivids (VVD) with better dimerization properties, enabling the formation of heterodimers initiated by electrostatic interactions. They have been used to recruit POIs to organelles or the cellular membrane. (C) Mini-singlet oxygen generator (miniSOG), a flavoprotein for correlative light and electron microscopy studies, enables tagging proteins genetically with a fluorophore and also plays a major role in ROS production *in vitro* and *in vivo*. It has been used to target the OMM and induce cellular ablation. (D) Cryptochrome 2/calcium- and integrin-binding protein 1 (CRY2/CIB1), a cryptochrome-based photoswitch, has been used to control Bcl-2-associated X protein (BAX) recruitment to the OMM. (E) KillerRed, the first intended reactive oxygen generator protein that generates reactive oxygen species (ROS) upon irradiation with green light, has been used to target the OMM for reversible mitochondrial fragmentation (see the "Reactive Oxygen Species Generators" section). (F) Bidirectional, cyanobacteriochrome-based light-inducible dimers (BICYCLs) can be used for controlling protein subcellular localization. (G) Spectral distribution of light specificities of photoswitches from ultraviolet (UV) to far-red light. (Table 9.1 complements the figure with further details and optogenetic applications.) (Created with BioRender.com.)

a target protein, the *As*LOV2 domain and Zdk sterically block the active site of the protein in the dark state. Illumination with blue light resolves the sterical block[65].

The light-inducible nuclear localization signal (LINuS), another tool based on *As*LOV2, uses blue light to reversibly activate the import of proteins into the cell nucleus. In this case, the *As*LOV2 domain cages a nuclear localization signal (NLS) within its C-terminal Jα-helix. LINuS facilitates nuclear translocation upon blue light illumination of fused POIs[63]. Controlling protein nuclear

TABLE 9.1
Non-Comprehensive List of Recent Non-Opsin-Based Optogenetic Tools for Mitochondrial Manipulation

Category	Light	Tool	Target	Description	Reference
Autophagy	Blue	CRY2-CIB1	OMM	BAX recruitment to OMM	(Hughes *et al.*, 2015[27], Godwin *et al.*, 2019[84])
	Blue	miniSOG	OMM	ROS-induced cell ablation	(Xu & Chisholm, 2016[121])
Mitophagy	Blue	iLID	OMM	AMBRA-1 recruitment to OMM	(D'Acunzo *et al.*, 2019[33])
	Green	KillerRed	OMM	Reversible mito fragmentation	(Williams *et al.*, 2013[122])
	Green	KillerRed	Matrix	ROS-induced Parkin-mediated autophagy	(Choubey *et al.*, 2011[123]) (Yang & Yang, 2011[124])
	Red/Far-red	MG2I-dl5***	COX4 COX8	ROS-induced neuronal mitochondria ablation	(Xie *et al.*, 2020[125])
Protein inactivation	Green/Yellow	Supernova	SDHB-1 SDHC-1	Inactivation of Complex II	(Trewin *et al.*, 2019[126])
	Blue	miniSOG	SDHC-1	Inactivation of Complex II	(Wojtovich *et al.*, 2016[127])
Protein localization	Blue	LOVTRAP	OMM	POI (mCherry) recruitment to OMM	(Wang *et al.*, 2016[15]) (Zhen *et al.*, 2020[128])
	Blue	cpLOV2	OMM	POI (mCherry) recruitment to OMM	(He *et al.*, 2021[128])
	Red	BYCICLE-green	OMM	POI (mCherry) recruitment to OMM	(Jang *et al.*, 2023[102])
	Green	BYCICLE-red	OMM	POI (mCherry) recruitment to OMM	(Jang *et al.*, 2023[102])
	Blue	iLID	OMM	Recruitment of mitotic spindle proteins to OMM	(Milas *et al.*, 2018[116])
Position and shape	Blue	iLID	OMM	ER-tethering	(Shi *et al.*, 2018[32])
	Blue	LOVpep	OMM	OMM attached to kinesin or dynein	(Harterink *et al.*, 2016[120])
	Blue	CRY2-CIB1	OMM	OMM attached to kinesin or dynein	(Duan *et al.*, 2015[82])
	Blue	CRY2/CIB1	OMM	OMM attached to kinesin	(Song *et al.*, 2022[119])
Membrane depolarization	Blue	CRY2/CIB1	IMM	Expressing K^+ channels Kesv and Kmp_{12T} targeted to IMM	(Engel *et al.*, 2020[129])

Notes: *** Approaches applied to mitochondria.

Abbreviations: IMM/OMM, inner/outer mitochondrial membrane; ROS, reactive oxygen species; SDHB-1/SDHC-1, succinate dehydrogenase subunits B/C; ER, endoplasmatic reticulum; POI, protein of interest.

export optogenetically has also been achieved with *As*LOV2, as exemplified with LEXY. In contrast to LINuS, LEXY harbors a nuclear export signal (NES) within the C-terminal Jα-helix of *As*LOV2. Blue light excitation leads to the exposure of the NES and hence to receptors that specifically assist active nuclear protein export through the nuclear pores[64].

In another study, a circular permutation technique was employed to re-engineer the *As*LOV2 domain and provide photoswitchable control of the C-terminus of a polypeptide[66]. As *As*LOV2 features the light-regulated Jα-helix at the C-terminus, it requires the fusion of target proteins to its C-terminus, which might affect the N-terminus of the target factor. Therefore, *As*LOV2 is often suitable for caging factors whose functions require a free N-terminus. However, caging the C-terminus might be the better choice for some target factors which, for example, harbor C-terminal residues that are crucial for activity. The circularly permuted *As*LOV2 domain (cpLOV) has reengineered termini which positions the Jα-helix close to the N-terminus. Therefore, cpLOV suits the latter caging requirements but may also be used in combination with *As*LOV2, enabling dual caging strategies[66,67]. cpLOV has been tested for various applications, thus broadening the scope of optogenetic tool designs[66].

EL222 is a transcription factor isolated from *Erythrobacter litoralis* HTCC2594 that can bind to DNA in a light-dependent way. It contains a C-terminal helix-turn-helix (HTH) DNA-binding domain in addition to the N-terminal LOV domain. The interactions between the LOV and HTH domains prevent DNA binding in the dark state, whereas blue light illumination resolves these interactions, releasing the LOV and HTH domains from each other and allowing EL222 dimerization to recognize its target DNA sequences. This conformational change is crucial for EL222 to bind DNA and control gene expression[68,69]. In one application, a VP-EL222 chimeric protein was created to activate gene expression by blue light, utilizing a VP16 transcriptional activation domain (AD) fused to EL222[70]. StrongLOV, a mutant version of EL222 developed in more recent applications in budding yeast, requires lower light intensities for the same degree of activity and may thereby result in lower phototoxicity[71]. As EL222 is widely used across kingdoms of life, such as in mammalian cells[72], bacteria[73], and plants[74], this mutation could be helpful for light-controlled experiments in various areas of synthetic biology.

Cryptochromes

Cryptochromes (e.g., plant CRY2) are photoreceptor proteins that play an important role in plant and animal circadian rhythms[75]. Although they were first discovered in Arabidopsis, cryptochromes were since found in many other organisms, ranging from bacteria to animals and even humans[76,77]. Cryptochromes evolved from photolyases that engage in DNA repair. Their evolution likely occurred multiple times independently leading to blue light perception through a conserved N-terminal photolyase-homologous region (PHR)[76]. Cryptochromes are composed of two domains: the N-terminal PHR domain that binds the FAD chromophore, and the CRY C-terminal extension (CCE) domain, which seems naturally unstructured but is crucial to cryptochrome function and control[78] (Figure 9.2D). Cryptochrome 2 (CRY2) from *Arabidopsis thaliana* uses FAD to sense light. The photoexcited CRY2 can both undergo homo-oligomerization and heterodimerization with its interaction partner calcium- and integrin-binding protein 1 (CIB1)[44,79,80]. Directional design and rational mutagenesis approaches yielded CRY2 variants with considerably enhanced[37,81,82], or suppressed[82], homo-oligomerization properties. The induction of clustering can result in the physical sequestration of CRY2-fused target proteins, inhibiting or activating relevant biological functions[38,39,49]. One of the most popular heterodimerization modules for inducing protein interactions is CRY2/CIB1. Upon blue light activation, photoexcited CRY2 changes conformation and can quickly heterodimerize with CIB1, within milliseconds of light exposure. The CRY2-CIB1 pair dissociates with a half-life time of around 5.5 min and this interaction can be repeatedly generated through numerous cycles[80]. This module requires no exogenous chromophore addition and is an advantageous tool for inducing protein translocation, transcription, and DNA recombination[44]. Capitalizing on CRY2 homo-oligomerization and CRY2/CIB1 heterodimerization, the photoswitch has found multiple applications, ranging from controlling organelle transport to positioning to the more recent induction of autophagy in cells[27,83,84].

Phytochromes

Phytochromes (Phys) are an important class of sensory (photochromic) biliproteins involved in regulating critical physiological processes across biological kingdoms. They can be found in all plants, the majority of cyanobacteria, most algae, some fungi, and many prokaryotes[85,86]. Phys are activated with red or far-red to near-infrared (NIR) light and rely on tetrapyrrole chromophores.

Typically, they are photochromic in the range of 600–750 nm, transitioning from the red light-absorbing ground state Pr to the far-red light absorbing Pfr state. Far-red light changes Pfr back to Pr. Plant Phys rely on phytochromobilin (PΦB) or phycocyanobilin (PCB) as chromophores, while bacterial phytochromes, such as bathy phytochromes, use biliverdin IXα (BV) instead[43,85,87,88]. In the presence of the cofactor PCB, the cyanobacterial phytochrome 1 (Cph1) from *Synechocystis* dimerizes under red light and monomerizes under far-red light, representing an immediate photoswitch for inducible homodimer formation[85].

The red light-induced binding capability of the plant photoreceptor phytochrome B (PhyB) to the interaction factor PIF3 and the reversal of this binding triggered by far-red light was used to engineer a split transcription factor for the control of gene expression in yeast cells[89]. Upon light activation, PIF3, a basic helix–loop–helix protein, interacts exclusively with the Pfr form (representing the active form of the photoreceptor) of both PhyA and PhyB[89]. Since then, many other Phy-based optogenetic applications ensued. In 2009, Levskaya *et al.* demonstrated how to deploy a new genetically encoded light-control system based on an improved, reversible PhyB/PIF6 protein-protein interaction switch derived from *Arabidopsis thaliana* components[48]. Plant PhyB covalently binds an exogenously supplied PCB chromophore in mammalian cells. This optoswitch can be utilized to precisely and reversibly translocate target proteins to the plasma membrane with high spatial resolution and within the timeframe of seconds to induce cell signaling responses. Shortly after, the first red/far-red light-triggered gene switch for mammalian cells was developed, which enables the spatiotemporal control of gene expression[48]. The authors used the PhyB/PIF6 red light-induced interaction and light illumination regimes at 660 and/or 740 nm to reversibly toggle the system between stable on and off states[90]. One year later, they adapted this concept into a red light-controlled synthetic gene expression switch for plant systems[91]. Plant phytochrome-based optogenetic tools have also been used in vertebrates *in vivo*, for instance, in 2015 Beyer *et al.* engineered a red light-inducible and far-red light-reversible synthetic system, based on the PhyB/PIF3 interaction, for controlling nucleo-cytoplasm shuttling of proteins in mammalian cells and zebrafish embryos[92].

*Dr*BphP, a bacterial phytochrome from *Deinococcus radiodurans*, forms a head-to-head parallel dimer that may alter conformation reversibly in milliseconds when illuminated with far-red or near-infrared light without the use of any exogenous cofactor, as BV is already endogenously present in bacteria and mammalian cells[93]. Under red light illumination (660 nm), bacterial phytochromes transmit the light signal from the photosensory module (PSM) to a C-terminal output domain. However, some bacterial phytochromes have an inverted photocycle with a far-red light-absorbing state (Pfr) as the resting state. *Dr*BphP, for instance, possesses a C-terminal histidine kinase domain, but other phytochromes employ, for example, protein domains called GGDEF-EAL with phosphodiesterase (PDE) or diguanylate cyclase (DGC) activity in engineered and truncated variants[94]. These domains approach each other in the dark or near-infrared light (740–780 nm), allowing for the homo-interaction of proteins fused to the C-termini[49,95–97].

Cyanobacteriochromes

Cyanobacteriochromes (CBCRs) constitute a class of photoreceptor proteins found in cyanobacteria with color sensitivities that extend over the full visible light spectrum from ultraviolet to far-red light[98–101]. The majority of these proteins consist of tandemly linked subdomains called GAF domains (cGMPphosphodiesterase/adenylate cyclase/FhlA). A conserved cysteine in the GAF domain forms a thioether bond with a bilin, usually phycocyanobilin (PCB), and upon radiation exposure, this chromophore undergoes E/Z isomerization of the C15 double bond[100].

The recently engineered bidirectional light-inducible dimers (BICYCLs) are the first CBCR-based devices that allow for reliable, adaptable control of protein-protein interactions *in vitro*, in yeast, and animal cells. They can be employed in regulating transcriptional processes, mitochondrial outer membrane protein localization, and can be used for multiplexing with the currently available blue-light tools LEXY, LINuS, and TULIPs[102].

BOX 9.1 OPTOGENETIC SYSTEMS AT THE PLASMA MEMBRANE

Due to similarities in applications between the cellular plasma membrane (PM) and the outer mitochondrial membrane (OMM), we provide here a short prospective introduction of optogenetic systems targeting the PM. The PM is a well-defined, two-dimensional subcellular unit engaged in highly regulated processes relevant to cell physiology. As a result, a wide range of optogenetic methods for dynamically controlling protein localization to the plasma membrane have been developed. These methods allow the optogenetic modulation of signaling cascades that start at the PM or change cell membrane dynamics, usually involving the recruitment of a protein-of-interest, or clustering at the membrane to regulate cell signaling. Previous reviews provide an extensive overview of these strategies[22,24,103,104], here we will only touch on a few relevant examples.

The most general strategy for modulating biological activities at the PM relies on the recruitment of various signaling components. For instance, the iLID system was recently used to enrich the red-fluorescent protein mCherry near macrophage plasma membranes[105]. A local increase in mCherry levels causes PM deformation, which in turn causes cells to move in the opposite direction. Another study reported the development of a blue light-inducible method for generating nanoscale membrane curvature in live cells. The authors used the Bin/Amphiphysin/Rvs (BAR) domain proteins to achieve a system that is controllable, reversible, and capable of precise spatial and temporal manipulation[106]. Cell motility is another process that could be modulated optogenetically. Various membrane structures important for cellular motility, such as lamellipodia, filopodia, and ruffles, arise from actin-dependent cell membrane motions[104]. Using the CRY2-CIB system, researchers showed that phosphatidylinositol-3,4,5-triphosphate (PIP3) production can be modulated to affect membrane dynamics. Phosphatidylinositol-3-kinase (PI3K) is recruited to the cell membrane by blue light-induced dimerization of CRY2 and CIBN, where it transforms local phosphatidylinositol-4,5-bisphosphate (PIP2) to PIP3, thereby increasing cellular motility and causing the development of apical ruffles[104]. Another approach is to use photoactivated binders for protein recruitment to the PM. Optogenetic nanobodies (OptoNBs) were used to modulate Ras activity *in vivo* by targeted recruitment of the SOS catalytic domain (SOScat) to the PM using both light-induced and light-inhibited OptoNBs[107]. SOS, which is mostly found at the plasma membrane, acts as a positive regulator of Ras[108]. Another study combined dual-wavelength control with actin-targeted intrabodies to attract actin to the plasma membrane and the nucleus. Here, they used BphP1/Q-PAS1, a phytochrome-based switch. The mCherry-tagged Q-PAS1 is recruited to the target protein bound to the membrane in a light-induced manner. Irradiation with far-red light boosted actin targeting to the PM, whereas illumination with blue light enhanced nuclear intrabodies targeted to actin levels. Their sequestration to the PM reduces cytoskeleton-driven cellular motility[95].

Employing such optogenetic techniques enables precise spatial and temporal control and modulation of large cellular structures without the use of pharmaceutical effectors[24]. Understanding how these systems work to control the plasma membrane can help us further understand how to employ optogenetic techniques on organelle membranes, such as the OMM discussed in this review.

NON-OPSIN-BASED OPTOGENETICS INVOLVING MITOCHONDRIA

Mitochondria play a pivotal role in metabolic homeostasis and the regulation of various cellular pathways, and research on functional and regulatory aspects of these organelles can certainly profit from implementing optogenetic approaches. Maintaining stable homeostasis allows mitochondria to regulate cellular health as well as death processes. Mitochondria are the main source of ATP and reactive oxygen species (ROS) production. They are also important for second messenger generation and production of biosynthetic precursors for macromolecules, some of which also function as second messengers involved in regulating vital processes like cellular aging[109,110]. The function and regulation of the cellular mitochondrial network is quite complex. Therefore, utilizing the spatial and temporal regulation possibilities of optogenetic approaches represents one of the most attractive strategies to gain access to the study of mitochondrial physiology. Optogenetically regulating diverse mitochondrial functions opens additional avenues to the study of intricate cellular programing and processes, such as autophagy, mitophagy, and apoptosis (regulated cell death), which can assist in revealing new insights into the function of the complex organelle. Optogenetics could in the future provide, for example, novel strategies to study the role of mitochondria in disease etiology and their key role in metabolic engineering[23,24]. In this chapter, we discuss optogenetic solutions, highlighting applications in mitochondria-related processes including autophagy, apoptosis, mitophagy or mitochondrial positioning, and shape modulation.

Autophagy

Autophagy is a crucial biological process in which cells sequester internal constituents such as organelles and proteins and transfer them to lysosomes for the breakdown and recycling of macromolecule precursors. It is a key cell survival mechanism that enables cells to respond to metabolic stress by degrading and recycling intracellular components[111]. Channelrhodopsins (ChRs), traditionally used to optogenetically modify the activity of neurons, now offer an interesting solution to change the potential of mitochondrial membranes. Disruption of the electrochemical gradient alters mitochondrial calcium mobilization and storage. Light stimulation of mitochondrial membrane-localized ChRs causes strong impacts on cell survival (a more detailed review on channelrhodopsins is presented in Chapter 2: Energetics/Opsin-Based Optogenetics)[25,26].

Apoptosis

Apoptosis is a form of programmed cell death with characteristic features, such as chromatin condensation, nuclear disintegration, and plasma membrane blebbing[112]. Before autophagic cell death and programmed necrosis (necroptosis) were described, the terms apoptosis and programmed cell death were frequently used interchangeably. Optogenetics also allows for the regulation of apoptosis *via* the targeted control of the localization of apoptosis-regulating proteins such as Bcl-2-associated X protein (BAX)[27]. Optogenetic recruitment of CRY2-BAX fusion proteins to CIB monomers attached to the outer mitochondrial membrane (OMM) *via* translocase of OMM 20 (TOM20) sequences could, for example, promote cell death in response to blue light. BAX targets the mitochondrial membrane and causes the release of cytochrome *c*, which in turn triggers apoptosis[24].

Mitophagy

Mitophagy, or autophagy-mediated degradation of mitochondria, is another process that has been opened for optogenetic manipulations. This specific kind of autophagy is responsible for the turnover of damaged mitochondria by selectively destroying them. Mitophagy acts often independently of the signaling that regulates basal autophagy and primarily serves as a quality control mechanism by identifying and removing malfunctioning mitochondria that could otherwise contribute to death signaling activation. The pathway is further responsible for the elimination of healthy mitochondria during cell differentiation[113]. D'Acunzo and his colleagues presented a bimodular system to reverse the light-dependent induction of mitophagy and rescue cells from apoptosis[33]. iLIDs have been implemented to attract autophagy and beclin-1 regulator 1 (AMBRA1) to the mitochondrial surface, thereby activating mitochondrial recycling and mitophagy in the presence of blue light. This design utilizes the fusion of SsrA to an OMM linker sequence from actin assembly-inducing protein (ActA). Furthermore, Shi *et al.* were able to employ light-triggered dimerization of iLIDs to induce mitochondrial-ER tethering with high spatiotemporal precision, which is required for mitophagy[32].

Protein Localization

Optogenetic tools employed for the control of subcellular protein localization are usually customized for the recruitment of proteins to the mitochondrial surface (Figure 9.2). In fact, some examples of protein localization have already been introduced before, for instance, the implementation of iLIDs in combination with AMBRA1 (application in mitophagy) uses protein localization to attract the protein to the mitochondrial surface. Furthermore, LOVTRAP-induced protein sequestration at the OMM has been employed to modulate intracellular protein accumulation at the OMM of mammalian cells[15].

Another tool with applications in mitochondrial protein localization are BICYCLs, the smallest green/red optogenetic tools known so far, which were used in combination with LEXY[64] and LINuS[63] to control subcellular protein localization and gene expression[102]. In the dark, LEXY-tagged Amg2 (the photoreceptor utilized in BICYCLs) was mostly localized in the nucleus and accumulated at the mitochondria upon red light illumination; green light reversed this mitochondrial accumulation. Exploiting the bidirectional switching behavior of phytochromes enhances spatial localization in contrast to, for example, LOV-based photoswitches, where dark reversion kinetics control the relaxation into the ground state[102,114,115]. Mitochondrial sequestration of Zdk fusion proteins and their optical release has been realized with *As*LOV2 anchored to the OMM using a sequence derived from TOM20. LOVTRAP, like other optogenetic tools, usually requires customization and optimization when transferred to other cell systems. In particular, portability to yeast cells where the high levels of mitochondrial *As*LOV2 required for successful protein trapping was challenging, as it affected cellular development[31]. Milas *et al.* introduced an optogenetic tool based on iLID for rapid and reversible translocation of proteins in human cells using blue light[116]. Proteins were removed from the mitotic spindle to the mitochondria and turned back to the mitotic spindle on a minute-by-minute scale, allowing regulation of protein activity in different periods of mitosis and in individual cells. Optogenetically exploring the complicated nature, dynamics, and function of mitotic proteins bears the potential to further investigate the mechanisms underlying mitosis[116].

Position and Shape

Organelle transport and distribution play a significant role in a variety of cellular processes. However, the paucity of tools to alter molecular motors within the complicated intracellular environment makes it difficult to draw a direct link between organelle distribution and cellular functions. The use of optogenetic strategies represents a potential solution to gain access to such methods, enabling a deeper investigation of these aspects. Duan *et al.* showed that it is possible to redistribute organelles like mitochondria, peroxisomes, and lysosomes in the cell in response to light[82]. Using a light-induced CRY2/CIB1 interaction, they recruited molecular motors to the outer membrane of organelles, thereby achieving optical control of directional transport and redistribution of organelles in living cells.

Mitochondria are also affected by mechanical stimuli, which directly impact and regulate their shape and function. Mechanical strains applied to cells have been shown to cause mitochondrial elongation, a decrease in mitochondrial membrane potential, and the generation of ROS[117,118]. Recently, Song *et al.* demonstrated the possibility of optogenetically controlling mitochondria shape and motility through light-activated protein-protein hetero-dimerization[119]. They were able to implement different optogenetic modules (CRY2/CIBN, iLID/SspB, and LOVpep/ePDZ) into their system to trigger mitochondria deformation by recruiting molecular motors to the OMM[119,120].

Reactive Oxygen Species (ROS) Generators

The processes involved in the oxidative phosphorylation at the mitochondrial respiratory chain give rise to a significant source of cellular ROS. ROS play a crucial role in cellular signaling, stress, and mitophagy where they engage as second messengers. The intended induction of ROS in a precise spatiotemporal manner may help to understand cellular signaling, as well as apoptotic and necroptotic pathways in whole organisms[130,131]. ROS generators (ROGs) can be used to investigate second messengers in the cytosol, ROS-induced ROS release[132], to mimic biological ROS production and mitophagy[121,123–125], as well as for cell ablation experiments[121]. Using genetically encoded light–activated ROGs has one main advantage over chemical light–activated ROGs. The production and thus the location of ROGs can be spatially controlled within the cellular compartments by genetically encoded regulative principles. Furthermore, it is possible to attach ROGs to proteins of interest *in vitro* and *in vivo.*

Formation of Reactive Oxygen Species

Activating a chromophore by the light of a specific wavelength within the excitation range leads to the molecule reaching a higher energetic state S_1 (Figure 9.3, blue arrow). The return to the basal S_0 state occurs *via* several mechanisms, including the non-radiative dissipation of energy (called internal conversion, gray wavy arrows) and emission of photons, fluorescence, which can occur from the S_1 state (green arrow). Another very rare and quantum mechanically "forbidden" state is the triplet state T_1, reached by intersystem crossing (ISC). Deactivation from this state may either occur *via* phosphorescence (yellow arrow) or the production of ROS that can occur through one of the two different types of reactions (Figure 9.3B). Since the triplet state has a lifetime of microseconds to minutes, molecules reaching this state can perform random energy transfer collisions. When colliding directly with substrates like lipids, proteins, or nucleic acids, the activated molecule tends to transfer an electron (type I reaction). The so-generated radicals themselves react with oxygen, forming ROS. The direct transfer of an electron to oxygen by collision is also possible, yielding the anion superoxide (O_2^{*-}). Superoxide dismutases generate H_2O_2 and O_2 while the hydrogen peroxide can be further catalyzed by copper or iron ions to the extremely reactive hydroxyl radical, HO^* (Fenton reaction). Type II reaction is the energy transfer from a triplet state chromophore to a ground state oxygen ($^3O_2$) yielding singlet oxygen, $^1O_2$. It has to be stated that ROS production does not always follow just one reaction type. Since singlet

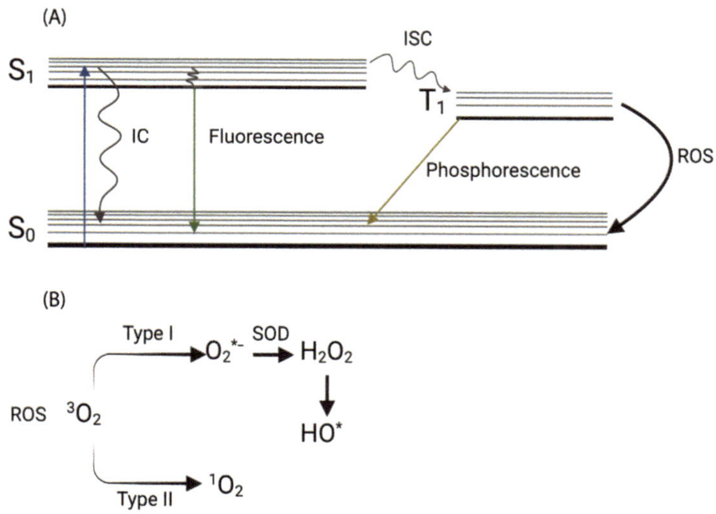

FIGURE 9.3 ROS production. (A) Jablonsky scheme, ROGs are excited to the high-energy singlet state S_1 upon photon absorption (blue arrow). Relaxation into the ground state follows either internal conversion (IC), fluorescence, or intersystem crossing (ISC), followed by phosphorescence or ROS generation. (B) ROS mechanisms, type I: Collision of the excited ROG leads to a single electron transfer. The superoxide dismutase (SOD) converts the superoxide radical O_2^{*-} to hydrogen peroxide. In the presence of metal cations, H_2O_2 can give rise to highly reactive hydroxyl radicals, HO^*, *via* the Fenton reaction. Type II: Energy transfer by a direct collision between ROS and oxygen molecules.

TABLE 9.2
Photophysical Properties of a Series of ROGs

Name	λ_{ex} (nm)	λ_{em} (nm)	$\Phi_{\Delta}{}^1O_2$ (i)	$\Phi_{\Delta}{}^1O_2$ (ii)	$\Phi_{\Delta}{}^1O_2$ (iii)
KillerRed	585	610	0[130]	n.d.	0.0076[136]
KillerOrange	512	555	n.d.	n.d.	n.d.
Supernova (red)	579	610	n.d.	n.d.	0.022[136]
SupernovaGreen	440/480	510/560	n.d.	n.d.	n.d.
mCherry	587	610	n.d.	n.d.	0.0057[136]
miniSOG	501	526	0.42[133]	0.0[137] 0.03[133,138]	n.d.
miniSOG Q103V	n.d.	n.d.	n.d.	0.39[139]	n.d.
miniSOG2	n.d.	n.d.	n.d.	n.d.	n.d.
SOPP	488	515	n.d.	0.25[137] 0.23[140] 0.19[138]	n.d.
SOPP2	491	513	n.d.	0.51[140]	n.d.
SOPP3	490	514	n.d.	0.61[140]	n.d.
FMN	444	531	n.d.	0.57[137] 0.51[141] 0.49[138]	n.d.
MG-2I-dl5***	666	693	0.13[130]	0.18[134]	n.d.

*** *Singlet oxygen quantum yields $\Phi_{\Delta}\ {}^1O_2$ determined by (i) use of anthracene dipropionic acid (ADPA), (ii) phosphorescence measurement at 1,275 nm, or (iii) singlet oxygen sensor green (SOSG). The wavelength of the excitation and emission maximum are indicated by λ_{ex} and λ_{em}, respectively.*

oxygen (type II) is highly reactive and plays a major role in phototoxic effects, the quantum yield of ${}^1O_2$ is the mainly used reference value to compare ROGs[133,134]. It has to be mentioned that the indirect measurements of singlet oxygen *via* loss of fluorescence signal due to photooxidation by scavengers like anthracene dipropionic acid (ADPA) or singlet oxygen sensor green (SOSG) is shown to be not ${}^1O_2$ exclusive[133,135]. Therefore, the direct measurement of ${}^1O_2$ phosphorescence can be seen as the most reliable detection method[133,134]. Table 9.2 shows the respective data for all mentioned ROGs.

KillerRed

KillerRed was the first intended ROG protein described, which was developed by Bulina *et al.* from the hydrozoan chromoprotein anm2CP, a homolog of the green fluorescent protein (GFP)[142]. Unlike GFP, KillerRed provides a water-filled channel to the chromophore which most likely is the reason for the photosensitizing effect. KillerRed generates ROS upon irradiation with green light mostly *via* the type I mechanism (Figures 9.2C and 9.3). Producing superoxide anions, which are the major ROS species in the cellular environment, makes KillerRed and all of its variants an attractive tool assisting in the study of ROS signaling and stress.

KillerOrange was described by Sarkisyan *et al.* in 2015[143]. The new variant is phototoxic upon green and blue light exposure. Furthermore, the ability of KillerRed and KillerOrange to form dimers affects the cellular localization and perturbed mitotic cell division. To face this problem, the monomeric variants Supernova red[144] and Supernova green[145] were subsequently developed.

MINISOG

MiniSOG was initially described as a new tool for correlative light and electron microscopy (CLEM) studies. Shu *et al.* used its ability to produce ROS for catalyzing the conversion of diaminobenzidine into a product stainable with osmium tetroxide[146]. Thus, miniSOG acts as a fluorophore tag and as an ROS production tag. This makes it possible to localize a protein of interest, which is genetically tagged with miniSOG and irradiated with the ROS-specific wavelength under the electron microscope when providing diaminobenzidine as substrate. MiniSOG not only revolutionized CLEM staining but also plays a major role in ROS production *in vitro* and *in vivo*. The 106-residue long miniSOG is the smallest ROG. It is derived from the *At*LOV2 domain and depends on the availability of intracellular FMN as a cofactor[146].

The mechanism of ROS production is not yet fully understood. Nevertheless, one may assume that miniSOGs can produce superoxide as well as singlet oxygen (type I and II mechanism, Figure 9.3). The production of $^1O_2$ likely causes cellular damage rather than ROS signaling. Therefore, miniSOG is well suited for loss of function and ablation experiments (Figure 9.3)[133]. In recent years, many improved variants were developed with increased triplet state lifetime and shifted ratios toward the production of $^1O_2$ (type II) rather than O_2^{*-} (type I), including SOPP[138], SOPP2[140], SOPP3[140], miniSOG Q103V[139], and miniSOG2[147].

THE USE OF ROG IN THE CONTEXT OF MITOCHONDRIA

Even though ROS production and investigation of mitochondria seem to be a perfect match, only a few mitochondrial targeting approaches are described in the literature. Here we will briefly describe some approaches exemplarily to state the possibilities provided by ROGs.

Mitophagy

Induction of mitophagy is by far the most widely used approach of ROGs in the context of mitochondria research, as exemplified by the activation and monitoring of the Parkin-mediated mitophagy pathway[123,124]. In 2011, Choubey and Yang & Yang addressed this pathway optogenetically by targeting KillerRed to the mitochondrial matrix and further fusing Parkin to a fluorophore of a different excitation color than the light used for the excitation of the ROG, to avoid bleaching. After excitation of the ROG, the impaired mitochondria membrane produces ROS, and Parkin-mediated autophagy is initiated through the assembly of Parkin at illuminated mitochondria. Later, Tan and Tu *et al.* fused miniSOG to PINK1, and mCherry to Parkin to investigate the role of RhoA[148] and HK-II[149] in Parkin-mediated mitophagy.

The choice of the ROG used affects the result not just quantitatively but also qualitatively. Additionally, the capacity to buffer/detoxify ROS in different cellular environments plays a major role in ROG experiments. The fusion of KillerRed to the OMM of muscle cells in *C. elegans* did not result in cell death. However, when fusing KillerRed to the plasma membrane, the cells effectively died. Since KillerRed produces superoxide, which is the most common natural ROS in cells, the mitochondria were able to buffer the effect and fully recover within 48 h[122]. In contrast, fusing miniSOG, a type II ROG, to the OMM of neuronal mitochondria in *C. elegans* led to apoptotic cells. Still, a direct comparison revealed an efficacy of just about 10% when miniSOG was fused to the OMM rather than to the plasma membrane, most likely due to the different ROS buffering capacities within the cellular sub-environments[121].

Chromophore-Assisted Light Inactivation

Chromophore-assisted light inactivation (CALI) relies on the oxidation of the protein of interest fused to the ROG. Since ROS are highly active and the intracellular environment is quite dense, the produced ROS have a very small reaction radius. Trewin *et al.* attached the Supernova-ROG to the Complex II succinate dehydrogenase subunits B and C, monitoring the effect of ROS on the inner mitochondrial membrane and in the intermembrane space[126]. Wojtovich *et al.* inactivated the SDHC subunit of the Complex II by fusing it to miniSOG[127].

Recent Chemo-Optogenetic Approaches

While setting the primary focus of this review article to genetically encoded gene optogenetic applications in mitochondria, we wish to mention one recent chemo-optogenetic hybrid system with a broad scope for use in mitochondria. Chemo-optogenetic hybrid systems utilize chemical compounds, e.g., a chemical dye, induced or activated by an optogenetic system. Optogenetic ROGs that have been used so far in mitochondria share a relatively low quantum yield in comparison to chemical ROGs. Furthermore, the spectral limitation to blue and green excitation wavelengths compromises their use for applications, that require a relatively deep tissue penetration, for example, in whole animals without causing cytotoxic effects by the excitation light. The chemo-optogenetic FAP-TAP (fluorogenic-activated protein-targeted activated photosensitizer) system addresses these issues[130,150]. Fluorogenic-activating proteins (FAP) are single-chain antibodies that activate and stabilize chemical dyes in their triplet state[125,130,151]. The here-used FAP, dl5***[103] activates TAP (targeted activated photosensitizer)-fluorescence approximately 1,000-fold[150]. Malachite green, modified by substitution with two iodines, acts as TAP. The heavy atom insertions (2I) additionally shift the emission spectra toward longer wavelengths and increase the probability of intersystem crossing (Figure 9.3), thus leading to a higher quantum yield. The ROS production follows mainly the type II mechanism and is, therefore, more destructive than KillerRed and its derivatives. Fusing the antibody genetically to a protein of interest leads to a highly specific ROS production tag. In comparison to KillerRed, the FAP-TAP system dl5*** – Malachite Green (MG2I) required a 60-fold lower light dose for effective cell ablation in *C. elegans* (90 s *vs.* 90 min of illumination)[130]. A recent study triggered mitophagy of neuronal mitochondria in zebrafish larvae, which led to a massive loss of neuronal function in combination with either apoptosis or necroptosis[125]. MG2I can penetrate membranes in its non-activated form so that very low concentrations suffice. ROS production of inactive TAP is usually not detectable; therefore, FAP-TAP systems are in principle well suited for studies in entire organisms[130].

Even though the actual singlet oxygen quantum yield might be lower than SOPP and its variants, MG-2I-dl5*** possesses a significantly stronger phototoxic effect on mitochondria than KillerRed[130,134]. Additionally, unlike miniSOG, SOPP and their derivatives do not depend on FMN, which is not evenly distributed over the different sub-environments. Furthermore, the absorption spectra do not overlap with free FMN which is itself known to have a quite high singlet oxygen quantum yield (Table 9.2) and provides a well-spread background in ROS production in every illuminated cell[134].

Since fully genetically encoded ROGs with activation wavelengths in red or far-red light are urgently needed, the excitation maximum of MG2I-dl5*** between 660 and 670 nm makes it a very promising tool for deep tissue experiments in red/far-red light conditions[125,130].

ENERGETICS/OPSIN-BASED OPTOGENETICS

During the past two decades, opsin-based optogenetics has increasingly developed and matured into a powerful research approach. For a long time, optogenetics was synonymous with neuronal optogenetics, used to control the plasma membrane potential in neurons. Nowadays, the expanding opsin-based optogenetics approaches strongly impact not only neurobiology but also cardiology for the control of the PM potential in myocytes and other related application areas[3,152,153]. Recently, opsins were also used to control the membrane potential in plants to study electrical signaling networks[154–157]. While opsin-based optogenetics is widely employed to modify the PM potential in the abovementioned environments, its use to influence the activity of mitochondria remained unexplored until very recently[25,26,28–30,158,159]. Conventionally, one would resort to the use of chemical solutions for altering the mitochondrial membrane potential ($\Delta\psi_m$). Carbonyl cyanide-*p*-trifluoromethoxyphenylhydrazone (FCCP) and carbonyl cyanide *m*-chloro-phenylhydrazone (CCCP), for instance, represent two of the most frequently used protonophores to control $\Delta\psi_m$. However, they also affect the PM potential and are cytotoxic[160]. Moreover, the difficulty of applying them in tissues and organs, the lack of spatiotemporal and quantitative control resolution, and their low reversibility hamper the use of these drugs for therapeutic purposes. Recent studies that successfully applied opsin-based optogenetics in mitochondria suggest an alternative and precise methodology which overcomes the mentioned bottlenecks[25,26,28–30,159].

Opsins are transmembrane proteins that respond to light. Their activation leads to ion transport, resulting in changes in the membrane potential. Their use as optogenetic tools offers the possibility to manipulate membrane potentials with high spatiotemporal resolution, on the millisecond scale, and with minimized toxicity[161,162]. Natural opsins generally belong to one of two types, microbial or animal opsins, also called type I and type II rhodopsins, respectively[163]. Microbial rhodopsins act as channels and pumps which, in response to light, control the potential across membranes by transporting ions. Animal opsins possess a different functional mechanism and act as G protein-coupled receptors (GPCRs)[164]. In this section, we review the use of microbial rhodopsins to artificially control membrane potentials, specifically in mitochondria.

Type I rhodopsins have been identified in algae, archaea, bacteria, and fungi[165]. All type I rhodopsins share a seven-transmembrane alpha-helical topology and require *all-trans*-retinal (ATR) as a cofactor[166]. Upon light illumination, ATR isomerizes to 13-*cis* retinal and promotes structural changes of the opsin leading to ion fluxes. Note that 13-*cis* retinal reversion to ATR in the dark inactivates the opsin, making the stimulation reversible[167]. Type I rhodopsins in mammalian neurons do not require ATR supplementation, although when used in other cell types and organisms, ATR supplementation may enable or boost opsin function[168]. In addition to the strong general light dependence of plants, the absence of retinal makes it difficult to use opsins as optogenetic tools. The recently engineered ATR production in *Arabidopsis thaliana*, obtained by expression of a β-carotene 15,15′-dioxygenase from a marine bacterium that converts β-carotene into retinal, might enable the generalized use of opsins for plant research in the future[157].

Over the past decades, several opsins with different properties have been discovered and further engineered to generate characteristic variants, which give rise to a broad and diverse opsin-based optogenetic toolbox. Table 9.3 summarizes some of the most used opsins, which have been described elsewhere[10,25,156,161,164,169–211], indicating the ones used for mitochondrial research.

Microbial rhodopsins further form two subgroups, light-gated ion channels and light-driven ion pumps[212]. Pumps actively transport ions in a unidirectional way, even against an existing electrochemical gradient, dependent on energy input. In this case, each absorbed photon results in the transport of a single ion. Therefore, high light intensities and expression levels are generally required to significantly alter the membrane potential[213]. Light-driven ion pumps belong to four main groups based on the nature of the transported ion and the direction of the transport: Inward chloride pumps, outward sodium pumps, and inward or outward hydrogen pumps[167] (Figure 9.4). Bacteriorhodopsin from *Halobacterium salinarum* (*Hs*BR) was the first discovered microbial rhodopsin[211]. It belongs to the outward hydrogen pumps; this subgroup induces membrane hyperpolarization by pumping protons into the extracellular space. Other bacteriorhodopsins and archaerhodopsins of this subgroup include archaerhodopsin-3 (Arch) from *Halorubrum sodomense*, which is the most frequently used pump for inducing hyperpolarization, and the fungal light-driven proton pump from *Leptosphaeria maculans* (MAC)[205], which is used for mitochondrial membrane potential control[28,29,159]. Contrasting the broad availability of outward proton pumps, very few inward proton pumps are known. *Po*XeR, from *Parvularcula oceani*, was the first one characterized[187]. Inward proton pumps could potentially be used for the depolarization of membranes. Although their capacity does not reach that of the light-gated ion channels, their activity does not depend on the ion gradient. Halorhodopsins are inward chloride pumps that trigger hyperpolarization by the intrusion of negative charges. Unlike proton pumps, halorhodopsins do not cause pH changes. *Np*HR, from *Natronomonas pharaonic*, for instance, is a chloride pump

TABLE 9.3
Commonly Used Natural Microbial Rhodopsins and Engineered Variants

	Name(s)	Description	Ion	Activation Wavelength (nm)	Function	Used in Mitochondria	Reference
PUMPS	**Arch/aR-3**	Archaerhodopsin-3 from *Halorubrum sodomense*. Widely used pump with rapid kinetics and high dynamic range	H^+ (outward)	552	Hyperpolarization	No	(Mukohata *et al.*, 1999)[197]
	eArch 3.0	Arch variant with high photocurrents				No	(Mattis *et al.*, 2012)[195]
	Arch M128A/ S151A/A225T	Arch blue-shifted variant (550 nm)				No	(Sudo *et al.*, 2013)[203]
	ArchT	Archaerhodopsin from *Halorubrum* strain TP009, higher light sensitivity than Arch	H^+ (outward)	566	Hyperpolarization	No	(Han *et al.*, 2011)[182]
	***Hs*BR**	Bacteriorhodopsin from *Halobacterium salinarum*. First-discovered microbial rhodopsin	H^+ (outward)	560	Hyperpolarization	No	(Oesterhelt *et al.*, 1971)[211]
	Ace/AR	BR-like rhodopsin from *Acetabularia acetabulum*. First eukaryotic proton pump reported	H^+ (outward)	518	Hyperpolarization	No	(Tsunoda *et al.*, 2006)[204]
	dR	Deltarhodopsin from *Haloterrigena* sp. Arg-4	H^+ (outward)	550	Hyperpolarization	Yes	(Ihara *et al.*, 1999)[185]
	MAC/LR/Ops	From *Leptosphaeria maculans*, lower photocurrent than Arch	H^+ (outward)	555	Hyperpolarization (also used for depolarization)	Yes	(Waschuk *et al.*, 2005)[205]
	***Po*XeR**	Xenorhodopsin from *Parvularcula oceani*	H^+ (inward)	567	Depolarization	No	(Inoue *et al.*, 2016)[187]
	***Ns*XeR**	Xenorhodopsin from *Nanosalina*	H^+ (inward)	565	Depolarization	No	(Shevchenko *et al.*, 2017)[202]
	***Hs*HR/HR**	Halorhodopsin from *Halobacterium salinarum*, poor heterologous expression	Cl^-(inward)	580	Hyperpolarization	No	(Zhang *et al.*, 2007)[209]
	***Np*HR/Halo**	Halorhodopsin from *Natronomas pharaonis* with low photocurrents	Cl^- (inward)	580	Hyperpolarization	No	(Zhang *et al.*, 2007)[209]
	*eNp*HR 3.0	*Np*HR variant with high photocurrents (20 times stronger than *Np*HR)				No	(Gradinaru *et al.*, 2010)[180]
	***Sal*HO**	Halorhodopsin-like chloride pump from *Salinibacter ruber*	Cl^- (inward)	582	Hyperpolarization	No	(Mongodin *et al.*, 2005)[196]

(Continued)

TABLE 9.3 (Continued)
Commonly Used Natural Microbial Rhodopsins and Engineered Variants

	Name(s)	Description	Ion	Activation Wavelength (nm)	Function	Used in Mitochondria	Reference
	Jaws	Cruxhalorhodopsin from *Haloarcula (Halobacterium) salinarum*	Cl^- (inward)	632	Hyperpolarization	No	(Chuong *et al.*, 2014)[172]
	KR2	Rhodopsin 2 from *Krokinobacter eikastus*	Na^+ (outward), H^+ in the absence of Na^+	524	Hyperpolarization	No	(Inoue *et al.*, 2013)[186]
CHANNELS	***Gt*CCR4**	Cation channelrhodopsin from *Guillardia theta* with high light sensitivity and rapid desensitization	CCR (H^+ and Na^+)	540	Depolarization	No	(Yamauchi *et al.*, 2017)[210]
	***Cr*ChR1**	Cation channelrhodopsin-1 from *Chlamydomonas reinhardtii*	CCR (highly permeable to H^+)	495	Depolarization	No	(Nagel *et al.*, 2002)[198]
	***V*ChR1**	Cation channelrhodopsin from *Volvox carteri*, poor expression in mammalian cells	CCR	570	Depolarization	No	(Zhang *et al.*, 2008)[208]
	***Ca*ChR1/ *Ca*CCR1**	Cation channelrhodopsin from *Chlamydomonas augustae*	CCR	520	Depolarization	No	(Hou *et al.*, 2012)[184]
	Chronos /*Sh*ChR	Cation channelrhodopsin from *Stigeoclonium helveticum* with fast kinetics and high light sensitivity	CCR	480	Depolarization	No	(Klapoetke *et al.*, 2014)[189]
	***Sd*ChR**	Cation channelrhodopsin from *Scherffelia dubia* with high light sensitivity	CCR	460	Depolarization	No	(Klapoetke *et al.*, 2014)[189]
	CheRiff	Sd*ChR* variant with fast kinetics than *sd*	ChR and high light sensitivity			No	(Hochbaum *et al.*, 2014)[183]
	***Ps*ChR/ *Ps*ChR2**	Cation channelrhodopsin from *Platymonas subcordiformis*. More blue-shifted variant available together with *Ts*ChR	CCR	445	Depolarization	No	(Govorunova *et al.*, 2013)[176]
	*Ps*ChR D139H	*Ps*ChR variant with five-fold enhanced photocurrent and enhanced	Na^+ and Ca^{2+} conductance			No	(Duan *et al.*, 2019)[174]
	***Ts*ChR**	Cation channelrhodopsin from *Tetraselmis striata*. More blue-shifted variant available together with *Ps*ChR	CCR	445	Depolarization	No	(Klapoetke *et al.*, 2014)[189]
	***D*ChR1**	Cation channelrhodopsin from *Dunaliella salina*, H^+ selective	CCR (H^+ selective)	475	Depolarization	No	(Zhang *et al.*, 2011)[164]

(Continued)

TABLE 9.3 *(Continued)*
Commonly Used Natural Microbial Rhodopsins and Engineered Variants

Name(s)	Description	Ion	Activation Wavelength (nm)	Function	Used in Mitochondria	Reference
Chrimson (*Cn*ChR1)	Cation channelrhodpsin from *Chlamydomonas noctigama,* highly H^+ selective	CCR (H^+ selective)	590	Depolarization	No	(Klapoetke *et al.*, 2014)[189]
ChrimsonR	Chrimson variant with fast off kinetics				No	(Klapoetke *et al.*, 2014)[189]
vf-Chrimson	Chrimson variant with very fast kinetics (comparable to Chronos)				No	(Mager *et al.*, 2018)[193]
Chrimson-SA	Chrimson mutant with red-shifted activation spectra (605 nm)				No	(Oda *et al.*, 2018)[224]
ChRmine (RlCCR1)	Cation channelrhodopsin from *Tiarina fusus* with high conductance and light sensitivity	CCR	600	Depolarization	No	(Marshel *et al.*, 2019)[194]
***Cr*Chr2**	Cation channelrhodopsin-2 from *Chlamydomonas reinhardtii* with fast recovery. Widely used	CCR ($H^+ > Na^+ > K^+ >> Ca^{2+}$)	470	Depolarization	Yes	(Nagel *et al.*, 2003)[199]
Chr2 C128X(X=T/A/S)	ChR2 variant with long-lived open states and enhanced light sensitivity. It remains open after activation with a short light pulse and can be closed with orange light				No	(Berndt *et al.*, 2011)[169]
ChR2 H134R	Most used ChR2 variant with slow channel closure, improved conductivity, and modified ion selectivity				Yes	(Nagel *et al.*, 2005)[200]
ChR2 C128A/H134R	ChR2 variant with slow inactivation				Yes	(Tkatch *et al.*, 2017)[25]
ChR2 XXL	ChR2 D156C mutant with large photocurrents and high light sensitivity (even at low retinal concentrations)				No	(Dawydow *et al.*, 2014)[173]
SSFO	ChR2 C128S/D156A mutant with extremely slow τ-off and high light sensitivity				Yes	(Yizhar *et al.*, 2011)[239]
SOUL	SSFO variant with ultra-high light sensitivity				No	(Gong *et al.*, 2020)[175]
ChETA	Fast-cycling ChR2-E123T/A mutant with low light sensitivity and photocurrent				No	(Gunaydin *et al.*, 2010)[181]
CatCh	ChR2 L132C mutant with weak inactivation, fast response kinetics, large photocurrents and light sensitivity, and increased Ca^{2+} and Mg^{2+} conductance				No	(Berndt *et al.*, 2011[169]; Kleinlogel *et al.*, 2011)[190]
ChR2 T159C	ChR2 variant with increased photocurrent and light sensitivity				No	(Berndt *et al.*, 2011)[169]
ChromeQ	ChR2 variant with improved photocurrent and reduced Ca^{2+} and H^+ conductance				No	(Cho *et al.*, 2019)[191]
ChloC	Chloride-conducting ChR2 variant (still some H^+ conductance)				No	(Wietek *et al.*, 2014)[206]
iChloC	Optimized ChloC with high Cl^- selectivity				No	(Wietek *et al.*, 2015)[207]

(Continued)

TABLE 9.3 *(Continued)*
Commonly Used Natural Microbial Rhodopsins and Engineered Variants

Name(s)	Description	Ion	Activation Wavelength (nm)	Function	Used in Mitochondria	Reference
C1V1	Chimera of ChR1 and VChR1 with high light sensitivity	CCR	540	Depolarization	No	(Yizhar *et al.*, 2011)[239]
ReaChR	Chimera of ChEF, VChR1, and VChR2	CCR	590	Depolarization	No	(Lin *et al.*, 2013)[192]
bReaChES	ReaChR variant with improved trafficking and kinetics				No	(Rajasethupathy *et al.*, 2015)[201]
C1C2	Chimera of *Cr*ChR1 and *Cr*ChR2 with large photocurrent and small desensitization	CCR	470	Depolarization	No	(Kato *et al.*, 2012)[161]
iC1C2	Chloride-conducting C1C2 variant (still some H^+ conductance)				No	(Berndt *et al.*, 2014)[170]
iC++	Optimized iC1C2 with high Cl^- selectivity				No	(Berndt *et al.*, 2016)[171]
SwiChR++	Optimized iC++ with higher photocurrents				No	(Berndt *et al.*, 2016)[171]
***Gt*ACR1**	Natural anion channelrhodopsin-1 from *Guillardia theta* with high photocurrent	ACR	515	Hyperpolarization/ Depolarization	No	(Govorunova *et al.*, 2015)[177]
FLASH	*Gt*ACR1 variant with fast kinetics and large photocurrent				No	(Kato *et al.*, 2018)[188]
ACR1 2.0	*Gt*ACR1 variant with large photocurrents and increased expression and plasma membrane trafficking in plants				No	(Zhou *et al.*, 2021b)[157]
***Gt*ACR2**	Natural anion channelrhodopsin-2 from *Guillardia theta* with faster kinetics than *Gt*ACR1	ACR	470	Hyperpolarization/ Depolarization	No	(Govorunova *et al.*, 2015)[177]
***Al*ACR1**	Natural anion channelrhodopsin from *Aurantiochytrium* sp. KH105 (rubyACR)	ACR	590	Hyperpolarization/ Depolarization	No	(Govorunova *et al.*, 2020)[178]
***Hc*KCR1**	Kalium channelrhodopsin from *Hyphochytrium catenoides* with rapid kinetics and robust photocurrent	CCR (K^+)	540	Hyperpolarization	No	(Govorunova *et al.*, 2021)[179]

Notes: Outward proton pumps are shaded in red, inward proton pumps in green, inward chloride pumps in yellow, outward sodium pumps in purple, cation channelrhodopsins in blue, and anion channelrhodopsins in orange.

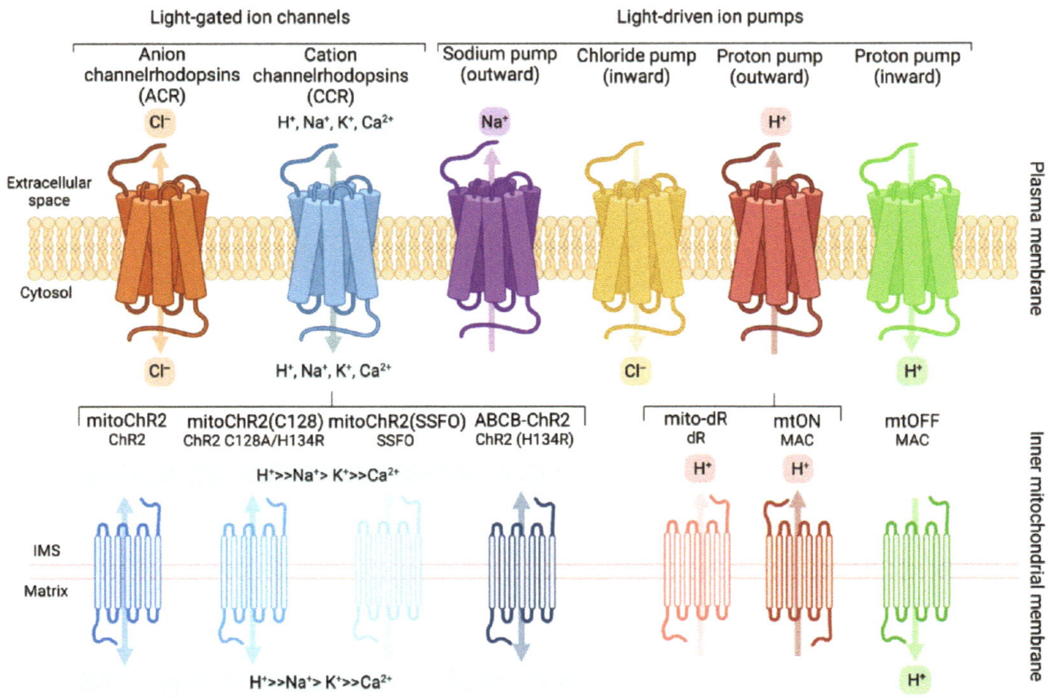

FIGURE 9.4 Types of microbial rhodopsins and variants used for mitochondrial optogenetic control. Upper part, a scheme of the different types of light-gated ion channels and light-driven ion pumps available. Lower part, channels and pumps used in mitochondrial applications: *Wild-type* and different variants of the cation channelrhodopsin 2 (ChR2) (ChR2 C128A/H134R, SSFO, ChR2 H134R) and two proton pumps (MAC in two different orientations and dR). (Created with BioRender.com.)

halorhodopsin (HR) that has been used *in vitro* and *in vivo* in primates for optogenetic control of neural activity[3,214,215]. Hyperpolarization can also be triggered by the use of the outward sodium pump of *Krokinobacter eikastus* KR2[186], as shown by its use in rat hippocampal neurons[216]. It is important to note that, depending on the orientation of the pump within the membrane, ion transport will take place in one or another direction. This means that the function of pumps employed for optogenetic approaches (hyperpolarization or depolarization) directly depends on their topology at the membrane. Therefore, preliminary experiments to analyze the pump orientation are highly recommended at the initial stages of an experiment[158].

Unlike proton pumps, channelrhodopsins (ChRs) passively transport ions along the electrochemical gradient, thereby dissipating energy[167]. Light absorption triggers the opening of the channel resulting in the bidirectional transport of ions. This multiple ion permeation makes channels more efficient than pumps. Two types of channelrhodopsins can be distinguished: Non-selective cation channels (CCR) and non-selective anion channels (ACR)[167]. Channelrhodopsin 1 (ChR1) and Channelrhodopsin 2 (ChR2) from *Chlamydomonas reinhardtii* were the first discovered cation-channelrhodopsins[198,199]. Not surprisingly, ChR2 has increasingly found application as cation-channelrhodopsins due to its fast response to blue light[217]. A great number of CCRs were since identified or engineered, displaying different ion selectivities, kinetics, photocurrent, and response spectra. However, *in vivo* applications would strongly benefit from activation using light of the red portion of the spectrum, as red light is far less toxic compared to light of shorter wavelength, and it also penetrates deeper into tissue than blue light, which is absorbed, for example, by hemoglobin or melanin[218]. Therefore, there is a quest to develop bathychromatically shifted opsins which absorb light of longer wavelengths than the original opsin. Engineering strategies to obtain "bathy-opsins" commonly involve mutagenesis and alteration of the proteinaceous chromophore environment, to match the therapeutic window of tissue light penetration. For instance, the recently engineered channelrhodopsin NeoR, with a maximum absorption peak at 690 nm, might enable non-invasive therapies[219]. Other red-shifted variants include, for example, the chimera ReaChR, C1V1, or ChrimsonR. The latter was recently used for partially restoring vision in a blind patient[220]. ChR2 kinetics were also modified to generate variants, like C128[169], which has long open states, or ChETA[181], with very fast response kinetics. Protein engineering has also yielded opsins with altered ion selectivity, as in the case of CatCh, where Ca^{2+} and Mg^{2+} conductance has been improved[190]. Another example of an engineered opsin is ChR2-XXL, which has lower retinal requirements, making it a good candidate for opsin use in plants, as shown in Reyer *et al.*[154].

Recently, kalium channelrhodopsins (KCR) were reported for the first time[179]. These light-gated cation channels have a 20-fold higher selectivity for K^+ than for Na^+. This new family of cation channelrhodopsins open up perspectives for the treatment of different diseases, such as epilepsy or mitochondrial channelopathies found in aging in which K^+ signaling is affected[221].

The opening of anion channelrhodopsins generates chloride fluxes that can result in hyperpolarization or depolarization depending on the existence of a high or low chloride gradient, respectively. These chloride gradients vary depending on the cell type and the organism. In 2014, two new ChR2 variants highly permeable to chloride were engineered, ChloC from ChR2[206] and iC1C2 from the C1C2 chimera[170]. Natural ACRs have some advantages in comparison to ACR derived from ChR2 since engineered ACRs still transport some cations and are sometimes pH-dependent. The natural *Gt*ACR1 and *Gt*ACR2 were discovered from *Guillardia theta* in 2015[177]. *Gt*ACR1 has been successfully used in plants for growth regulation[156,157].

Despite the difference in the number of ions transported by channels and pumps, one needs to consider further aspects when selecting the ideal candidate for an optogenetic experiment from the extensive toolkit of available opsins. Characteristics that deem particular attention involve their function, spectral sensitivity, response kinetics, and photocurrent[195,222]. Molecular engineering efforts supported by structural information enabled the tuning of opsins for their ion selectivity. Depending on the transported ion and the opsin orientation within the membrane (in the case of pumps), opsin activation either leads to hyperpolarization or depolarization. An inward flow of cations or an outward flow of anions causes depolarization, whereas the opposite transport will trigger hyperpolarization of the membranes.

The maximum excitation peak of the different opsins ranges from 445 to 690 nm. Utilizing opsins with different spectra allows multichannel optical control, as demonstrated by Vierock *et al.*[223], who combined a blue anion channel, *Gt*ACR2, with a red cation channel, Chrimson, to bidirectionally control the activity of a population of neurons. When combining different rhodopsins, one has to consider that red-shifted variants respond with a significant activity to blue light. One of the best choices, when using more than one opsin, is currently Chrimson-SA, a red-shifted mutant of Chrimson (*Cn*ChR1)[224]. Moreover, the different rhodopsins also vary in light sensitivity, which influences the photocurrent of the opsin[195].

Another important aspect that has to be contemplated is the kinetics of the light-activated proteins, which determines how fast the membrane potential changes, which therefore directly influences the possible temporal resolution.

The repertoire of opsins continues to expand as the redshift engineering is still in progress and increasingly peculiar rhodopsins are discovered, as for instance OLPVRII, a pentameric light-gated ion channel encoded by a virus[225], or enzyme-rhodopsins, opsins that modulate, e.g., cyclic nucleotide levels due to their fusion to one of the following catalytic domains: histidine kinases, guanylyl cyclases, or phosphodiesterases[226]. The exciting field of opsin-based optogenetics can look forward to the utilization of an increasing number of opsins with particular properties. These will likely expand the potential of the field beyond their use in regulating neuronal activities in the future.

Opsin-Based Optogenetics in Mitochondria

ATP and ROS production, mitochondrial protein import, calcium homeostasis, oxygen consumption, and many signaling pathways depend on the state of the mitochondrial membrane potential[227,228]. For this reason, precise manipulation of $\Delta\psi_m$ by the use of optogenetic tools opens up the possibility to control cell metabolism and study mitochondrial function in great detail.

Until six years ago, the possibility of using optogenetics to precisely regulate the mitochondrial membrane potential had not been explored. Nowadays, opsins represent a straightforward tool for use within the inner mitochondrial membrane to modify the membrane potential. The first step in developing such a system requires targeting the opsin to the inner mitochondrial membrane rather than to the plasma membrane, their natural homing environment. The following targeting peptides have been used for this purpose: one or four repeats of the mitochondria-targeting sequence of subunit VIII of human cytochrome *c* oxidase (cox-8)[25,30], the mitochondrial leading sequence (MLS) of ABCB10[26,229], 144 N-terminal amino acids of the rat SDHC1 protein[29], and the mitochondrial targeting sequence (MTS) of IMMT1[28,159].

Once the channel or pump has been correctly targeted to the inner mitochondrial membrane, changes in mitochondrial membrane potential due to opsin activation can be monitored, for example, with $\Delta\psi_m$-dependent fluorescent dyes, such as tetramethylrhodamine methyl ester perchlorate (TMRM), tetramethylrhodamine ethyl ester perchlorate (TMRE), MitoView 633, JC-1, or Rhod123. Up to date, TMRE and TMRM have been used in opsin-based optogenetic studies in mitochondria[25,26,28–30,159,229]. These two lipophilic cationic dyes undergo a potential-dependent accumulation in mitochondria: intact or hyperpolarized mitochondria accumulate more dye than depolarized mitochondria. The behavior of these dyes directly depends on their concentration. Under quenching mode conditions, when high concentrations (50–100 nM) are used, the dye forms aggregates in the matrix and inner membrane of intact or hyperpolarized mitochondria, resulting in the quenching of fluorescence (auto-quenching). Similarly, depolarization causes the dye to diffuse into the cytoplasm, lowering the concentration within the mitochondria and preventing aggregate formation, thus increasing the fluorescence signal. Low, non-quenching concentrations (0.5–30 nM) are not sufficient to produce dye aggregation, even in hyperpolarized mitochondria.

In conclusion, non-quenching mode fluorescence intensity increases upon polarization, since it is proportional to dye accumulation in the inner membrane of mitochondria, whereas quenching mode fluorescence signals increase upon depolarization. The addition of uncouplers that dissipate $\Delta\psi_m$ at the end of the experiment, such as FCCP or CCCP, assists in reliably determining the behavior of the dye[230,231].

Opsin activation and assays of dynamic mitochondrial membrane potential changes have been performed either in mammalian and *Drosophila* cells[25,26,30,31] or isolated mitochondria[28,29,159]. Dye accumulation can be detected by fluorescence microscopy and bulk-quantified using fluorescent plate readers. Dye excitation profiles often overlap with the spectra of opsins. Therefore, decreasing the energy of the excitation source and the frequency of image acquisition might be necessary in these cases to avoid undesired opsin activation.

Mitochondrial Hyperpolarization and Energization

Mitochondria not only provide energy to the cell but also engage in regulating key cell functions, such as apoptosis. However, the lack of reversible tools for modulating mitochondrial function has severely hampered the study of this organelle and its associated diseases in the past. Optogenetics offers ways to increase the electrochemical gradient at membranes reversibly. During the last few years, opsins have increasingly been used to artificially generate a proton motive force, leading to an "energized-mitochondrium". Two different proton pumps were used for the hyperpolarization of this organelle: the deltarhodopsin (dR) from *Haloterrigena* sp. *Arg-4*[30] and the fungal light-driven proton pump MAC[28,159].

The first mitochondria hyperpolarizing tool, mito-dR, was generated by Imai[30]. It is based on the genetic fusion of a single COX8A targeting sequence to the dR proton pump, which, upon green light illumination, transports protons out of the mitochondrion. They reported that activation of mito-dR boosts mitochondrial functions such as ATP production or Ca^{2+} metabolism while decreasing the generation of ROS. One of the main causes of Parkinson's disease is a defective mitochondrial respiratory chain. In a Parkinson's disease *Drosophila* sp. model, activation of the mito-dR pump decreased the effects of the disease (reduced alpha-synuclein accumulation, dopaminergic neuron loss, degeneration of mitochondrial cristae, and brain lipid peroxidation were reported), resulting in improved locomotor activity.

A second strategy to hyperpolarize the mitochondrial membrane was devised by Berry *et al.* using the light-driven proton pump from *Leptosphaeria maculans* (MAC)[159]. The authors fused the proton pump to a mitochondrial targeting sequence (MTS) to localize it to mitochondria. They made use of one transmembrane domain of MIC60/IMMT1, resulting in the so-called mtON tool. mtON transports protons out of the mitochondria upon 555 nm light absorption, which leads to an increase in the proton motive force (PMF) and thus, the pH of the mitochondrial matrix. As expected, the activation of mtON increased ATP synthesis independent of oxygen consumption in *C. elegans*. mtON could drive the synthesis of ATP even when mitochondrial electron-transport-chain inhibitors of Complex I, III, or IV were added. In *C. elegans*, they showed that mtON can alter metabolic pathways such as the sensing of food availability through AMP-activated protein kinase (AMPK) signaling[29]. Amitrano *et al.* further confirmed the performance of mtON in mammalian cells, where it was used to improve $CD8^+$ T cell-based immunotherapy in solid tumors[28]. The immunosuppressive tumor microenvironment has detrimental effects on $CD8^+$ T cells by impeding their energy production. The authors showed that supporting mitochondrial energy metabolism by activation of mtON using 590 nm light increased $CD8^+$ T cell migration and effector function.

In conclusion, optogenetic manipulations of mitochondria hyperpolarization enable the regulation of the PMF and matrix pH, and improve ATP synthesis and calcium metabolism.

Mitochondria Depolarization and "De-Energization"

Light-dependent energization of mitochondria through innovative optogenetic solutions represents a promising starting point for the study and treatment of diseases derived from mitochondrial dysfunction. However, optogenetic dissipation of the PMF likewise serves as a tool for gaining a deeper understanding of mitochondrial activity and function.

Mitochondrial membrane depolarization through optogenetic methods that initiate PMF dissipation has been successfully achieved by Tkatch *et al.*[25], Ernst *et al.*[26], and Berry *et al.*[159] by using the light-gated ion channel ChR2[25,26] or the proton pump MAC[29]. Tkatch *et al.* used ChR2 (mitoChR2) and two ChR2 variants with long open states: ChR2 C128A/H134R (mitoChR2 C128) and ChR2 SSFO (mitoChR2(SSFO)[25]. Targeted to the inner mitochondrial membrane *via* four repeats of the COX8A-derived MTS and the deletion of the first 24 residues of ChR2, both variants effectively decreased the mitochondrial membrane potential of mammalian cells quantitatively. Pulses of high light intensity (10 or 22 mW/mm^2) caused a stronger depolarization than lower light intensities (0.18 or 3.1 mW/mm^2). Moreover, they described how the channel can be switched off with long pulses of orange light (595 nm), suggesting that mitoChR2 causes a weaker depolarization and has a faster recovery than the two ChR2 variants. The activation of the channel by 10 s of blue light illumination of 2 mW/mm^2 caused a decrease in Ca^{2+} uptake and an increase in O_2 consumption. The use of mitoChR2(SSFO) could further suppress the spontaneous beating of cardiomyocytes in cell culture and prevent ATP production in pancreatic β-cells that were exposed to high glucose concentrations. Tkatch and collaborators also demonstrated the high spatial resolution achieved by the optogenetic approach. They depolarized a restricted group of mitochondria within a single cell by using a laser illumination source[25].

A similar approach was devised by Ernst *et al.* with the ChR2(H134R) variant fused to the ABCB10 MTS to generate ABCB-ChR2, which is capable of regulating cell fate decision and function[26,229]. Sustained light illumination for 24 h with 0.5 mW/mm^2 blue light caused cell death *via* apoptosis and mitophagy. Due to these findings, the authors concluded that IMM permeabilization itself can trigger apoptosis. Although increasing light intensities and exposure time have a detrimental effect on cell viability, mitochondrial pretreatment with 2 h of low-intensity (0.2 mW/mm^2) blue light showed a cytoprotective effect and could prevent cell death.

Berry and colleagues reused the MAC proton pump placed in the inverted direction to generate a depolarization tool called mtOFF. They fused MAC to the 144 N-terminal residues of the rat SDHC1 protein, which includes the mitochondrial targeting sequence and two transmembrane regions achieving the opposite topology to the mtON version[29]. mtOFF thus adopts an orientation that pumps protons into mitochondria, which induces depolarization upon light illumination. The authors showed, in *C. elegans*, that a light-induced decrease in the proton motive force reduces oxidative damage during oxygen starvation (hypoxia), thereby increasing the survival rate of the worm. In addition, they found that, upon activation of the opsin, oxygen consumption increased by around 70% while ATP production decreased. Food depletion, which represents a high-energy-demand condition in *C. elegans*, increased the AMPK activity, resulting in increased locomotion. In this regard, PMF dissipation through mtOFF activation can increase AMPK activity and locomotion in *C. elegans* independently of food availability.

Unlike hyperpolarizing tools, optogenetic depolarization of mitochondria decreases PMF, ATP production, and calcium uptake and increases oxygen consumption.

PERSPECTIVES AND DISCUSSION

Optogenetic technologies, once devised out of the need to differentially excite neuronal cells in order to study their function, developed at a rapid pace after the first reports appeared. The concept instantly found application in a broad range of cell biological disciplines, likewise benefitting from the excellent precision light stimulation offers. Light-regulated molecular switches hence enable versatile applications not limited to single-cell control but even extending to the regulation of cellular subcompartments. After gene-expression control in the nucleus and signaling pathway modulations at the plasma membrane and in the cytosol saw an extensive asset of optogenetic solutions[22], it is now less surprising to see the technology supporting the field of mitochondria research[158,229]. However, mitochondrial optogenetics is still at its earliest stages, as we are just starting to see how optogenetic switches and strategies transition to the study of mitochondria biology. We expect that new optogenetic approaches will be applied in mitochondria, for example, utilizing multi-color control of specific functions, or tracing and stimulating individual mitochondria within the cellular mitochondrial network. Thanks to these new strategies, complex pathways that, for example, lead to mitophagy can now be studied at high spatiotemporal resolution. By multiplexing the orthogonal optogenetic color palette, one may assist in detangling sequential physiological events at the mitochondrial membrane. Furthermore, by providing temporal control over enzymatic clustering, light-sensitive programming of synthetic mitochondria could boost metabolic engineering efforts.

Throughout this review, we have seen the different optogenetic technologies so far applied to mitochondria. The approaches include both opsin- and non-opsin-based tools, together aiming to better study the metabolism of mitochondria and the various cellular processes mitochondria engage in, such as mitophagy and apoptosis. Opsin-based tools constantly evolve and increasingly mature, enabling precise control of membrane potential. During the last decade, inward proton pumps, outward sodium pumps, potassium channelrhodopsins, and anion channelrhodopsins were identified, enriching the rapidly growing opsin toolbox. These newly identified variants, together with the engineered ones, have expanded the applications of opsin-based optogenetics. Moreover, *in vivo* limitations of blue light were overcome by the engineering of red-shifted variants, such as NeoR[219]. Opsin-based mitochondrial optogenetics successfully applied proton pumps and channelrhodopsins to hyperpolarize or depolarize the inner mitochondrial membrane and assisted in demonstrating the role of the mitochondrial membrane potential in Parkinson's disease[30] or in apoptosis[26,229], among other processes[25,28,29,159]. However, many strategies remain yet to be explored, such as the combination of opsins with different excitation wavelengths to hyper or depolarize mitochondria in a light-dependent manner with a millisecond-resolution, or the use of non-proton selective pumps or channels, such as anion and potassium channels, that do not cause pH changes and could help to shed light on the role of mitochondrial channelopathies in aging[221].

Reports of enzyme rhodopsins open another avenue for developing novel strategies of opsin-based optogenetics. Application of these opsins could lead to the control of biological functions by converting light inputs into enzyme functions. For instance, a fungi rhodopsin-guanylyl cyclase (RhGC) described by Scheib *et al.* is capable of producing cyclic guanosine monophosphate (cGMP) in mammalian neurons[232]. One could speculate about the potential uses of these enzyme rhodopsins to treat diseases such as hypertension, upon rebalancing the cGMP pathway at the outer mitochondrial membrane[233]. In terms of light-assisted protein inactivation, mitochondria research is still in its early stages. The promising ROS generators SOPP, SOPP2, and SOPP3 with their outstanding high $^1O_2$ quantum yields are not yet applied in CALI studies.

Finally, optogenetic transcriptional control and genome editing technologies, both targeting nuclear and mitochondrial genes, represent methods that have not yet been applied to the study of mitochondria. While genome editing using TALEN and CRISPR/Cas9-based methods has already yielded new insights into mitophagy and hypoxia[234], optogenetic control would enable the study of individual cells and mitochondria, offering a venue for controlling complex processes pertaining to the organelle. Together with the potential of ROS generators and the steady evolution of versatile methods such as OptoNBs, we foresee a bright future for the field.

BOX 9.2 WHAT'S HOLDING BACK OPTOGENETICS?

Optogenetic techniques are constantly shedding fresh light on numerous dynamic cell processes, even though they have only been available for a relatively short time. However, implementing optogenetic approaches requires sophisticated instruments and a suitable light-shielded experimental environment. Most optogenetic experiments can be performed simply by using fluorescent light microscopes. However, to achieve true spatial control, equipment such as laser-scanning or digital micromirror devices are needed. Another experimental limitation of using optogenetic tools is that the spatial precision with which an area of interest can be exposed to light in a microscope is physically limited by diffraction. Photochemistry kinetics further restrict the accuracy of intracellular photoactivation[2]. Thermal decay causes blue light–sensitive regions to revert to the dark state, and the pace at which this happens is crucial for generating a crisp photoactivation boundary. Slow dark state reversal permits photoactivated molecules to diffuse deeper, allowing the boundary between activated and non-activated regions to be widened. The *As*LOV2 photoactivated state has a half-life of about 5–50 s, which is long enough to form intracellular gradients of easily diffusible molecules[235]. On the other hand, the half-life of photoactivated CRY2 is in the range of minutes, which is still fast enough to form gradients if diffusion is limited. Furthermore, fast dark state reversal requires frequent exposure to photoactivation light, which can cause phototoxicity even at low intensities. Bennedetti *et al.* provide a detailed comparison of the three main blue light–sensitive dimerizers and their performance in live cells. Magnets produced the best spatial confinement in comparison with iLID/nano and CRY2/CIBN. The authors imply that this is because both parts of the Magnet dimerizer must be photoactivated, restricting interaction within the light-exposed region[236]. A more "stable" system is PhyB, which photoconverts between two very stable states in mammalian cells using either red or infrared light. Longer wavelengths are less harmful to cells and penetrate deeper into biological samples. However, perfect spatial control of the PhyB/PIF interaction necessitates complementary 650–670 and 740–760 nm illumination patterns[13,235].

The recently published BICYCL systems enable tight spatial and temporal control due to their bi-stable nature. However, the requirement for a green light may be a drawback for *in vivo* applications. The switch also requires the external addition of chromophores (PCB) to function, which could be improved by further engineering to enhance phycocyanobilin and/or biliverdin absorption or by simply using endogenous biliverdin[101].

Another point to keep in mind is that the inactive and active states of photosensors are in a chemical equilibrium, thus dark is never completely dark. For instance, even in the dark, a very low percentage of LOV2 molecules (1.5%) are still activated, with a dissociated Jα-helix[237], which means individual LOV2 molecules have a limited dynamic range. This is to a greater degree the case for Phys, due to the spectral overlap of the two forms (Pfr/Pr) in the red region.

A major challenge arises when combining optogenetics and fluorescence microscopy because of overlapping excitation spectra, as this may cause unwanted photoactivation. For example, blue light–sensitive domains can often only be combined with red or far-red fluorescent proteins[235]. For PhyB, both blue and red/far-red light cause its photoconversion, making YFP one of the only possible combinations[2,238].

ACKNOWLEDGMENTS

This study was supported in part by the Deutsche Foschungsgemeinschaft (DFG, German Research Foundation) under the Germany's Excellence Strategy (CEPLAS – EXC-2048 Project No. 390686111), NEXTPlant (Project ID 391465903/GRK 2466), and by the Collaborative Research Centers SFB1208 (Project No. 267205415) and SFB1535 (Project No. 458090666) to MDZ, by Human Frontiers Scientific Program HFSP RGP 0067 Grant to MDZ, the European Commission – Research Executive Agency (H2020 Future and Emerging Technologies [FET-Open]) Project ID 801041 CyGenTiG to SS, HMB, and MDZ, and MN and HMB were supported by the "Freigeist" Fellowship of the Volkswagen Foundation.

REFERENCES

1. Deisseroth, K. Optogenetics. *Nat. Methods* **8**, 26–29 (2011).
2. Wittmann, T., Dema, A. & Van Haren, J. Lights, Cytoskeleton, Action: Optogenetic Control of Cell Dynamics. *Curr. Opin. Cell Biol.* **66**, 1–10 (2020).
3. Rost, B. R., Schneider-Warme, F., Schmitz, D. & Hegemann, P. Optogenetic Tools for Subcellular Applications in Neuroscience. *Neuron* **96**, 572–603 (2017).
4. Ankenbruck, N., Courtney, T., Naro, Y. & Deiters, A. Optochemical Control of Biological Processes in Cells and Animals. *Angew. Chem. Int. Ed.* **57**, 2768–2798 (2018).
5. Losi, A., Gardner, K. H. & Möglich, A. Blue-Light Receptors for Optogenetics. *Chem. Rev.* **118**, 10659–10709 (2018).
6. Caldwell, R. M. *et al.* Optochemical Control of Protein Localization and Activity within Cell-like Compartments. *Biochemistry* **57**, 2590–2596 (2018).
7. Bartelt, S. M., Steinkühler, J., Dimova, R. & Wegner, S. V. Light-Guided Motility of a Minimal Synthetic Cell. *Nano Lett.* **18**, 7268–7274 (2018).
8. Glantz, S. T. *et al.* Functional and Topological Diversity of LOV Domain Photoreceptors. *Proc. Natl. Acad. Sci.* **113** (11) E1442–E1451, (2016).
9. Glantz, S. T. *et al.* Directly Light-Regulated Binding of RGS-LOV Photoreceptors to Anionic Membrane Phospholipids. *Proc. Natl. Acad. Sci.* **115** (33), E7720–E7727, (2018).
10. Govorunova, E. G. *et al.* The Expanding Family of Natural Anion Channelrhodopsins Reveals Large Variations in Kinetics, Conductance, and Spectral Sensitivity. *Sci. Rep.* **7**, 43358 (2017).
11. Conrad, K. S., Bilwes, A. M. & Crane, B. R. Light-Induced Subunit Dissociation by a Light–Oxygen–Voltage Domain Photoreceptor from *Rhodobacter sphaeroides*. *Biochemistry* **52**, 378–391 (2013).

12. Ohlendorf, R., Schumacher, C. H., Richter, F. & Möglich, A. Library-Aided Probing of Linker Determinants in Hybrid Photoreceptors. *ACS Synth. Biol.* **5**, 1117–1126 (2016).
13. Guntas, G. *et al.* Engineering an Improved Light-Induced Dimer (iLID) for Controlling the Localization and Activity of Signaling Proteins. *Proc. Natl. Acad. Sci.* **112**, 112–117 (2015).
14. Kawano, F., Suzuki, H., Furuya, A. & Sato, M. Engineered Pairs of Distinct Photoswitches for Optogenetic Control of Cellular Proteins. *Nat. Commun.* **6**, 6256 (2015).
15. Wang, H. *et al.* LOVTRAP: An Optogenetic System for Photoinduced Protein Dissociation. *Nat. Methods* **13**, 755–758 (2016).
16. Dagliyan, O. *et al.* Engineering Extrinsic Disorder to Control Protein Activity in Living Cells. *Science* **354**, 1441–1444 (2016).
17. Diensthuber, R. P., Bommer, M., Gleichmann, T. & Möglich, A. Full-Length Structure of a Sensor Histidine Kinase Pinpoints Coaxial Coiled Coils as Signal Transducers and Modulators. *Structure* **21**, 1127–1136 (2013).
18. Kim, Y. S. *et al.* Crystal Structure of the Natural Anion-Conducting Channelrhodopsin GtACR1. *Nature* **561**, 343–348 (2018).
19. Polesskaya, O. *et al.* Optogenetic Regulation of Transcription. *BMC Neurosci.* **19**, 12 (2018).
20. Gangemi, C. G. & Janovjak, H. *Modular Light-Emitting Diode Shelving Systems for Scalable Optogenetics.* https://www.researchsquare.com/article/rs-2801275/v1 (2023). https://doi.org/10.21203/rs.3.rs-2801275/v1.
21. Harmer, Z. P. & McClean, M. N. Lustro: High-Throughput Optogenetic Experiments Enabled by Automation and a Yeast Optogenetic Toolkit. *ACS Synth. Biol.* 12, 1943–1951 (2023).
22. Hannanta-Anan, P., Glantz, S. T. & Chow, B. Y. Optically Inducible Membrane Recruitment and Signaling Systems. *Curr. Opin. Struct. Biol.* **57**, 84–92 (2019).
23. Abate, M. *et al.* Mitochondria as Playmakers of Apoptosis, Autophagy and Senescence. *Semin. Cell Dev. Biol.* **98**, 139–153 (2020).
24. Kichuk, T. C., Carrasco-López, C. & Avalos, J. L. Lights up on Organelles: Optogenetic Tools to Control Subcellular Structure and Organization. *WIREs Mech. Dis.* **13**, e1500 (2021). https://doi.org/10.1002/wsbm.1500.
25. Tkatch, T. *et al.* Optogenetic Control of Mitochondrial Metabolism and Ca^{2+} Signaling by Mitochondria-Targeted Opsins. *Proc. Natl. Acad. Sci.* **114** (26), E5167–E5176 (2017).
26. Ernst, P. *et al.* Precisely Control Mitochondria with Light to Manipulate Cell Fate Decision. *Biophys. J.* **117**, 631–645 (2019).
27. Hughes, R. M. *et al.* Optogenetic Apoptosis: Light-Triggered Cell Death. *Angew. Chem. Int. Ed.* **54**, 12064–12068 (2015).
28. Amitrano, A. M. *et al.* Optical Control of $CD8^+$ T Cell Metabolism and Effector Functions. *Front. Immunol.* **12**, 666231 (2021).
29. Berry, B. J., Baldzizhar, A., Nieves, T. O. & Wojtovich, A. P. Neuronal AMPK Coordinates Mitochondrial Energy Sensing and Hypoxia Resistance in *C. elegans*. *FASEB J.* **34**, 16333–16347 (2020).
30. Imai, Y. *et al.* Light-Driven Activation of Mitochondrial Proton-Motive Force Improves Motor Behaviors in a Drosophila Model of Parkinson's Disease. *Commun. Biol.* **2**, 424 (2019).
31. Chen, S. Y. *et al.* Optogenetic Control Reveals Differential Promoter Interpretation of Transcription Factor Nuclear Translocation Dynamics. *Cell Syst.* **11**, 336–353.e24 (2020).
32. Shi, F. *et al.* Optogenetic Control of Endoplasmic Reticulum–Mitochondria Tethering. *ACS Synth. Biol.* **7**, 2–9 (2018).
33. D'Acunzo, P. *et al.* Reversible Induction of Mitophagy by an Optogenetic Bimodular System. *Nat. Commun.* **10**, 1533 (2019).
34. Ye, H., Baba, M. D.-E., Peng, R.-W. & Fussenegger, M. A Synthetic Optogenetic Transcription Device Enhances Blood-Glucose Homeostasis in Mice. *Science* **332**, 1565–1568 (2011).
35. Airan, R. D., Thompson, K. R., Fenno, L. E., Bernstein, H. & Deisseroth, K. Temporally Precise In Vivo Control of Intracellular Signalling. *Nature* **458**, 1025–1029 (2009).
36. Zheng, W. *et al.* Spatiotemporal Control of GPR37 Signaling and Its Behavioral Effects by Optogenetics. *Front. Mol. Neurosci.* **11**, 95 (2018).
37. Taslimi, A. *et al.* An Optimized Optogenetic Clustering Tool for Probing Protein Interaction and Function. *Nat. Commun.* **5**, 4925 (2014).
38. Bugaj, L. J., Choksi, A. T., Mesuda, C. K., Kane, R. S. & Schaffer, D. V. Optogenetic Protein Clustering and Signaling Activation in Mammalian Cells. *Nat. Methods* **10**, 249–252 (2013).
39. Lee, S. *et al.* Reversible Protein Inactivation by Optogenetic Trapping in Cells. *Nat. Methods* **11**, 633–636 (2014).

40. Zhou, X. X., Chung, H. K., Lam, A. J. & Lin, M. Z. Optical Control of Protein Activity by Fluorescent Protein Domains. *Science* **338**, 810–814 (2012).
41. Wu, Y. I. *et al.* A Genetically Encoded Photoactivatable Rac Controls the Motility of Living Cells. *Nature* **461**, 104–108 (2009).
42. Schmidt, D., Tillberg, P. W., Chen, F. & Boyden, E. S. A Fully Genetically Encoded Protein Architecture for Optical Control of Peptide Ligand Concentration. *Nat. Commun.* **5**, 3019 (2014).
43. Kaberniuk, A. A., Shemetov, A. A. & Verkhusha, V. V. A Bacterial Phytochrome-Based Optogenetic System Controllable with Near-Infrared Light. *Nat. Methods* **13**, 591–597 (2016).
44. Kennedy, M. J. *et al.* Rapid Blue-Light–Mediated Induction of Protein Interactions in Living Cells. *Nat. Methods* **7**, 973–975 (2010).
45. Pudasaini, A. *et al.* Kinetics of the LOV Domain of ZEITLUPE Determine its Circadian Function in Arabidopsis. *eLife* **6**, e21646 (2017).
46. O'Neill, P. R., Kalyanaraman, V. & Gautam, N. Subcellular Optogenetic Activation of Cdc42 Controls Local and Distal Signaling to Drive Immune Cell Migration. *Mol. Biol. Cell* **27**, 1442–1450 (2016).
47. Strickland, D. *et al.* TULIPs: Tunable, Light-Controlled Interacting Protein Tags for Cell Biology. *Nat. Methods* **9**, 379–384 (2012).
48. Levskaya, A., Weiner, O. D., Lim, W. A. & Voigt, C. A. Spatiotemporal Control of Cell Signalling Using a Light-Switchable Protein Interaction. *Nature* **461**, 997–1001 (2009).
49. Huang, P., Zhao, Z. & Duan, L. Optogenetic Activation of Intracellular Signaling Based on Light-Inducible Protein-Protein Homo-Interactions. *Neural Regen. Res.* **17**, 25 (2022).
50. Kolar, K., Knobloch, C., Stork, H., Žnidarič, M. & Weber, W. OptoBase: A Web Platform for Molecular Optogenetics. *ACS Synth. Biol.* **7**, 1825–1828 (2018).
51. Herrou, J. & Crosson, S. Function, Structure and Mechanism of Bacterial Photosensory LOV Proteins. *Nat. Rev. Microbiol.* **9**, 713–723 (2011).
52. Briggs, W. R., Christie, J. M. & Salomon, M. Phototropins: A New Family of Flavin-Binding Blue Light Receptors in Plants. *Antioxid. Redox Signal.* **3**, 775–788 (2001).
53. Crosson S., Rajagopal S., & Moffat K. The LOV Domain Family: Photoresponsive Signaling Modules Coupled to Diverse Output Domains. *Biochemistry* **42**, 2–10 (2003). https://doi.org/10.1021/bi026978l
54. Strickland, D., Moffat, K. & Sosnick, T. R. Light-Activated DNA Binding in a Designed Allosteric Protein. *Proc. Natl. Acad. Sci.* **105**, 10709–10714 (2008).
55. Mills, E., Chen, X., Pham, E., Wong, S. & Truong, K. Engineering a Photoactivated Caspase-7 for Rapid Induction of Apoptosis. *ACS Synth. Biol.* **1**, 75–82 (2012).
56. Hope, J. M., Liu, A., Calvin, G. J. & Cui, B. Construction of Light-Activated Neurotrophin Receptors Using the Improved Light-Induced Dimerizer (iLID). *J. Mol. Biol.* **432**, 3739–3748 (2020).
57. Grusch, M. *et al.* Spatio-Temporally Precise Activation of Engineered Receptor Tyrosine Kinases by Light. *EMBO J.* **33**, 1713–1726 (2014).
58. Humphreys, P. A. *et al.* Optogenetic Control of the BMP Signaling Pathway. *ACS Synth. Biol.* **9**, 3067–3078 (2020).
59. Gama, A. R., Miller, T., Lange, J. J., Unruh, J. & Halfmann, R. A Nucleation Barrier Spring-Loads the CBM Signalosome for Binary Activation. eLife 11, e79826 (2022).
60. Berlew, E. E. *et al.* Designing Single-Component Optogenetic Membrane Recruitment Systems: The Rho-Family GTPase Signaling Toolbox. *ACS Synth. Biol.* **11**, 515–521 (2022).
61. Zoltowski, B. D. *et al.* Conformational Switching in the Fungal Light Sensor Vivid. *Science* **316**, 1054–1057 (2007).
62. Wang, H. & Hahn, K. M. LOVTRAP: A Versatile Method to Control Protein Function with Light. *Curr. Protoc. Cell Biol.* **73**, (2016).
63. Wehler, P., Niopek, D., Eils, R. & Di Ventura, B. Optogenetic Control of Nuclear Protein Import in Living Cells Using Light-Inducible Nuclear Localization Signals (LINuS). *Curr. Protoc. Chem. Biol.* **8**, 131–145 (2016).
64. Niopek, D., Wehler, P., Roensch, J., Eils, R. & Di Ventura, B. Optogenetic Control of Nuclear Protein Export. *Nat. Commun.* **7**, 10624 (2016).
65. Stone, O. J. *et al.* Optogenetic Control of Cofilin and αTAT in Living Cells Using Z-Lock. *Nat. Chem. Biol.* **15**, 1183–1190 (2019).
66. Geng, L., Shen, J. & Wang, W. Circularly Permuted AsLOV2 as an Optogenetic Module for Engineering Photoswitchable Peptides. *Chem. Commun.* **57**, 8051–8054 (2021).
67. He, L. *et al.* Circularly Permuted LOV2 as a Modular Photoswitch for Optogenetic Engineering. *Nat. Chem. Biol.* **17**, 915–923 (2021).
68. Nash, A. I. *et al.* Structural Basis of Photosensitivity in a Bacterial Light-Oxygen-Voltage/Helix-Turn-Helix (LOV-HTH) DNA-Binding Protein. *Proc. Natl. Acad. Sci.* **108**, 9449–9454 (2011).

69. Rivera-Cancel, G., Motta-Mena, L. B. & Gardner, K. H. Identification of Natural and Artificial DNA Substrates for Light-Activated LOV–HTH Transcription Factor EL222. *Biochemistry* **51**, 10024–10034 (2012).
70. Motta-Mena, L. B. *et al.* An Optogenetic Gene Expression System with Rapid Activation and Deactivation Kinetics. *Nat. Chem. Biol.* **10**, 196–202 (2014).
71. Gligorovski, V., Sadeghi, A. & Rahi, S. J. Multidimensional Characterization of Inducible Promoters and a Highly Light-Sensitive LOV-Transcription Factor. *Nat. Commun.* **14**, 3810 (2023).
72. Blomeier, T. *et al.* Blue Light-Operated CRISPR/Cas13b-Mediated mRNA Knockdown (Lockdown). *Adv. Biol.* **5**, 2000307 (2021).
73. Jayaraman, P. *et al.* Blue Light-Mediated Transcriptional Activation and Repression of Gene Expression in Bacteria. *Nucleic Acids Res.* **44**, 6994–7005 (2016).
74. Ochoa-Fernandez, R. *et al.* Optogenetic Control of Gene Expression in Plants in the Presence of Ambient White Light. *Nat. Methods* **17**, 717–725 (2020).
75. Cashmore, A. Cryptochromes: Enabling Plants and Animals to Determine Circadian Time. *Cell* **114**, 537–543 (2003).
76. Yu, X., Liu, H., Klejnot, J. & Lin, C. The Cryptochrome Blue Light Receptors. *Arab. Book* **8**, e0135 (2010).
77. Partch, C. L. & Sancar, A. Cryptochromes and Circadian Photoreception in Animals. in *Methods in Enzymology* vol. **393** 726–745 (Elsevier, 2005).
78. Lin, C. *et al.* Association of Flavin Adenine Dinucleotide with the *Arabidopsis* Blue Light Receptor CRY1. *Science* **269**, 968–970 (1995).
79. Shao, K. *et al.* The Oligomeric Structures of Plant Cryptochromes. *Nat. Struct. Mol. Biol.* **27**, 480–488 (2020).
80. Che, D. L., Duan, L., Zhang, K. & Cui, B. The Dual Characteristics of Light-Induced Cryptochrome 2, Homo-oligomerization and Heterodimerization, for Optogenetic Manipulation in Mammalian Cells. *ACS Synth. Biol.* **4**, 1124–1135 (2015).
81. Park, H. *et al.* Optogenetic Protein Clustering through Fluorescent Protein Tagging and Extension of CRY2. *Nat. Commun.* **8**, 30 (2017).
82. Duan, L. *et al.* Optogenetic Control of Molecular Motors and Organelle Distributions in Cells. *Chem. Biol.* **22**, 671–682 (2015).
83. Van Bergeijk, P., Adrian, M., Hoogenraad, C. C. & Kapitein, L. C. Optogenetic Control of Organelle Transport and Positioning. *Nature* **518**, 111–114 (2015).
84. Godwin, W. C., Hoffmann, G. F., Gray, T. J. & Hughes, R. M. Imaging of Morphological and Biochemical Hallmarks of Apoptosis with Optimized Optogenetic Tools. *J. Biol. Chem.* **294**, 16918–16929 (2019).
85. Strauss, H. M., Schmieder, P. & Hughes, J. Light-Dependent Dimerisation in the N-Terminal Sensory Module of Cyanobacterial Phytochrome 1. *FEBS Lett.* **579**, 3970–3974 (2005).
86. Tang, K., Beyer, H. M., Zurbriggen, M. D. & Gärtner, W. The Red Edge: Bilin-Binding Photoreceptors as Optogenetic Tools and Fluorescence Reporters. *Chem. Rev.* **121**, 14906–14956 (2021).
87. Ulijasz, A. T. & Vierstra, R. D. Phytochrome Structure and Photochemistry: Recent Advances Toward a Complete Molecular Picture. *Curr. Opin. Plant Biol.* **14**, 498–506 (2011).
88. Piatkevich, K. D., Subach, F. V. & Verkhusha, V. V. Engineering of Bacterial Phytochromes for Near-Infrared Imaging, Sensing, and Light-Control in Mammals. *Chem. Soc. Rev.* **42**, 3441 (2013).
89. Shimizu-Sato, S., Huq, E., Tepperman, J. M. & Quail, P. H. A Light-Switchable Gene Promoter System. *Nat. Biotechnol.* **20**, 1041–1044 (2002).
90. Müller, K. *et al.* A Red/Far-Red Light-Responsive Bi-stable Toggle Switch to Control Gene Expression in Mammalian Cells. *Nucleic Acids Res.* **41**, e77–e77 (2013).
91. Müller, K. *et al.* A Red Light-Controlled Synthetic Gene Expression Switch for Plant Systems. *Mol BioSyst* **10**, 1679–1688 (2014).
92. Beyer, H. M. *et al.* Red Light-Regulated Reversible Nuclear Localization of Proteins in Mammalian Cells and Zebrafish. *ACS Synth. Biol.* **4**, 951–958 (2015).
93. Chernov, K. G., Redchuk, T. A., Omelina, E. S. & Verkhusha, V. V. Near-Infrared Fluorescent Proteins, Biosensors, and Optogenetic Tools Engineered from Phytochromes. *Chem. Rev.* **117**, 6423–6446 (2017).
94. Gomelsky, M. Special Issue on Synthetic Photobiology. *ACS Synth. Biol.* **3**, 780–781 (2014).
95. Redchuk, T. A. *et al.* Optogenetic Regulation of Endogenous Proteins. *Nat. Commun.* **11**, 605 (2020).
96. Leopold, A. V., Chernov, K. G., Shemetov, A. A. & Verkhusha, V. V. Neurotrophin Receptor Tyrosine Kinases Regulated with Near-Infrared Light. *Nat. Commun.* **10**, 1129 (2019).
97. Redchuk, T. A., Omelina, E. S., Chernov, K. G. & Verkhusha, V. V. Near-Infrared Optogenetic Pair for Protein Regulation and Spectral Multiplexing. *Nat. Chem. Biol.* **13**, 633–639 (2017).

98. Fushimi, K. & Narikawa, R. Cyanobacteriochromes: Photoreceptors Covering the Entire UV-to-Visible Spectrum. *Curr. Opin. Struct. Biol.* **57**, 39–46 (2019).
99. Ikeuchi, M. & Ishizuka, T. Cyanobacteriochromes: A New Superfamily of Tetrapyrrole-Binding Photoreceptors in Cyanobacteria. *Photochem. Photobiol. Sci.* **7**, 1159–1167 (2008).
100. Jang, J., McDonald, S., Uppalapati, M. & Woolley, G. A. *Green, Orange, Red, and Far-red Optogenetic Tools Derived from Cyanobacteriochromes.* (2019). https://doi.org/10.1101/769422.
101. Rockwell, N. C., Martin, S. S. & Lagarias, J. C. Red/Green Cyanobacteriochromes: Sensors of Color and Power. *Biochemistry* **51**, 9667–9677 (2012).
102. Jang, J. *et al.* Engineering of Bidirectional, Cyanobacteriochrome-Based Light-Inducible Dimers (BICYCL)s. *Nat. Methods* **20**, 432–441 (2023).
103. Mühlhäuser, W. W. D., Fischer, A., Weber, W. & Radziwill, G. Optogenetics - Bringing Light into the Darkness of Mammalian Signal Transduction. *Biochim. Biophys. Acta BBA — Mol. Cell Res.* **1864**, 280–292 (2017).
104. Ueda, Y. & Sato, M. Cell Membrane Dynamics Induction Using Optogenetic Tools. *Biochem. Biophys. Res. Commun.* **506**, 387–393 (2018).
105. Meshik, X., O'Neill, P. R. & Gautam, N. Physical Plasma Membrane Perturbation Using Subcellular Optogenetics Drives Integrin-Activated Cell Migration. *ACS Synth. Biol.* **8**, 498–510 (2019).
106. Jones, T., Liu, A. & Cui, B. Light-Inducible Generation of Membrane Curvature in Live Cells with Engineered BAR Domain Proteins. *ACS Synth. Biol.* **9**, 893–901 (2020).
107. Gil, A. A. *et al.* Optogenetic Control of Protein Binding Using Light-Switchable Nanobodies. *Nat. Commun.* **11**, 4044 (2020).
108. Prior, I. A. & Hancock, J. F. Ras Trafficking, Localization and Compartmentalized Signalling. *Semin. Cell Dev. Biol.* **23**, 145–153 (2012).
109. Dakik, P., Medkour, Y., Mohammad, K. & Titorenko, V. I. Mechanisms Through Which Some Mitochondria-Generated Metabolites Act as Second Messengers That Are Essential Contributors to the Aging Process in Eukaryotes Across Phyla. *Front. Physiol.* **10**, 461 (2019).
110. Spinelli, J. B. & Haigis, M. C. The Multifaceted Contributions of Mitochondria to Cellular Metabolism. *Nat. Cell Biol.* **20**, 745–754 (2018).
111. Galluzzi, L. *et al.* Molecular Definitions of Autophagy and Related Processes. *EMBO J.* **36**, 1811–1836 (2017).
112. Fan, Y.-J. & Zong, W.-X. The Cellular Decision between Apoptosis and Autophagy. *Chin. J. Cancer* **32** (3), 121–129 (2012).
113. Ashrafi, G. & Schwarz, T. L. The Pathways of Mitophagy for Quality Control and Clearance of Mitochondria. *Cell Death Differ.* **20**, 31–42 (2013).
114. Hallett, R. A., Zimmerman, S. P., Yumerefendi, H., Bear, J. E. & Kuhlman, B. Correlating *in Vitro* and *in Vivo* Activities of Light-Inducible Dimers: A Cellular Optogenetics Guide. *ACS Synth. Biol.* **5**, 53–64 (2016).
115. Toettcher, J. E., Gong, D., Lim, W. A. & Weiner, O. D. Light Control of Plasma Membrane Recruitment Using the Phy–PIF System. in *Methods in Enzymology* vol. **497** 409–423 (Elsevier, 2011).
116. Milas, A., Jagrić, M., Martinčić, J. & Tolić, I. M. Optogenetic Reversible Knocksideways, Laser Ablation, and Photoactivation on the Mitotic Spindle in Human Cells. in *Methods in Cell Biology* vol. **145** 191–215 (Elsevier, 2018).
117. Bartolák-Suki, E., Imsirovic, J., Nishibori, Y., Krishnan, R. & Suki, B. Regulation of Mitochondrial Structure and Dynamics by the Cytoskeleton and Mechanical Factors. *Int. J. Mol. Sci.* **18**, 1812 (2017).
118. Feng, Q. & Kornmann, B. Mechanical Forces on Cellular Organelles. *J. Cell Sci.* **131**, jcs218479 (2018).
119. Song, Y. *et al.* Light-Inducible Deformation of Mitochondria in Live Cells. *Cell Chem. Biol.* **29**, 109–119.e3 (2022).
120. Harterink, M. *et al.* Light-Controlled Intracellular Transport in *Caenorhabditis elegans*. *Curr. Biol.* **26**, R153–R154 (2016).
121. Xu, S. & Chisholm, A. D. Highly Efficient Optogenetic Cell Ablation in *C. elegans* Using Membrane-Targeted miniSOG. *Sci. Rep.* **6**, 21271 (2016).
122. Williams, D. C. *et al.* Rapid and Permanent Neuronal Inactivation In Vivo via Subcellular Generation of Reactive Oxygen with the Use of KillerRed. *Cell Rep.* **5**, 553–563 (2013).
123. Choubey, V. *et al.* Mutant A53T α-Synuclein Induces Neuronal Death by Increasing Mitochondrial Autophagy. *J. Biol. Chem.* **286**, 10814–10824 (2011).
124. Yang, J.-Y. & Yang, W. Y. Spatiotemporally Controlled Initiation of Parkin-Mediated Mitophagy within Single Cells. *Autophagy* **7**, 1230–1238 (2011).
125. Xie, W. *et al.* Chemoptogenetic Ablation of Neuronal Mitochondria In Vivo with Spatiotemporal Precision and Controllable Severity. *eLife* **9**, e51845 (2020).

126. Trewin, A. J. *et al.* Mitochondrial Reactive Oxygen Species Generated at the Complex-II Matrix or Intermembrane Space Microdomain Have Distinct Effects on Redox Signaling and Stress Sensitivity in *Caenorhabditis elegans. Antioxid. Redox Signal.* **31**, 594–607 (2019).
127. Wojtovich, A. P., Wei, A. Y., Sherman, T. A., Foster, T. H. & Nehrke, K. Chromophore-Assisted Light Inactivation of Mitochondrial Electron Transport Chain Complex II in *Caenorhabditis elegans. Sci. Rep.* **6**, 29695 (2016).
128. Zhen, Y. *et al.* ESCRT-Mediated Phagophore Sealing during Mitophagy. *Autophagy* **16**, 826–841 (2020).
129. Engel, A. J. *et al.* Light-Regulated Transcription of a Mitochondrial-Targeted K^+ Channel. *Cells* **9**, 2507 (2020).
130. He, J. *et al.* A Genetically Targetable Near-Infrared Photosensitizer. *Nat. Methods* **13**, 263–268 (2016).
131. Das, S., Tiwari, M., Mondal, D., Sahoo, B. R. & Tiwari, D. K. Growing Tool-Kit of Photosensitizers for Clinical and Non-Clinical Applications. *J. Mater. Chem. B* **8**, 10897–10940 (2020).
132. Zorov, D. B., Juhaszova, M. & Sollott, S. J. Mitochondrial ROS-induced ROS Release: An Update and Review. *Biochim. Biophys. Acta BBA—Bioenerg.* **1757**, 509–517 (2006).
133. Ruiz-González, R. *et al.* Singlet Oxygen Generation by the Genetically Encoded Tag miniSOG. *J. Am. Chem. Soc.* **135**, 9564–9567 (2013).
134. Dichmann, L. *et al.* Photophysics of a Protein-Bound Derivative of Malachite Green that Sensitizes the Production of Singlet Oxygen. *Photochem. Photobiol. Sci.* **20**, 435–449 (2021).
135. Liu, B., Marston, D. J. & Hahn, K. M. Engineering Optogenetic Protein Analogs. in *Photoswitching Proteins* (ed. Niopek, D.) vol. **2173** 113–126 (Springer US, 2020).
136. Onukwufor, J. O. *et al.* Quantification of Reactive Oxygen Species Production by the Red Fluorescent Proteins KillerRed, SuperNova and mCherry. *Free Radic. Biol. Med.* **147**, 1–7 (2020).
137. Endres, S. *et al.* An Optogenetic Toolbox of LOV-Based Photosensitizers for Light-Driven Killing of Bacteria. *Sci. Rep.* **8**, 15021 (2018).
138. Westberg, M., Holmegaard, L., Pimenta, F. M., Etzerodt, M. & Ogilby, P. R. Rational Design of an Efficient, Genetically Encodable, Protein-Encased Singlet Oxygen Photosensitizer. *J. Am. Chem. Soc.* **137**, 1632–1642 (2015).
139. Rodríguez-Pulido, A. *et al.* Assessing the Potential of Photosensitizing Flavoproteins as Tags for Correlative Microscopy. *Chem. Commun.* **52**, 8405–8408 (2016).
140. Westberg, M., Bregnhøj, M., Etzerodt, M. & Ogilby, P. R. No Photon Wasted: An Efficient and Selective Singlet Oxygen Photosensitizing Protein. *J. Phys. Chem. B* **121**, 9366–9371 (2017).
141. Bäumler, W., Regensburger, J., Knak, A., Felgenträger, A. & Maisch, T. UVA and Endogenous Photosensitizers — the Detection of Singlet Oxygen by Its Luminescence. *Photochem. Photobiol. Sci.* **11**, 107–117 (2012).
142. Bulina, M. E. *et al.* A Genetically Encoded Photosensitizer. *Nat. Biotechnol.* **24**, 95–99 (2006).
143. Sarkisyan, K. S. *et al.* KillerOrange, a Genetically Encoded Photosensitizer Activated by Blue and Green Light. *PLOS ONE* **10**, e0145287 (2015).
144. Takemoto, K. *et al.* SuperNova, a Monomeric Photosensitizing Fluorescent Protein for Chromophore-Assisted Light Inactivation. *Sci. Rep.* **3**, 2629 (2013).
145. Riani, Y. D., Matsuda, T., Takemoto, K. & Nagai, T. Green Monomeric Photosensitizing Fluorescent Protein for Photo-Inducible Protein Inactivation and Cell Ablation. *BMC Biol.* **16**, 50 (2018).
146. Shu, X. *et al.* A Genetically Encoded Tag for Correlated Light and Electron Microscopy of Intact Cells, Tissues, and Organisms. *PLoS Biol.* **9**, e1001041 (2011).
147. Makhijani, K. *et al.* Precision Optogenetic Tool for Selective Single- and Multiple-Cell Ablation in a Live Animal Model System. *Cell Chem. Biol.* **24**, 110–119 (2017).
148. Tu, Y. *et al.* Light-Induced Reactive Oxygen Species (ROS) Generator for Tumor Therapy through an ROS Burst in Mitochondria and AKT-Inactivation-Induced Apoptosis. *ACS Appl. Bio Mater.* **4**, 5222–5230 (2021).
149. Tan, V. P. *et al.* Dissociation of Mitochondrial HK-II Elicits Mitophagy and Confers Cardioprotection against Ischemia. *Cell Death Dis.* **10**, 730 (2019).
150. Szent-Gyorgyi, C. *et al.* Fluorogen-Activating Single-Chain Antibodies for Imaging Cell Surface Proteins. *Nat. Biotechnol.* **26**, 235–240 (2008).
151. Telmer, C. A. *et al.* Rapid, Specific, No-wash, Far-red Fluorogen Activation in Subcellular Compartments by Targeted Fluorogen Activating Proteins. *ACS Chem. Biol.* **10**, 1239–1246 (2015).
152. Entcheva, E. & Kay, M. W. Cardiac Optogenetics: A Decade of Enlightenment. *Nat. Rev. Cardiol.* **18**, 349–367 (2021).
153. Ferenczi, E. A., Tan, X. & Huang, C. L.-H. Principles of Optogenetic Methods and Their Application to Cardiac Experimental Systems. *Front. Physiol.* **10**, 1096 (2019).

154. Reyer, A. *et al.* Channelrhodopsin-Mediated Optogenetics Highlights a Central Role of Depolarization-Dependent Plant Proton Pumps. *Proc. Natl. Acad. Sci.* **117**, 20920–20925 (2020).
155. Zhou, Y., Ding, M., Nagel, G., Konrad, K. R. & Gao, S. Advances and Prospects of Rhodopsin-Based Optogenetics in Plant Research. *Plant Physiol.* **187**, 572–589 (2021).
156. Zhou, Y. *et al.* Extending the Anion Channelrhodopsin-Based Toolbox for Plant Optogenetics. *Membranes* **11**, 287 (2021).
157. Zhou, Y. *et al.* Optogenetic Control of Plant Growth by a Microbial Rhodopsin. *Nat. Plants* **7**, 144–151 (2021).
158. Berry, B. J. & Wojtovich, A. P. Mitochondrial Light Switches: Optogenetic Approaches to Control Metabolism. *FEBS J.* **287**, 4544–4556 (2020).
159. Berry, B. J. *et al.* Optogenetic Control of Mitochondrial Protonmotive Force to Impact Cellular Stress Resistance. *EMBO Rep.* **21**, e49113 (2020).
160. Demine, Renard, & Arnould. Mitochondrial Uncoupling: A Key Controller of Biological Processes in Physiology and Diseases. *Cells* **8**, 795 (2019).
161. Kato, H. E. *et al.* Crystal Structure of the Channelrhodopsin Light-Gated Cation Channel. *Nature* **482**, 369–374 (2012).
162. Ernst, O. P. *et al.* Microbial and Animal Rhodopsins: Structures, Functions, and Molecular Mechanisms. *Chem. Rev.* **114**, 126–163 (2014).
163. Spudich, J. L., Yang, C.-S., Jung, K.-H. & Spudich, E. N. Retinylidene Proteins: Structures and Functions from Archaea to Humans. *Annu. Rev. Cell Dev. Biol.* **16**, 365–392 (2000).
164. Zhang, F. *et al.* The Microbial Opsin Family of Optogenetic Tools. *Cell* **147**, 1446–1457 (2011).
165. Kandori, H. History and Perspectives of Ion-Transporting Rhodopsins. in *Optogenetics* (eds. Yawo, H., Kandori, H., Koizumi, A. & Kageyama, R.) vol. **1293** 3–19 (Springer Singapore, 2021).
166. Zabelskii, D. *et al.* Structure-based insights into evolution of rhodopsins. *Commun. Biol.* **4**, 821 (2021).
167. Kato, H. E. Structure–Function Relationship of Channelrhodopsins. in *Optogenetics* (eds. Yawo, H., Kandori, H., Koizumi, A. & Kageyama, R.) vol. **1293** 35–53 (Springer Singapore, 2021).
168. Srinivasan, S., Schelhaas, B., Maimon, B., Song, H. & Herr, H. Retinal Supplementation Augments Optogenetic Stimulation Efficacy *In Vivo. J. Neural Eng.* **16**, 054002 (2019).
169. Berndt, A. *et al.* High-Efficiency Channelrhodopsins for Fast Neuronal Stimulation at Low Light Levels. *Proc. Natl. Acad. Sci.* **108**, 7595–7600 (2011).
170. Berndt, A., Lee, S. Y., Ramakrishnan, C. & Deisseroth, K. Structure-Guided Transformation of Channelrhodopsin into a Light-Activated Chloride Channel. *Science* **344**, 420–424 (2014).
171. Berndt, A. *et al.* Structural Foundations of Optogenetics: Determinants of Channelrhodopsin Ion Selectivity. *Proc. Natl. Acad. Sci.* **113**, 822–829 (2016).
172. Chuong, A. S. *et al.* Noninvasive Optical Inhibition with a Red-Shifted Microbial Rhodopsin. *Nat. Neurosci.* **17**, 1123–1129 (2014).
173. Dawydow, A. *et al.* Channelrhodopsin-2–XXL, a Powerful Optogenetic Tool for Low-Light Applications. *Proc. Natl. Acad. Sci.* **111**, 13972–13977 (2014).
174. Duan, X., Nagel, G. & Gao, S. Mutated Channelrhodopsins with Increased Sodium and Calcium Permeability. *Appl. Sci.* **9**, 664 (2019).
175. Gong, X. *et al.* An Ultra-Sensitive Step-Function Opsin for Minimally Invasive Optogenetic Stimulation in Mice and Macaques. *Neuron* **107**, 38–51.e8 (2020).
176. Govorunova, E. G., Sineshchekov, O. A., Li, H., Janz, R. & Spudich, J. L. Characterization of a Highly Efficient Blue-shifted Channelrhodopsin from the Marine Alga *Platymonas subcordiformis. J. Biol. Chem.* **288**, 29911–29922 (2013).
177. Govorunova, E. G., Sineshchekov, O. A., Janz, R., Liu, X. & Spudich, J. L. Natural Light-Gated Anion Channels: A Family of Microbial Rhodopsins for Advanced Optogenetics. *Science* **349**, 647–650 (2015).
178. Govorunova, E. G. *et al.* RubyACRs, Nonalgal Anion Channelrhodopsins with Highly Red-Shifted Absorption. *Proc. Natl. Acad. Sci.* **117**, 22833–22840 (2020).
179. Govorunova, E. G. *et al. Kalium rhodopsins: Natural light-gated potassium channels.* (2021). https://doi.org/10.1101/2021.09.17.460684.
180. Gradinaru, V. *et al.* Molecular and Cellular Approaches for Diversifying and Extending Optogenetics. *Cell* **141**, 154–165 (2010).
181. Gunaydin, L. A. *et al.* Ultrafast Optogenetic Control. *Nat. Neurosci.* **13**, 387–392 (2010).
182. Han, X. *et al.* A High-Light Sensitivity Optical Neural Silencer: Development and Application to Optogenetic Control of Non-Human Primate Cortex. *Front. Syst. Neurosci.* **5,** 18 (2011). https://doi.org/10.3389/fnsys.2011.00018.

183. Hochbaum, D. R. *et al.* All-Optical Electrophysiology in Mammalian Neurons Using Engineered Microbial Rhodopsins. *Nat. Methods* **11**, 825–833 (2014).
184. Hou, S.-Y. *et al.* Diversity of Chlamydomonas Channelrhodopsins: Photochemistry and Photobiology. *Photochem. Photobiol.* **88**, 119–128 (2012).
185. Ihara, K. *et al.* Evolution of the Archaeal Rhodopsins: Evolution Rate Changes by Gene Duplication and Functional Differentiation. *J. Mol. Biol.* **285**, 163–174 (1999).
186. Inoue, K. *et al.* A Light-Driven Sodium Ion Pump in Marine Bacteria. *Nat. Commun.* **4**, 1678 (2013).
187. Inoue, K. *et al.* A Natural Light-Driven Inward Proton Pump. *Nat. Commun.* **7**, 13415 (2016).
188. Kato, H. E. *et al.* Structural Mechanisms of Selectivity and Gating in Anion Channelrhodopsins. *Nature* **561**, 349–354 (2018).
189. Klapoetke, N. C. *et al.* Independent Optical Excitation of Distinct Neural Populations. *Nat. Methods* **11**, 338–346 (2014).
190. Kleinlogel, S. *et al.* Ultra Light-Sensitive and Fast Neuronal Activation with the Ca^{2+}-permeable Channelrhodopsin CatCh. *Nat. Neurosci.* **14**, 513–518 (2011).
191. Cho, Y. K. *et al.* Multidimensional Screening Yields Channelrhodopsin Variants Having Improved Photocurrent and Order-of-Magnitude Reductions in Calcium and Proton Currents. *J. Biol. Chem.* **294**, 3806–3821 (2019).
192. Lin, J. Y., Knutsen, P. M., Muller, A., Kleinfeld, D. & Tsien, R. Y. ReaChR: A Red-Shifted Variant of Channelrhodopsin Enables Deep Transcranial Optogenetic Excitation. *Nat. Neurosci.* **16**, 1499–1508 (2013).
193. Mager, T. *et al.* High Frequency Neural Spiking and Auditory Signaling by Ultrafast Red-Shifted Optogenetics. *Nat. Commun.* **9**, 1750 (2018).
194. Marshel, J. H. *et al.* Cortical Layer–Specific Critical Dynamics Triggering Perception. *Science* **365**, eaaw5202 (2019).
195. Mattis, J. *et al.* Principles for Applying Optogenetic Tools Derived from Direct Comparative Analysis of Microbial Opsins. *Nat. Methods* **9**, 159–172 (2012).
196. Mongodin, E. F. *et al.* The Genome of *Salinibacter ruber*: Convergence and Gene Exchange Among Hyperhalophilic Bacteria and Archaea. *Proc. Natl. Acad. Sci.* **102**, 18147–18152 (2005).
197. Mukohata, Y., Ihara, K., Tamura, T. & Sugiyama, Y. Halobacterial Rhodopsins. *J. Biochem. (Tokyo)* **125**, 649–657 (1999).
198. Nagel, G. *et al.* Channelrhodopsin-1: A Light-Gated Proton Channel in Green Algae. *Science* **296**, 2395–2398 (2002).
199. Nagel, G. *et al.* Channelrhodopsin-2, a Directly Light-Gated Cation-Selective Membrane Channel. *Proc. Natl. Acad. Sci.* **100**, 13940–13945 (2003).
200. Nagel, G. *et al.* Light Activation of Channelrhodopsin-2 in Excitable Cells of *Caenorhabditis elegans* Triggers Rapid Behavioral Responses. *Curr. Biol.* **15**, 2279–2284 (2005).
201. Rajasethupathy, P. *et al.* Projections from Neocortex Mediate Top-Down Control of Memory Retrieval. *Nature* **526**, 653–659 (2015).
202. Shevchenko, V. *et al.* Inward H^+ Pump Xenorhodopsin: Mechanism and Alternative Optogenetic Approach. *Sci. Adv.* **3**, e1603187 (2017).
203. Sudo, Y. *et al.* A Blue-shifted Light-driven Proton Pump for Neural Silencing. *J. Biol. Chem.* **288**, 20624–20632 (2013).
204. Tsunoda, S. P. *et al.* H^+-Pumping Rhodopsin from the Marine Alga Acetabularia. *Biophys. J.* **91**, 1471–1479 (2006).
205. Waschuk, S. A., Bezerra, A. G., Shi, L. & Brown, L. S. *Leptosphaeria* Rhodopsin: Bacteriorhodopsin-like Proton Pump from a Eukaryote. *Proc. Natl. Acad. Sci.* **102**, 6879–6883 (2005).
206. Wietek, J. *et al.* Conversion of Channelrhodopsin into a Light-Gated Chloride Channel. *Science* **344**, 409–412 (2014).
207. Wietek, J. *et al.* An Improved Chloride-Conducting Channelrhodopsin for Light-Induced Inhibition of Neuronal Activity In Vivo. *Sci. Rep.* **5**, 14807 (2015).
208. Zhang, F. *et al.* Red-Shifted Optogenetic Excitation: A Tool for Fast Neural Control Derived from *Volvox carteri*. *Nat. Neurosci.* **11**, 631–633 (2008).
209. Zhang, F. *et al.* Multimodal Fast Optical Interrogation of Neural Circuitry. *Nature* **446**, 633–639 (2007).
210. Yamauchi, Y. *et al.* Molecular Properties of a DTD Channelrhodopsin from *Guillardia theta*. *Biophys. Physicobiology* **14**, 57–66 (2017).
211. Oesterhelt, D. & Stoeckenius, W. Rhodopsin-like Protein from the Purple Membrane of *Halobacterium halobium*. *Nature. New Biol.* **233**, 149–152 (1971).

212. Rozenberg, A., Inoue, K., Kandori, H. & Béjà, O. Microbial Rhodopsins: The Last Two Decades. *Annu. Rev. Microbiol.* **75**, 427–447 (2021).
213. Kandori, H. Ion-pumping Microbial Rhodopsins. *Front. Mol. Biosci.* **2**, (2015).
214. Diester, I. *et al.* An Optogenetic Toolbox Designed for Primates. *Nat. Neurosci.* **14**, 387–397 (2011).
215. Chow, B. Y. *et al.* High-Performance Genetically Targetable Optical Neural Silencing by Light-Driven Proton Pumps. *Nature* **463**, 98–102 (2010).
216. Tsunoda, S. P. *et al.* Functional Characterization of Sodium-Pumping Rhodopsins with Different Pumping Properties. *PLOS ONE* **12**, e0179232 (2017).
217. Boyden, E. S., Zhang, F., Bamberg, E., Nagel, G. & Deisseroth, K. Millisecond-Timescale, Genetically Targeted Optical Control of Neural Activity. *Nat. Neurosci.* **8**, 1263–1268 (2005).
218. Lehtinen, K., Nokia, M. S. & Takala, H. Red Light Optogenetics in Neuroscience. *Front. Cell. Neurosci.* **15**, 778900 (2022).
219. Broser, M. *et al.* NeoR, a Near-Infrared Absorbing Rhodopsin. *Nat. Commun.* **11**, 5682 (2020).
220. Sahel, J.-A. *et al.* Partial Recovery of Visual Function in a Blind Patient after Optogenetic Therapy. *Nat. Med.* **27**, 1223–1229 (2021).
221. Strickland, M., Yacoubi-Loueslati, B., Bouhaouala-Zahar, B., Pender, S. L. F. & Larbi, A. Relationships Between Ion Channels, Mitochondrial Functions and Inflammation in Human Aging. *Front. Physiol.* **10**, 158 (2019).
222. Lin, J. Y. A user's Guide to Channelrhodopsin Variants: Features, Limitations and Future Developments: A User's Guide to Channelrhodopsin Variants. *Exp. Physiol.* **96**, 19–25 (2011).
223. Vierock, J. *et al.* BiPOLES is an Optogenetic Tool Developed for Bidirectional Dual-Color Control of Neurons. *Nat. Commun.* **12**, 4527 (2021).
224. Oda, K. *et al.* Crystal Structure of the Red Light-Activated Channelrhodopsin Chrimson. *Nat. Commun.* **9**, 3949 (2018).
225. Bratanov, D. *et al.* Unique Structure and Function of Viral Rhodopsins. *Nat. Commun.* **10**, 4939 (2019).
226. Tsunoda, S. P., Sugiura, M. & Kandori, H. Molecular Properties and Optogenetic Applications of Enzymerhodopsins. in *Optogenetics* (eds. Yawo, H., Kandori, H., Koizumi, A. & Kageyama, R.) vol. **1293** 153–165 (Springer Singapore, 2021).
227. Vercellino, I. & Sazanov, L. A. The Assembly, Regulation and Function of the Mitochondrial Respiratory Chain. *Nat. Rev. Mol. Cell Biol.* **23**, 141–161 (2022).
228. Zorova, L. D. *et al.* Mitochondrial Membrane Potential. *Anal. Biochem.* **552**, 50–59 (2018).
229. Chen, K., Ernst, P., Liu, X. M. & Zhou, L. Optogenetic Studies of Mitochondria. in *Rhodopsin* (ed. Gordeliy, V.) vol. **2501** 311–324 (Springer US, 2022).
230. Esteras, N., Adjobo-Hermans, M. J. W., Abramov, A. Y. & Koopman, W. J. H. Visualization of Mitochondrial Membrane Potential in Mammalian Cells. in *Methods in Cell Biology* vol. **155** 221–245 (Elsevier, 2020).
231. Perry, S. W., Norman, J. P., Barbieri, J., Brown, E. B. & Gelbard, H. A. Mitochondrial Membrane Potential Probes and the Proton Gradient: A Practical Usage Guide. *BioTechniques* **50**, 98–115 (2011).
232. Scheib, U. *et al.* The Rhodopsin–Guanylyl Cyclase of the Aquatic Fungus *Blastocladiella emersonii* Enables Fast Optical Control of cGMP Signaling. *Sci. Signal.* **8** (389), rs8 (2015).
233. Li, D. *et al.* Abnormal Cyclic Nucleotide Signaling at the Outer Mitochondrial Membrane In Sympathetic Neurons During the Early Stages of Hypertension. *Hypertension* **79**, 1374–1384 (2022).
234. Lazarou, M. *et al.* The Ubiquitin Kinase PINK1 Recruits Autophagy Receptors to Induce Mitophagy. *Nature* **524**, 309–314 (2015).
235. Van Haren, J. *et al.* Local Control of Intracellular Microtubule Dynamics by EB1 Photodissociation. *Nat. Cell Biol.* **20**, 252–261 (2018).
236. Benedetti, L. *et al.* Light-Activated Protein Interaction with High Spatial Subcellular Confinement. *Proc. Natl. Acad. Sci.* **115**, (2018).
237. Yao, X., Rosen, M. K. & Gardner, K. H. Estimation of the Available Free Energy in a LOV2-Jα Photoswitch. *Nat. Chem. Biol.* **4**, 491–497 (2008).
238. Burgie, E. S. *et al.* Photosensing and Thermosensing by Phytochrome B Require Both Proximal and Distal Allosteric Features within the Dimeric Photoreceptor. *Sci. Rep.* **7**, 13648 (2017).
239. Yizhar, O. *et al.* Neocortical Excitation/Inhibition Balance in Information Processing and Social Dysfunction. *Nature* **477**, 171–178 (2011).

10 Mitochondrial Biology

A Vibrating and Dynamic Research Field Still Breaking Boundaries

Andreas S. Reichert

MITOCHONDRIAL RESEARCH OVER THE LAST ~150 YEARS – LESSONS FROM THIS FASCINATING ORGANELLE

When asked 'What is so unique about mitochondria and mitochondrial research?', everyone presumably has a different picture in mind. The answer depends on many things: your teachers at school or university, your family, your career stage and the talks and paper you may have read, of course also your current main field of research, and many more. Is it about one of the many metabolic processes located there, or the unique role of mitochondria in eukaryotic evolution, or the complexity of bioenergetics and its OXPHOS complexes involved, or the import of proteins and lipids, or the extraordinary dynamic nature of this organelle able to adapt rapidly to changing conditions, or the amazing structural variability and dynamics of inner membrane architecture, or the strong link to a broad range of human disorders, or the role in regulating apoptosis or immunity, or the interplay with other organelles via contact sites, or that it is linked to stem cell maintenance or differentiation, or that there is still no therapy for diseases summarized as 'mitochondriopathies', or is it the complex interplay of some of these selected possibilities? Let me give you some thoughts that I feel are of particular relevance.

There is the evolutionary aspect placing the endosymbiotic origin of this organelle at the very basis of the appearance of eukaryotes about two billion years ago. This event was very special as it fundamentally changed the biology of living organisms, rapidly allowing unprecedented complexity and an endomembrane system generating cellular compartmentalization at an entirely new level. The symbiotic installation of mitochondria within an ancestral archezoan organism can be seen as adding a powerplant as well as hundreds of batteries in an organism that was presumably highly efficient with the usage of limited energy sources but now faced a new situation, namely having a vast surplus of chemical energy provided by oxidative phosphorylation and the ability to store this energy in multiple membrane-rich compartments (Gray et al., 1999; Lane and Martin, 2010; Wallace, 2011). This surplus opened an evolutionary race with unthought possibilities – the energy barrier of prokaryotic life and genome content was overcome so that more genes, proteins, membrane surface, compartments, cell volume, and functions could evolve. Rough estimates suggest that this allowed the estimated 1000- to 100,000-fold increase in energy cost per gene accompanied with this symbiosis (Lane, 2020). Certainly, there are many unknown details of this process. However, it is clear that major steps in evolution are often inherently linked to steps in optimizing energy conversion.

With the critical role of mitochondria in energy conversion, it is not surprising that mitochondria also play central regulatory and signalling roles in many processes, such as innate immunity, cell differentiation, cell death, ion homeostasis, thermogenesis, and multiple stress responses including autophagy. Moreover, this interplay has to be highly adaptive and dynamic as any cell and all organisms have to be able to adapt rapidly either to changing environmental conditions, altered nutrient supply, and external and internal stresses or to developmental programs. The dynamic

DOI: 10.1201/9781003221180-15

interplay of mitochondrial energy conversion to the extra-mitochondrial environment was evolutionary shaped directly by central metabolites such as ATP, $NAD(P)^+$, Acetyl-CoA, fumarate, succinate, and many others. They are well known cofactors of enzymes and which in part are able to chemically modify proteins, RNA and DNA (e.g. acetylation and deacetylation of DNA). Many cofactors required for various enzymatic activities are directly linked to the energy- and the redox-status of a cell. One such example is that the levels of Acetyl-CoA and NAD^+ are strongly determined by mitochondrial energy metabolism which in turn regulate Histone-binding to DNA via and interplay between Acetyl-CoA-dependent acetylases and NAD^+-dependent deacetylases (Sirtuins). This is a nice example of how gene expression is directly linked to certain metabolites. Further, recent findings highlight that numerous metabolites directly bind or modify proteins or protein activity and by that modulate numerous cellular functions (Hicks et al., 2023). In the light of evolution, this makes perfect sense as controlling energy conversion and metabolism is of fundamental importance for survival. Thus, very ancient mechanisms regulating, for example, gene expression epigenetically via metabolites (and other processes) do not come as a surprise. Of course, regulation of efficient energy supply and distribution occurs at different levels and sophisticated signalling pathways (e.g. hypoxia or AMPK signalling) have evolved to ensure this. Yet, what I want to emphasize is that many of these biochemical facts we have learned need to be seen also in the light of how energy flow shaped evolution and vice versa. We still do not fully understand this interplay, yet mitochondria are now more and more recognized as signalling hubs controlling a pleiotropy of functions in eukaryotes.

Mitochondria are involved in many fundamental processes that go beyond energy conversion and ATP production. Mitochondria are essential for Fe-sulfur biogenesis, which actually is a truly essential function exerted by mitochondria (Lill and Freibert, 2020). Moreover, they are required, for example, for heme biosynthesis, amino acid and lipid metabolism, one carbon metabolism, detoxification of ammonia in the urea cycle, Ca^{2+} buffering, and, not to forget, also for thermogenesis. Mitochondria control the flux of amazingly many metabolites across the inner and the outer membrane and were, for example, reported to provide phospholipids in large amount for the biogenesis of autophagosomal membranes under starvation (Hailey et al., 2010). There is strong evidence that the entire endomembrane system in ancestral eukaryotes has evolved from vesicles derived from the outer membrane of the mitochondrial endosymbiont (Gould et al., 2016). In this regard, it is amazing to see that mitochondria still form mitochondria-derived vesicles (MDVs) and also show other ways to release mitochondrial material into the cytosol and also extracellular compartments (see also Chapter 6 by **Carla Lopes** and **Nuno Raimundo** and Chapter 7 by **Andrea Irazoki** and **David Pla-Martín**).

If you look at the history of mitochondrial research, it is quite clear that major discoveries went hand in hand with major technological advances, and mitochondria were often in the centre of interest (Figure 10.1). More than hundred years ago, researchers were fascinated by the possibilities light microscopes opened and mitochondria ('*mitos*' Greek for 'thread' and '*chondros*' Greek for 'grain') – also termed 'bioblasts' earlier – were visualized for the first time around the 1840s (reviewed in Zick et al., 2009). Although not appreciated at that time, early studies already revealed the dynamic nature of mitochondria undergoing constantly fusion and fission events (Lewis and Lewis, 1914). The next eras established the essential role of mitochondria in cellular metabolism and bioenergetics driven by the discovery and analysis of cytochromes, heme, flavin nucleotides, and ATP, and was further established with the formulation of the chemiosmotic theory by Mitchell (Mitchell, 1961) and the discovery and sequencing of mitochondrial DNA (mtDNA) (Anderson, 1981; Nass and Nass, 1963a; Nass and Nass, 1963b) (Figure 10.1). The later was the basis for the emergence of 'molecular evolution' as a new field, allowing for the first time to use mtDNA sequence variations in populations to track the evolution and migration routes of humans, primates, and other species. This topic actually was instrumental for me to become interested in mitochondria. In grad school during a one-year visit at the University of Delaware (USA), my professor Harold B White III, a wonderful, motivating teacher, gave me an assignment on different topics in biochemical evolution including to write about the so-called 'Mitochondrial Eve Hypothesis' (Cann et al., 1987). It was fascinating to

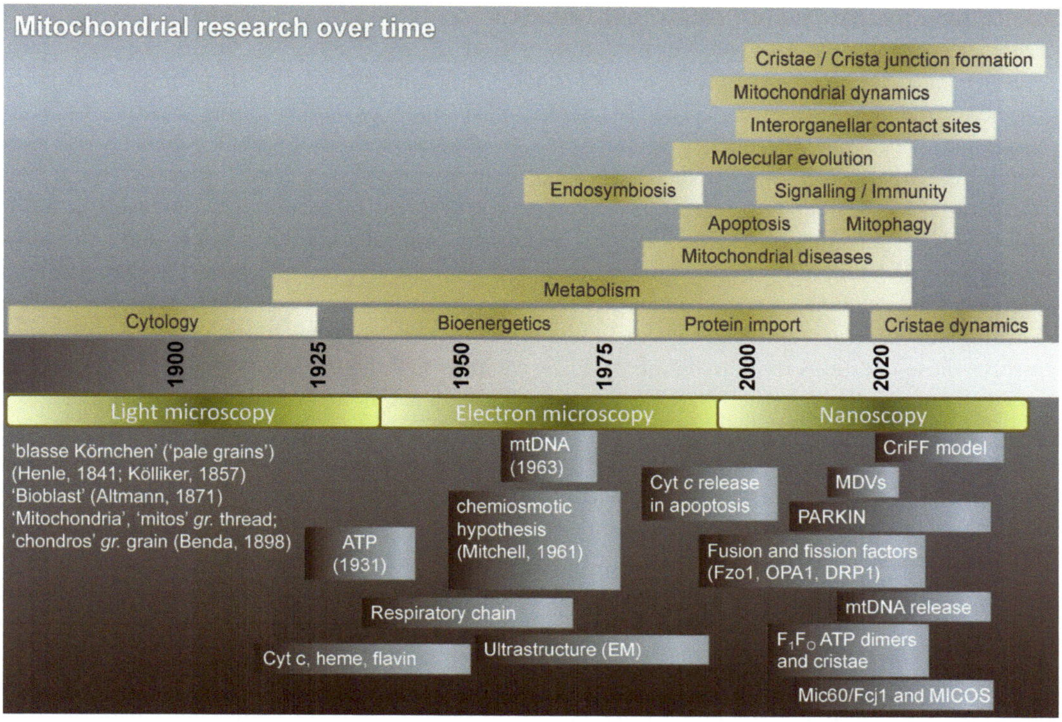

FIGURE 10.1 Simplified historical overview of mitochondrial research over time. Time spans of selected fields of research (above time axis) and technologies (below time axis) are only roughly depicted based on major discoveries or developments. Some highlights/discoveries linked to above fields are shown in the bottom part. First scientific descriptions of this organelle, later termed mitochondria, started around 1841.

learn how from mtDNA sequencing of individuals across the world, Allan C. Wilson and colleagues derived the conclusion that an ancestral female populating the world started from Africa a few hundred thousand years ago. This certainly influenced my decision to join the lab of Svante Pääbo, an earlier student of Allan C. Wilson, to do my PhD in the group of Mario Mörl on tRNA editing in human mitochondria and later to join the lab of Walter Neupert, a pioneer in studying protein import into mitochondria, where I decided to decipher the molecular mechanisms shaping cristae. Without these inspiring mentors and working together on these diverse aspects of mitochondrial biology, my research would not have gone from molecular evolution and RNA biochemistry to the dynamic nature of mitochondrial architecture and its role in health in disease.

But let's get back to important technological advances driving the field. With the development of electron microscopy and the analysis of biological specimens, our view on mitochondria and their ultrastructure was revolutionized. The inner membrane shows extensive infoldings, termed cristae, and it was quite clear from early on that certain cells or tissues harbour a certain shape of cristae and that, depending on the bioenergetic state of cells, the surface area is positively correlated with the amount of OXPHOS complexes and ATP generation. In the 1950s and 1960s, several competing hypotheses about the internal structure of the inner membrane have been discussed. The long-time prevailing and most accepted model was the so-called 'baffle' model proposed by Palade (1952; 1953). According to this model, cristae represent broad infoldings of the inner membrane with broad openings. An alternative, less known view was the 'Septa-model' by Sjöstrand who suggested that the inner membrane forms sheets/septa, forming distinct subcompartments within mitochondria (Sjöstrand, 1953). Analysing carefully performed electron microscopy on very thin serial sections, Daems and Wisse (1966) proposed the existence of so-called 'pediculi cristae' (*'little cristae feet'*) which were rediscovered in 1990 with the development of electron tomography, allowing the 3D reconstruction of mitochondrial ultrastructure in great detail. These pore- or slit-like very narrow (12–40 nm in diameter) structures are now termed 'crista junctions' (CJs). They subdivide the inner boundary membrane from the cristae membrane. The simple existence of this structure has fascinated many researchers, including myself, as it imposed many possibilities and allowed to ask many questions in the field. For example, do CJs form a physical barrier for lipids, proteins, metabolites, and protons? How does this determine assembly of mitochondrial membrane complexes including the OXPHOS machinery? Do they help to coordinate replication, transcription, and segregation of mtDNA? What determines the flux in and out of these newly formed subcompartments? Is this regulated and dependent on the bioenergetic status of the cell? And of course, what are the proteins forming this structure? Many groups including ourselves have contributed in the last 10–15 years to resolve these questions in more detail. In 2009, we identified the first subunit of a high-molecular-weight complex (later termed MICOS complex) located at CJs and required for CJ formation in baker's yeast, which we termed 'Formation of crista junction protein 1' (Fcj1), and its antagonistic interplay with the F_1F_O-ATP synthase (Rabl et al., 2009). The field has grown tremendously in the meantime and the characterization of the MICOS complex as well as its interplay with other cristae-shaping factors and its regulation has revealed major insights (see Chapter 1 by **Patrick Horten** and **Heike Rampelt**, and Chapter 2 by **Gunjan Purohit** and **Oleh Khalimonchuk**).

New technologies further accelerated our scientific understanding and a beautiful example is highlighted by very recent findings revealing, for the first time, that cristae are much less static and rather undergo remodelling events, and even fission and fusion processes, at least at a timescale of seconds (Kondadi et al., 2020a). Studying cristae membrane dynamics within individual mitochondria at a high spatiotemporal resolution became only possible with advanced super-resolution nanoscopy techniques in living cells (see Chapter 3 by **Ruchika Anand** and **Arun Kondadi** and Kondadi et al., 2020b). Our own studies further revealed that the MICOS complex is involved in determining this dynamic behaviour of cristae (Kondadi et al., 2020a). Lipids and their dynamic distribution within organellar membranes as well as dynamic interactions with protein complexes is not well explored so far. Still, for Complex I a number of specific lipid proteins interactions have been described and it is likely that the role of lipid dynamics will play an emerging role in the future (see Chapter 4 by **Nanami Senoo** and **Steven Claypool**).

The '*Dynamic-Trap-and-Flux*' ('DynaTrux') Model – a Possible Rational of Cristae Membrane Dynamics

After observing cristae membrane dynamics by STED-nanoscopy in living cells for the first time, we even recognized letters such as A, E, R, X, Y (see example in Figure 10.2) and we joked about mitochondria aiming to text us. We also thought about the question 'What is the main physiological function of cristae membrane dynamics?' One possibility that I would like to put forward is related to the role of mitochondria in energy conversion discussed at the beginning of this chapter. Maybe cristae dynamics has evolved to optimize energy conversion and at the same time allowing mitochondria to dynamically switch modes, for example, to optimize other functions depending on the current need. Under physiological conditions, ADP is the metabolite that usually is limiting for ATP synthesis via OXPHOS. Thus, a sufficiently high flux of ADP from the cytosol to the matrix is mandatory under high ATP consumption conditions, whereas this is less critical in resting conditions. Depending on the metabolite, analogous switching conditions can be formulated (e.g. for fatty acid oxidation or electron transport from cytosolic NADH to matrix-localized NADH). On the other hand, it is a long discussion in bioenergetics how protons are efficiently trapped within the intermembrane space to be used for ATP synthesis by the F_1F_O-ATP synthase. Loss of protons to the cytosol would reduce ATP yield per molecule of oxygen. Several mechanisms for proton trapping are proposed to act, including the role of cardiolipin (Haines and Dencher, 2002; Kates et al., 1993), the formation of CJs separating the intracristal lumen from the intermembrane space compartment that is near the outer

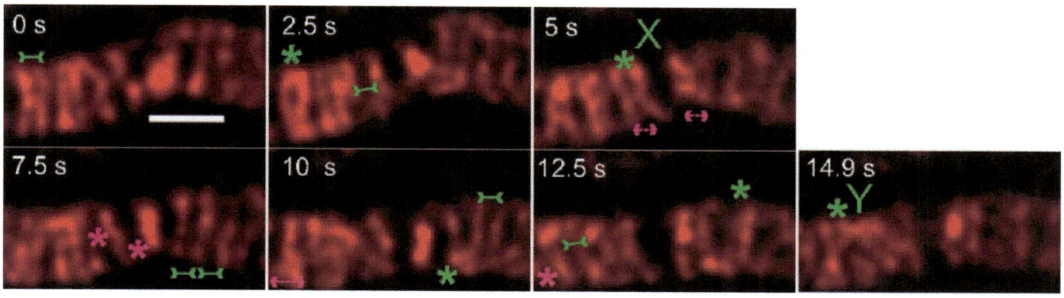

FIGURE 10.2 Cristae membrane dynamics in mammalian cells revealed by live cell STED nanoscopy. Time-lapse image series of a mitochondrion expressing MIC13-SNAP imaged at a time interval of 2.5 s/frame. Green and magenta asterisks show cycles of cristae merging and splitting marked by MIC13-SNAP. Green arrows pointing inward connected by solid line and magenta arrows pointing outward connected by dotted line show sites of imminent merging and splitting events, respectively. Cristae mergence events that form structure resembling letter 'X' or 'Y' are marked at 5 or 14.9 s, respectively. Scale bar 500 nm. (From Fig. 6 in Kondadi et al., 2020a.)

membrane (Zick et al., 2009), and the localization of the F_1F_O-ATP synthase oligomers at the rim of cristae membranes (Strauss et al., 2008). Our studies provide evidence that cristae can undergo fission and fusion and can even exist transiently as isolated vesicles (Kondadi et al., 2020a). This is also supported by findings that cristae can act as independent bioenergetic units and can adopt different levels of membrane potentials (Wolf et al., 2019). The transient formation of cristae vesicles is predicted to trap protons very efficiently, allowing those protons to be used efficiently for ATP synthesis via the F_1F_O-ATP synthase (Figure 10.3). Nevertheless, it is important to note that there is a clear trade-off to expect. This membrane architecture, if it would be permanent, would impair the entry of metabolites from the cytosol to both the intercristal lumen and the matrix as the directly accessible surface area would be reduced. Conversely, the exit to the cytosol is equally impaired. Important metabolites that may help to explain the concept better are ADP and ATP. Cytosolic $ADP3^-$ can be exchanged for matrix-located $ATP4^-$ via adenine nucleotide transporters (ANTs), proteins that are highly abundant and present in the inner boundary membrane and the cristae membrane (Vogel et al., 2006). When cristae are detached as in the 'proton trapping' mode, there is less accessible surfaces area for direct exchange between the cytosol and the matrix. ADP/ATP exchange between the matrix and the intracristal lumen may occur to a very limited extent for a limited time but would overall not be productive in promoting ATP replenishing of the cytosol. Furthermore, import of new proteins and lipids into the cristae membrane would be impaired. On the contrary, an optimal ADP/ATP exchange is expected when mitochondria are in the 'metabolite exchange' mode with a high extent of accessible cristae membrane surface. Thus, having the ability to switch between the 'proton trapping' mode and the 'metabolite exchange' mode is likely to be crucial for mitochondrial flexibility and OXPHOS efficiency. Based on these thoughts, I suggest to term this model the '*Dyna*mic-*Tr*ap-and-Fl*ux*' ('DynaTrux') model (Figure 10.3). The model implies that mitochondria have the ability to alter between a mode primarily optimized for ATP synthesis and a mode optimized for replenishing limiting metabolites. The 'proton trapping' mode is also predicted to prevail for thermogenesis. The notion that ADP can induce changes in cristae morphology is actually quite old. It was observed over five decades ago by Hackenbrock (1966; 1968) that isolated rat liver mitochondria treated with ADP (state III) adopt a matrix condensed appearance. Depletion of ADP (state IV) resulted in the formation of an 'orthodox', matrix-expanded appearance (Mannella, 2006; Mannella et al., 1994; Mannella et al., 2001). The intracristal compartments were extensively enlarged and interconnected in state III compared to state IV. Interestingly, also our recent data in living cells show that inhibiting ADP/ATP change across the inner membrane impaired cristae dynamics, although this was not observed when complexes I, III, or V were inhibited (Golombek et al., 2023).

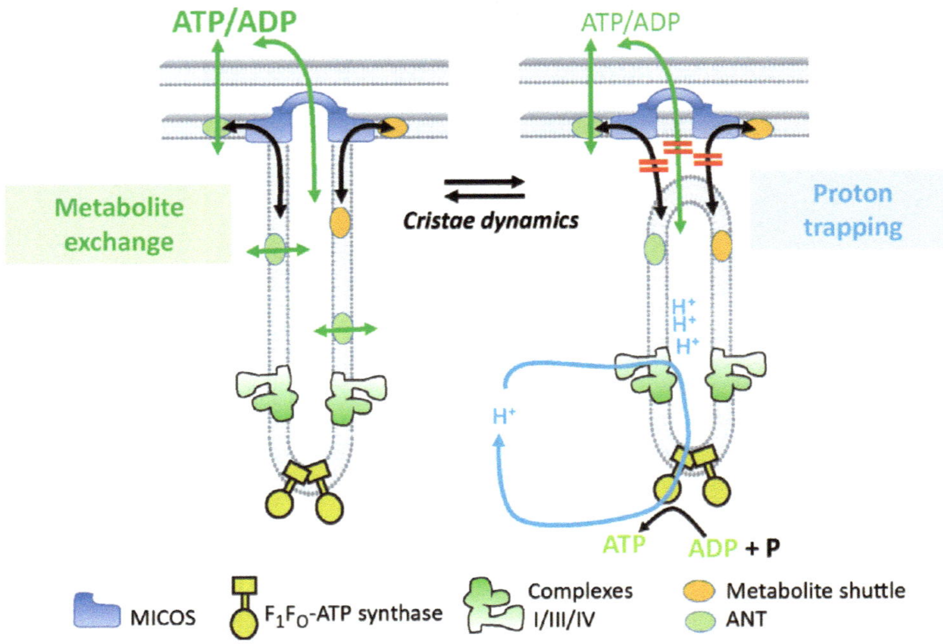

FIGURE 10.3 '*Dyna*mic-*Tr*ap-and-Fl*ux*' ('DynaTrux') model. The notion of cristae undergoing fission and fusion events implies a transient formation of cristae vesicles that can later re-fuse with the inner boundary membrane (IBM) and form crista junctions (CJs). Thus, two cristae morphologies exist and can dynamically interconvert: one is the 'metabolite exchange' mode with high extent of accessible membrane surface (left); and one is the 'proton trapping' mode in which formation of an entirely enclosed or separated compartment, the intracristal lumen, is formed (right). In the latter mode, the protons pumped during electron transport from reducing equivalents (NADH and $FADH_2$) to oxygen are efficiently trapped in this compartment and cannot escape to the cytosol. A high proton motive force can be generated to drive ATP generation via the F_1F_O-ATP synthase enriched in cristae. Protons can flow in a unidirectional manner and drive ATP generation. In this mode, the dynamic redistribution of membrane complexes to the IBM is not possible. However, upon re-fusion and formation of CJs, which occurs in a MICOS-dependent manner, metabolites can be exchanged more efficiently (higher surface) and membrane complexes including metabolite carriers/shuttles can redistribute, further adjusting dynamically metabolic and other needs.

Another level linked to this model not discussed yet is the aspect how certain metabolite transporters/carriers or ion pumps are distributed between the inner boundary membrane and the cristae membrane and how this can change dynamically. Let's assume that a certain metabolite shuttle/carrier is highly enriched or even exclusively located in the cristae membrane. It will not be accessible for its substrate coming from the cytosol when the 'trapping' mode prevails. Further, in this mode also no dynamic redistribution of protein complexes is possible (see Figure 10.3, right). Thus, cristae dynamics is expected to have major influences on metabolic fluxes and this, on the other hand, depends on the (dynamic) distribution of relevant protein complexes. Future studies will have to experimentally address the validity and the implications of the 'DynaTrux' model, and it will be interesting to see how it affects other mitochondrial functions such as Ca^{2+} buffering, cell death or immune signalling, or thermogenesis.

The Three Levels of Mitochondrial Dynamics

There are different levels of mitochondrial dynamics and it should be emphasized that they are connected on a mechanistic level as they also partly share molecular players: (1) intramitochondrial dynamics, (2) intermitochondrial fusion and fission dynamics, and (3) interorganellar dynamics (for review, see Kondadi and Reichert, 2023). The intramitochondria level includes the dynamic movement of proteins, lipids, metabolites, ions, RNAs, and mtDNA and also cristae membrane dynamics. The putative major physiological relevance of this level is discussed in the DynaTrux model above (Figure 10.3). The fact that mitochondria constantly undergo fusion and fission events is known for long time by pioneering work of Jürgen Bereiter-Hahn (1994) and also by Peter Walter and Jodi Nunnari (Nunnari et al., 1997) and many others. This process is linked to a vast number of metabolic and physiological processes such as mitochondrial quality control via mitophagy, apoptosis, mtDNA inheritance, OXPHOS, fatty acid oxidation, and metabolic switching during stem cell differentiation and in immune cells. Impairments of fusion and fission, not surprisingly, are linked to ageing and diseases such as various neuropathies, diabetes, cancer, and altered immunity (see Chapter 3 by **Ruchika Anand** and **Arun Kondadi** and Kondadi and Reichert, 2023). Interorganellar contacts, another level of mitochondrial dynamics, have been reported for various organelles, and mitochondria are known to form contacts to the endoplasmatic reticulum (ER), peroxisomes, lysosomes, the plasma membrane, and lipid droplets (see Chapter 6 by **Carla Lopes** and **Nuno Raimundo** and Kondadi and Reichert, 2023). It is more and more evident that organelles have to be seen as a dynamically adapting network rather than isolated organelles. Mitochondrial fission is governed by the fission factor DRP1 as well as by the close interaction to the ER. Mitochondria-ER contacts can be mediated by several molecular players but one kind depends on the fusion factors MFN1 and MFN2 (Chan, 2020). This is one example how intermitochondrial dynamics and interorganellar contacts between different organelles share factors and how these levels are interconnected. All levels described above allow one important aspect of cell biology that is of major importance, namely cellular flexibility. It allows the rapid exchange of metabolites and signalling molecules as well as adaptation to changing conditions and stress cues. Impairment of mitochondrial functions and dynamics is now well established to cause a pleiotropy of human diseases.

Outlook

Overall, we now have a basic understanding about mechanisms that shape and regulate the architecture of the mitochondrial inner membrane (see Chapters 1 to 3). We also have knowledge on certain steps in the assembly pathways of major OXPHOS complexes and also about their interplay with lipids (Chapters 4 and 8). Yet, we have only little insight into how all these are coordinated – in particular with respect to the spatiotemporal distribution of assembly intermediates within the distinct membrane compartments and what is the role of cristae morphology. Moreover, with the new era of cristae dynamics, additional questions arise. For example, how is the assembly of OXPHOS complexes, the flux of metabolites including lipids, influenced by the dynamic nature of mitochondria (Chapters 4 and 8)? These questions have to be answered also in the context that few OXPHOS subunits are encoded in mtDNA and many are encoded in the nuclear genome. How and where do transient assembly intermediates form and what determines the final steps in the assembly of OXPHOS complexes and supercomplexes? Is the 'DynaTrux' model correct and how can it be tested? What happens to damaged, misfolded or aggregated proteins, complexes? How is mtDNA and mtRNA release mediated (Chapter 7)? Is there any sort of active sorting of damaged material within mitochondria allowing a dynamic quality control, for example, by mitochondrial proteases (Chapter 2) and/or by selective mitophagy in combination with cristae remodelling (Chapter 3)? What is the role of redox signalling in integrating mitochondria and cellular homeostasis (Chapter 5)? How is selective mtDNA segregation ensured and how is it connected to the different levels of mitochondrial dynamics and human health (Chapters 7 and 8)? What are novel developments in mitochondrial research with respect to optogenetics (Chapter 9)?

In conclusion, the field of mitochondrial research has always been full of surprises and has helped science to advance tremendously. It developed hand in hand with landmark developments in technology and is influencing an increasing number of scientific disciplines now. There are many challenges ahead of us, many are related to basic research which I have discussed here in part, but many are highly relevant in medicine, for example, to find therapies for the treatment of mitochondriopathies. I am convinced that there is still much 'new' to find out about this 'old' organelle and that mitochondria will continue to break boundaries for the next 150 years.

ACKNOWLEDGEMENTS

I am very grateful to the entire team of the Institute of Biochemistry and Molecular Biology I in Düsseldorf, Germany, to all former lab members in Frankfurt am Main and Munich, to my mentors, and to my family – in particular, to my wife Karin. This work was supported by the Deutsche Forschungsgemeinschaft (DFG) CRC 1208 ID 267205415 – Project B12 and CRC 1535 ID 458090666 - Project A04.

REFERENCES

Anderson, L. 1981. Identification of mitochondrial proteins and some of their precursors in two-dimensional electrophoretic maps of human cells. *Proc Natl Acad Sci USA*. 78:2407–2411.

Bereiter-Hahn, J., and M. Vöth. 1994. Dynamics of mitochondria in living cells: shape changes, dislocations, fusion, and fission of mitochondria. *Microsc Res Tech*. 27:198–219.

Cann, R.L., M. Stoneking, and A.C. Wilson. 1987. Mitochondrial DNA and human evolution. *Nature*. 325:31–36.

Chan, D.C. 2020. Mitochondrial dynamics and its involvement in disease. *Annu Rev Pathol*. 15:235–259.

Daems, W.T., and E. Wisse. 1966. Shape and attachment of the cristae mitochondriales in mouse hepatic cell mitochondria. *J Ultrastruct Res*. 16:123–140.

Golombek, M., T. Tsigaras, Y. Schaumkessel, S. Hänsch, S. Weidtkamp-Peters, R. Anand, A.S. Reichert, and A.K. Kondadi. 2023. Live-cell super-resolution nanoscopy reveals modulation of cristae dynamics in bioenergetically compromised mitochondria. *bioRxiv*. https://doi.org/10.1101/2023.04.27.538553

Gould, S.B., S.G. Garg, and W.F. Martin. 2016. Bacterial vesicle secretion and the evolutionary origin of the eukaryotic endomembrane system. *Trends Microbiol*. 24:525–534.

Gray, M.W., G. Burger, and B.F. Lang. 1999. Mitochondrial evolution. *Science*. 283:1476–1481.

Hackenbrock, C.R. 1966. Ultrastructural bases for metabolically linked mechanical activity in mitochondria. I. Reversible ultrastructural changes with change in metabolic steady state in isolated liver mitochondria. *J Cell Biol*. 30:269–297.

Hackenbrock, C.R. 1968. Chemical and physical fixation of isolated mitochondria in low-energy and high-energy states. *Proc Natl Acad Sci USA*. 61:598–605.

Hailey, D.W., A.S. Rambold, P. Satpute-Krishnan, K. Mitra, R. Sougrat, P.K. Kim, and J. Lippincott-Schwartz. 2010. Mitochondria supply membranes for autophagosome biogenesis during starvation. *Cell*. 141:656–667.

Haines, T.H., and N.A. Dencher. 2002. Cardiolipin: a proton trap for oxidative phosphorylation. *FEBS Lett*. 528:35–39.

Hicks, K.G., A.A. Cluntun, H.L. Schubert, S.R. Hackett, J.A. Berg, P.G. Leonard, M.A. Ajalla Aleixo, Y. Zhou, A.J. Bott, S.R. Salvatore, F. Chang, A. Blevins, P. Barta, S. Tilley, A. Leifer, A. Guzman, A. Arok, S. Fogarty, J.M. Winter, H.C. Ahn, K.N. Allen, S. Block, I.A. Cardoso, J. Ding, I. Dreveny, W.C. Gasper, Q. Ho, A. Matsuura, M.J. Palladino, S. Prajapati, P. Sun, K. Tittmann, D.R. Tolan, J. Unterlass, A.P. VanDemark, M.G. Vander Heiden, B.A. Webb, C.H. Yun, P. Zhao, B. Wang, F.J. Schopfer, C.P. Hill, M.C. Nonato, F.L. Muller, J.E. Cox, and J. Rutter. 2023. Protein-metabolite interactomics of carbohydrate metabolism reveal regulation of lactate dehydrogenase. *Science*. 379:996–1003.

Kates, M., J.Y. Syz, D. Gosser, and T.H. Haines. 1993. pH-dissociation characteristics of cardiolipin and its 2'-deoxy analogue. *Lipids*. 28:877–882.

Kondadi, A.K., R. Anand, S. Hänsch, J. Urbach, T. Zobel, D.M. Wolf, M. Segawa, M. Liesa, O.S. Shirihai, S. Weidtkamp-Peters, and A.S. Reichert. 2020a. Cristae undergo continuous cycles of membrane remodelling in a MICOS-dependent manner. *EMBO Rep*. 21:e49776.

Kondadi, A.K., R. Anand, and A.S. Reichert. 2020b. Cristae membrane dynamics - a paradigm change. *Trends Cell Biol*. 30:923–936.

Kondadi, A.K., and A.S. Reichert. 2023. Mitochondrial dynamics at different levels: from cristae dynamics to inter-organellar crosstalk. *Annu Rev Biophys*. Vol 53. https://doi.org/10.1146/annurev-biophys-030822-020736

Lane, N. 2020. How energy flow shapes cell evolution. *Curr Biol.* 30:R471–R476.

Lane, N., and W. Martin. 2010. The energetics of genome complexity. *Nature.* 467:929–934.

Lewis, M.R., and W.H. Lewis. 1914. Mitochondria in tissue culture. *Science.* 39:330–333.

Lill, R., and S.A. Freibert. 2020. Mechanisms of mitochondrial iron-sulfur protein biogenesis. *Annu Rev Biochem.* 89:471–499.

Mannella, C.A. 2006. Structure and dynamics of the mitochondrial inner membrane cristae. *Biochim Biophys Acta.* 1763:542–548.

Mannella, C.A., M. Marko, P. Penczek, D. Barnard, and J. Frank. 1994. The internal compartmentation of rat-liver mitochondria: tomographic study using the high-voltage transmission electron microscope. *Microsc Res Tech.* 27:278–283.

Mannella, C.A., D.R. Pfeiffer, P.C. Bradshaw, Moraru, II, B. Slepchenko, L.M. Loew, C.E. Hsieh, K. Buttle, and M. Marko. 2001. Topology of the mitochondrial inner membrane: dynamics and bioenergetic implications. *IUBMB Life.* 52:93–100.

Mitchell, P. 1961. Coupling of phosphorylation to electron and hydrogen transfer by a chemi-osmotic type of mechanism. *Nature.* 191:144–148.

Nass, M.M., and S. Nass. 1963a. Intramitochondrial fibers with DNA characteristics. I. fixation and electron staining reactions. *J Cell Biol.* 19:593–611.

Nass, S., and M.M. Nass. 1963b. Intramitochondrial fibers with DNA characteristics. II. Enzymatic and other hydrolytic treatments. *J Cell Biol.* 19:613–629.

Nunnari, J., W.F. Marshall, A. Straight, A. Murray, J.W. Sedat, and P. Walter. 1997. Mitochondrial transmission during mating in *Saccharomyces cerevisiae* is determined by mitochondrial fusion and fission and the intramitochondrial segregation of mitochondrial DNA. *Mol Biol Cell.* 8:1233–1242.

Palade, G. 1953. An electron microscope study of the mitochondrial structure. *J Histochem Cytochem.* 1:188–211.

Palade, G.E. 1952. The fine structure of mitochondria. *Anat Rec.* 114:427–451.

Rabl, R., V. Soubannier, R. Scholz, F. Vogel, N. Mendl, A. Vasiljev-Neumeyer, C. Korner, R. Jagasia, T. Keil, W. Baumeister, M. Cyrklaff, W. Neupert, and A.S. Reichert. 2009. Formation of cristae and crista junctions in mitochondria depends on antagonism between Fcj1 and Su e/g. *J Cell Biol.* 185: 1047–1063.

Sjöstrand, F.S. 1953. Electron microscopy of mitochondria and cytoplasmic double membranes. *Nature.* 171:30–32.

Strauss, M., G. Hofhaus, R.R. Schroder, and W. Kuhlbrandt. 2008. Dimer ribbons of ATP synthase shape the inner mitochondrial membrane. *EMBO J.* 27:1154–1160.

Vogel, F., C. Bornhövd, W. Neupert, and A.S. Reichert. 2006. Dynamic subcompartmentalization of the mitochondrial inner membrane. *J Cell Biol.* 175:237–247.

Wallace, D.C. 2011. Bioenergetic origins of complexity and disease. *Cold Spring Harb Symp Quant Biol.* 76:1–16.

Wolf, D.M., M. Segawa, A.K. Kondadi, R. Anand, S.T. Bailey, A.S. Reichert, A.M. van der Bliek, D.B. Shackelford, M. Liesa, and O.S. Shirihai. 2019. Individual cristae within the same mitochondrion display different membrane potentials and are functionally independent. *EMBO J.* 38:e101056.

Zick, M., R. Rabl, and A.S. Reichert. 2009. Cristae formation-linking ultrastructure and function of mitochondria. *Biochim Biophys Acta.* 1793:5–19.

Index

Note: Locators in *italics* represent figures and **bold** indicate tables in the text.

D

E

F

G

H

I

K

L

M

Q

R

S

T

V

Y